ASTR Y
MANUAL

THE PRACTICAL GUIDE TO THE NIGHT SKY

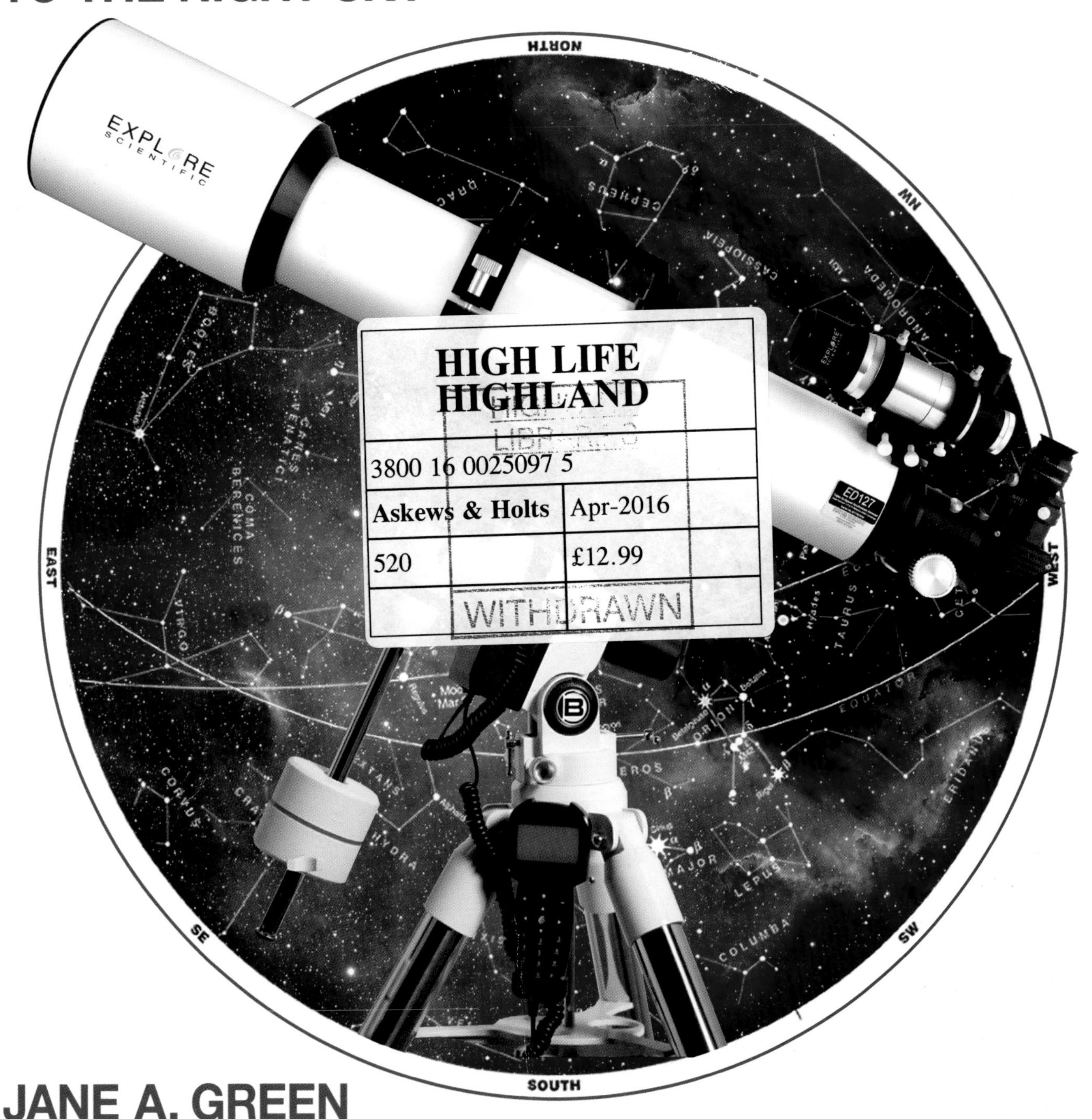

JANE A. GREEN

FOREWORD BY DR BRIAN MAY INTRODUCTION BY SIR PATRICK MOORE

To Cheryl
My dear friend – a dazzling light amidst universal darkness

First published in October 2010
First reprint January 2011
Second reprint April 2011

Reprinted in paperback with revisions in April 2016

A catalogue record for this book is available from the British Library

ISBN 978 0 85733 850 1

Library of Congress catalog card no. 2015935804

Published by Haynes Publishing,
Sparkford, Yeovil, Somerset BA22 7JJ, UK

Tel: +44 (0) 1963 440635
Website: www.haynes.co.uk

Haynes North America, Inc.,
861 Lawrence Drive, Newbury Park,
California 91320, USA

Design and layout: Richard Parsons
Illustrations: Dominic Stickland

Printed in the USA by Odcombe Press LP,
1299 Bridgestone Parkway, La Vergne, TN 37086

Front Cover Image: *An ED Triplet Essential Series 127mm Apochromatic Refractor with Bresser Exos-II GoTo EQ Mount*
(Daniel Cerda/Explore Scientific/Bresser)

Contents

Foreword

By Dr Brian May

In the 1950s, when I was a boy, I was inspired by the work of Patrick Moore, who is very much recognised by today's British astronomers, amateur and professional alike, as the father of English astronomy, and the genial uncle of all of us who today feel called to study the mysteries of the boundless universe. His writings made the state of astronomical knowledge at the time understandable and enticing to us all.

More than 60 years later, our knowledge of astronomy and astrophysics has been transformed almost beyond recognition. A constant clamour of exciting new discoveries hits the journals every week, making it hard even for astronomers themselves to keep up with the developments in every part of the subject. Jane Green has, in this luxurious *Astronomy Manual*, taken up the challenge of doing for a new generation what Patrick, now Sir Patrick, did for us all those years ago. Summarising and elucidating current knowledge in all areas of astronomy, she opens a new door for potential future astronomers to walk through. The *Astronomy Manual* encompasses all the newest discoveries in observational astronomy, and boldly tackles many of the most important current theories of the way the Universe works, including how it was created and how it evolves. Every page is graced with the latest astronomical photographs from the Hubble Space Telescope, the Spitzer Infrared Telescope and many others; and each new concept is accompanied by clear and beautiful explanatory diagrams.

Significantly, this book devotes large sections to the current observational techniques of both professional and amateur astronomers. In an age where any amateur can summon up more computing power on his laptop than was used in the entire first moon landing, the lines between amateur and professional have become blurred, and many new discoveries are being made daily by lone amateurs, especially in the field of exosystems – the planetary families of other suns. This book introduces the constellations, and discusses ways to enjoy and study the night sky, from naked-eye observations, through binoculars to the awesome new giant, land-based telescopes which are probing ever deeper into the echoes of our universe in the distant past. The newest techniques of astrophotography are also introduced, again, many accessible to both professional and amateur.

I have had the luxury of seeing Jane Green's project in its development, and can bear witness to the passion and care that she has put into it, which leap from every page. So I'm honoured to contribute these few words to encourage anyone who feels a curiosity for the subject to take this book home. I suggest keeping it around, to delve into, in those moments when we wonder what is the current state of knowledge of, say, the planets, or our Sun, or the local part of the Milky Way galaxy where our Solar System resides, or … way out in deepest space … where we look back in time, and the history of the universe is spread out before our eyes – if we have the skills to read and interpret it. This book will give many future astronomers their first insights into the secrets of the cosmos, and will inspire many to take up the challenge, and start their own explorations where we, the current generation, leave off.

And perhaps they will inch closer to answering the fundamental questions that we all eventually dare to ask – "Who are we?" and "Why are we here?".

Dr Brian May
August 2010

Introduction

By Sir Patrick Moore

Astronomy is a science open to everybody. Jane Green, a skilled amateur astronomer who has given lectures in many parts of the world, clearly appreciates this, so that her book will be of value to readers of all kinds, from complete beginners to highly-qualified experts. It is extremely well written, and the illustrations are beautiful.

Many new books on astronomy are now published every year, but I am emphatic in saying that Jane Green's is outstanding. As well as covering the usual topics – descriptions of the Sun, Moon, planets and so on, together with clear maps – there are features that are not often found in what may be termed "popular" books. There is a tremendous amount of information in these sections, but it is easy to absorb because the text is so clear. Obviously the author's lecturing experience stands her in good stead. There are also some fascinating asides about old beliefs and legends. For example, did you know that to the Yoingu people of North Australia, the creator planet Venus "still remains a doorway to heaven, a portal for reaching ancestors of the great beyond"? I have to admit that before reading this paragraph I had never heard of the Yoingu!

As well as being carefully chosen, the illustrations are excellently produced; they include images from the world's best telescopes. The diagrams are of equally high standard. All in all, it is clear that a great deal of research has gone into the preparation and production of this book, but the result shows that the effort has been well worthwhile.

Whether you are a complete novice or an experienced astronomer, you will benefit from reading and owning Jane Green's book. It deserves an honoured place in every scientific library.

Patrick Moore, CBE FRS
March 1923 – December 2012

The Witch Head Nebula, IC 2118: a reflection nebula where tenuous clouds of gas and dust are illuminated by neighbouring stars

Preface

The cosmos is all there ever was and all there ever will be. Yet we look down, not up, inward, not outward. We do not know that we need to know nor that a journey awaits us, one of revelation from infinitesimal to infinite, manifest to mystery, certainty to uncertainty, and relevant to remote. Yet its gallery beckons, as it has always beckoned. It waits; the giant glass doors are closed, but unlocked. Entrance is free. We need only nudge them, gaze inside … and greet history, science, drama, and art engraved on the grandest canvas of all.

Humanity has always sought to know the 'out there'; that world beyond the sky. Nowadays, many look but do not question, instead glancing at myriad pinpricks of light without context or comprehension, appreciating prettiness, but little more. Google 'star', and, invariably, a plethora of celebrities cascade forth along with words like 'megastar', 'superstar', and 'meteoric rise'. Bedazzled by glitz and glamour, sensationalism and spin, we miss the greatest show of all. Why, when observation, experimentation, innovation, and exploration has made the cosmos so accessible? Those pinpricks are stars, that is to say, suns, like our Sun, some 'close', many immeasurably distant, some 'small', others mind-numbingly immense. We are enabled because, in the past, humankind has wondered and enquired, because we have sacrificed or been courageous. We have journeyed, despite ourselves ... and we journey still…

But what could the Aborigines imagine when, perhaps 40,000 years ago, they gathered around campfires and saw the vast arm of stars arc overhead? What was it to them? A backbone bracing the sky without which the heavens would collapse? Some saw in its clouds a giant emu, its head recognisable today as the dark, starless patch of sky now known as the Coal Sack near the eponymous Southern Cross constellation. For most, the cosmos was simply a timeless synergy of earth and sky; an amalgam of myth and magic derived from Dreaming – when creator spirits conceived the world. Such beliefs engendered an enduring oneness with nature. For the Yolngu people in north Australia, the creator planet Venus still remains a doorway to heaven, a portal for reaching ancestors in the great beyond. The Sun was worshipped as a goddess, the Moon as a god, and it was these gods who policed their moral meanderings. The orange star Karambal – Aldebaran in the constellation of Taurus the bull – warned of the dangers of adultery. Others were more benign, like the famous three stars of Orion's belt seen as men paddling a canoe. For many indigenous peoples, star patterns were of more practical use: eggs could be collected when Lyra rose or meat could be eaten with the advent of the Pleiades. Stars affirmed structure, were an inseparable anchor servicing everyday life.

Not so for the ancient Polynesians seeking expansion. They looked to the stars for navigation as they plied their canoes across the Pacific carrying families, livestock, and crops. Incredibly, using only a cerebral star-compass – dividing the horizon into 32 directions that correlated with rising and setting stars – they were able to cross the equator to Hawaii, course south-east to Easter Island, and finally return to New Zealand, colonising islands as they went. If Arcturus was on the zenith, they knew they were in Hawaii. If Sirius blazed overhead, home was Tahiti. Although stars were not intrinsically understood, they were a calendar, their motion dividing the year. They had purpose and function.

For the agrarian society of Egypt the night skies were for worship, propitiation, and timekeeping rather than being physically understood. Star patterns were painted on pyramid tombs to assist their Sun god, Ra, in his underworld journey. He would descend at dusk and, assisted by their light, surface at dawn. Their brightest star, Sothis – Sirius – also signalled the Nile's seasonal flooding when rejuvenated soil would enable crop cultivation. Their studies of sequential stellar risings and settings led to today's 24-hour clock. Yet there was no serious astronomical investigation and no immanent understanding. Astronomy was still astrology: the belief that the positions of the planets and the Sun influenced human behaviour.

Astronomer priests of Ancient Egypt: take advantage of the unique facilities of the Great Pyramid of Khufu (Cheops), using it as an observatory.

Incredibly, elsewhere on the planet, great edifices, like the solar shrine of southern England's Stonehenge, were being erected. Fifty-tonne blocks were assembled in circles,

Planets align over stonehenge: the massive stone structure dates from around 2000 BC.

perhaps forming a 'cathedral' for the cosmos or a timepiece for the heavens; its purpose is still not fully known. But the stones bestow a global symbol, a reminder that mankind worshipped and wondered, had *need* of the heavens ... and the *need* continues.

Unwittingly, it was the Babylonians with their desire for order in the cosmos who left a fundamental legacy still in use today. From around 747 BC their astronomers kept regular ephemerides – tables recording the movements of the Moon and planets. Cuneiform script inscribed on thousands of clay tablets became astronomical diaries and contained such periodic phenomena as the lunar eclipse 18-year cycle. To provide a framework for planetary motion, zodiacal constellations were also created: Orion (known as the warrior, Gilgamesh), Gemini, and Taurus all originate in Mesopotamia. The first predictions for planetary positions had arrived. Commensurate with observation was meticulous mathematics: the sexagesimal system (counting in 60s) left its legacy in our 60 minutes in an hour and 360 (6 x 60) degrees in a circle. As with other ancients, the stars were still gods but were now bestowed characters: Jupiter was Marduk, head of the pantheon; Mars, Nergal, the god of war; and Venus, Ishtar, the goddess of love. Such astronomical and astrological enquiry embodied a thousand-year tradition that would travel the Tigris and Euphrates rivers to Greece, India, and beyond. Even in today's Middle East, the Babylonian names of the months are familiar to speakers of modern Hebrew or Arabic. But *what were* the stars, *what were* the heavens?

Statue of Atlas bearing the globe depicting 41 constellations: known as the Farnese Atlas, in the Museo Nazionale of Napoli.

Stone tablet: detailing the cosmos according to Babylonian astronomers, *c*500 BC.

The Greeks sought answers with their invention of science, and the first scientist was Thales of Miletus (624–546 BC). Drawing on Babylonian findings, he accurately predicted the first total solar eclipse – 28 May 585 BC. Greek intellectual life thrived, fuelled by prosperous trading between regional city states – 'polis'. Knowledge spread and seeded like a great interplanetary pan-spermia. Egyptian geometry passed to Pythagoras of Samos (*c*580–*c*500 BC); his preoccupation with circles, simple observations of the Earth's curved shadow on the Moon, and stellar observations south of the equator, led to mankind's giant intellectual leap: the planet was spherical and floated in space. Around this sphere, the Sun, Moon, and planets sailed in circular paths. He thought he understood. Plato and his pupil, Eudoxus of Cnidus (*c*410–*c*347 BC), followed, each speculating complex planetary motion with copious complex spheres. But it was Aristotle (384–322 BC), the grounded philosopher, who offered simplicity: life comprised four elements – earth, water, air, and fire. And it was the courage of Aristarchus (*c*310–230 BC) that took mankind forward. He dared to ask, 'Why, if these objects circle Earth, is the Sun ten times bigger?' Could not the Sun be at the centre? It was revolutionary – and rejected for almost 2,000 years – but not forgotten. There were others: the engineering genius, Archimedes of Syracuse, perhaps the mentor of the famous Antikythera Mechanism. Hipparchus of Rhodes (190–120 BC), the discoverer of precession, who measured the length of the year to within an amazing seven minutes in 365 days, inscribed the positions of 850 of the brightest stars on a globe, and improved the accuracy of predicting lunar eclipses.

Egypt, too, was an eager participant. In 240 BC, Eratosthenes (*c*276–194 BC) – a Libyan polymath and third librarian of the great Alexandrian Library – used sticks and imagination to study Earth's shadow. In so doing he accurately measured the circumference of the planet. Apollonius of Perga, the mathematician, painstakingly pioneered the geometry of ellipses. And, finally, in AD 150, Ptolemy gathered Greek knowledge in his 13-volume *Amalgest* – the astronomical 'bible' that would survive 14 centuries and bequeath 48 constellations – followed by the occult *Tetrabiblos*, and cemented with his *Geographia* comprising everything then known about the world.

Yet the Universe was still deemed geocentric and such thinking would fossilise for 1,400 years, mired in religionist theories of an ethereal heaven. Enter the Muslim Qu'ran, in which Allah's stars gave direction to Mecca, where followers were told where and when to pray and whence the Arabic month would start. The Persian star-gazer Abd Al-Rahman Al Sufi (903–86) glamorised the constellations in his *Book of Fixed Stars* and was the first to notice galaxies: the giant Andromeda and the Large Magellanic Cloud. Muslim scholars invented trigonometry and algebra, including Omar Khayyám's (1048–1122) quadratic equations and his accurate calendar year. Ulugh Beg's stunning Samarkand observatory soared over the Silk Road. Yet still elusive understanding of 'out there' remained unperceived, beyond the astronomical 'event horizon'.

Elsewhere, other great temples, pyramids, and observatories had arisen. The Chinese were watching and recording, a total eclipse on 5 June 1302 BC being the oldest written astronomical account. As with the Babylonians, astronomy assumed precision. Orderly events in the sky translated to order on Earth. A Chinese Astronomical Bureau was created, its bequest – the longest continuous set of sky records – and the Ancient Observatory of Beijing, where 283 constellations were monitored, still stands! If the skies were quiescent, there was peace for the Emperor and China; recorded disturbances, such as comets, eclipses or the 1054 supernova, now known to have been an exploding star, were portents of doom. For modern-day astronomers, the Chinese records are bookmarks, references for refining calculations.

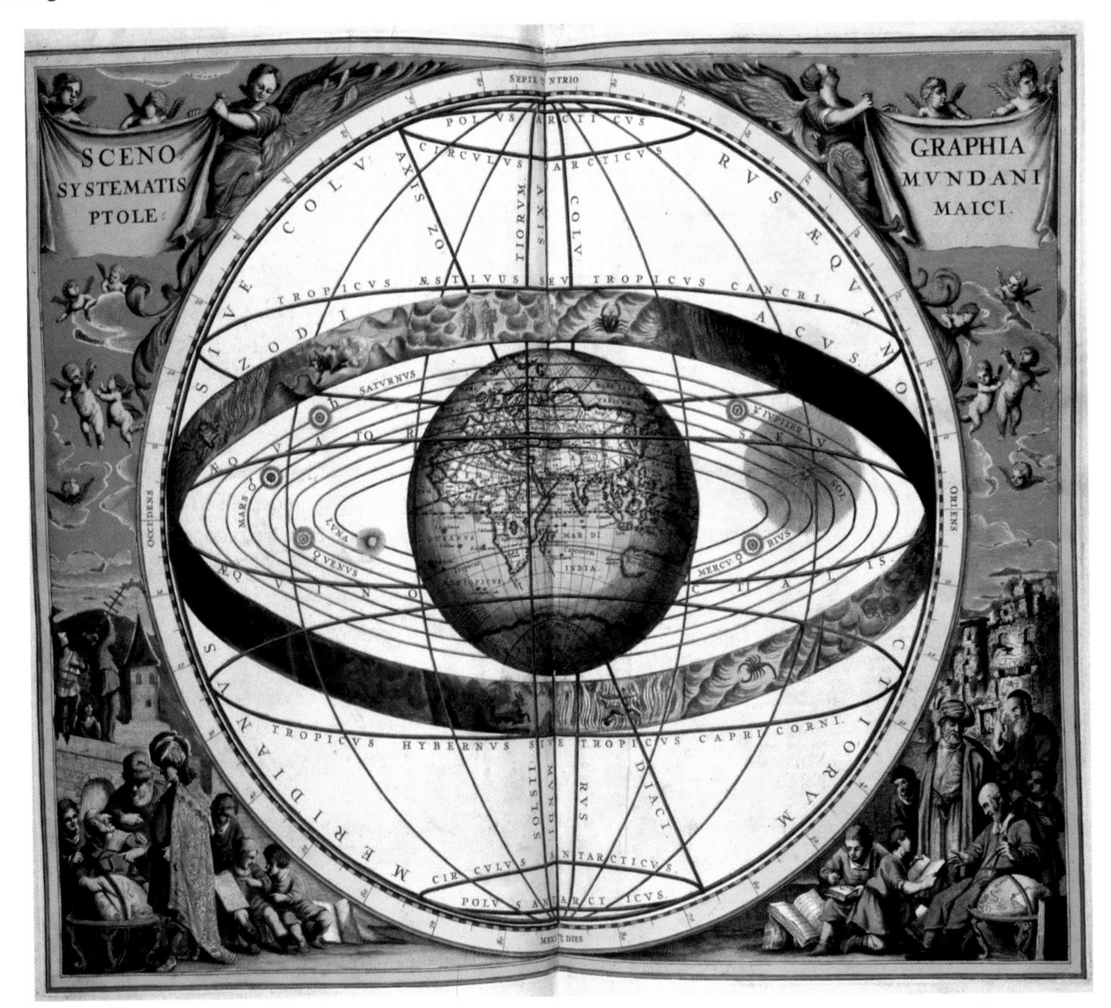

Scenographia Systematis Mundani: Geocentric System, Andreas Cellarius. (The British Library Board/ Maps.C.6.c.2, after 12)

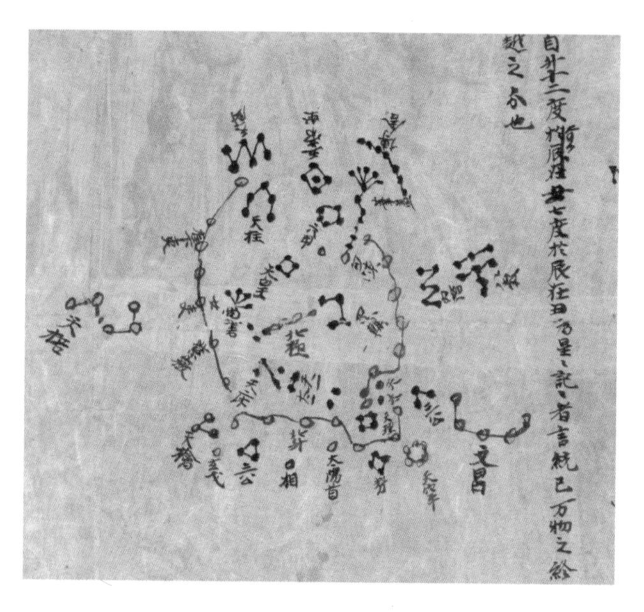

Chinese Star Map: China, Tang Dynasty 618–906.
(The British Library Board/Or.8210/S.3326)

Whilst the Chinese watched Jupiter in its 12-year orbit, the Maya in their Caracol observatory watched Venus. From this Chichen Itza site they sought similar stellar stillness and significance. Some records survive on bark – codices – but most were cremated in 1519 by Spanish missionaries, knowledge, like the great fire in the Library of Alexander, lost forever. Contrastingly, others still gazed and worshipped, measured and mused, like the Sun Chiefs in America's south-western Chaco Canyon or the Anasazi cave-dwellers of Arizona's Mesa Verde. Their lunar and solar ceremonies are silenced in sandstone but some traditions still perpetuate in the Pueblo today.

It was in the 12th century, with the liberation of Spain from the Muslims, that solid scientific exploration passed to Christianity and scholars were united. Buried knowledge was exhumed. With the fall to the Muslims in 1453 of Constantinople – a fortress of Roman and Greek philosophy – came an exodus of wisdom that would enflame a renaissance of astronomy. Astrology matured from medieval mysticism to mathematical fact. Regiomontanus (Johannes Müller), a Bavarian (1436–76), recalculated Ptolemy's *Amalgest* and built Germany's first observatory, later capitalising on the newly invented printing press by producing copies of his *Epherimides*, subsequently used by Christopher Columbus himself. Indeed, Columbus' prediction of a lunar eclipse, interpreted as the disappearance of a god, sufficed to threaten natives into providing food to ensure the Moon's return.

Cliff Palace in Mesa Verde National Park, USA: built in the AD 1200s by the Anasazi who sought to understand the cosmos.

But it was the Polish canon Nicolaus Copernicus (1473–1543), a reader of Regiomontanus and admirer of Columbus, who truly turned the astronomical tide. Frustrated by confusing epicycles and religious-controlled crystalline spheres, the 16th century was a time for untangling the heavens. Copernicus would concentrate on planetary distance, speed, size, and ratio, and it soon became spectacularly clear: all planetary paths revolved around but one central source ... the Sun. It was both revelation ... and revolution. An astronomical atlas was born ... but remained unpublished, for years. When *De Revolutionibus* was finally printed, it sparked a century of argument and deliberation. It confounded, was at odds with skies that appeared perfect and preordained. Many clung to the certainties of Aristotle and Ptolemy. But when a comet ripped through the heavens in November 1572, and subsequently decayed, its repercussions resounded and eroded the shores of certainty. All was not perfect. The heliocentric theory tantalised.

Inspired by a comet, a 1560 eclipse, and another exploding supernova, it was a golden-nosed Danish aristocrat and wealthy alchemist, Tycho Brahe (1546–1601), who took up the cudgel in the form of a large wooden sextant. By studying parallax, he proved the Universe was inconceivably distant, that current dogma was wrong, and he would correct it. Utilising giant 'celestial castles' at Uraniborg and Stjerneborg he filled them with instrumentation including compasses, a steel quadrant and his giant celestial globe. Yet, despite calculation and observation, his Tychonic system still set Earth at the centre. But now the spheres were no longer solid; a new understanding was needed.

In an age of crippling Reformation and stifling Catholicism, that epiphany would come after a fateful meeting in February 1600 between Brahe and the mathematical mind of Johannes Kepler (1571–1630). With the help of Brahe's observations and contributions on magnetism from an Englishman, physician and philosopher William Gilbert (1544–1603), the introvert German became the first in human history to comprehend the motion of the planets and how the Solar System worked. His famous three laws form the basis of space exploration today. Through his courage and dedication astronomy made a physical connection not seen since Hellenisation. Planets orbited the Sun in ellipses, not circles, at varying speeds and were driven by an understandable force. The science of physics was resumed. But now a threshold for naked eye viewing had been reached. Where to next, and why?

To Holland, and a spectacle-maker, Hans Lippershey (or Lipperhey), in the summer of 1608. Even though Venetian

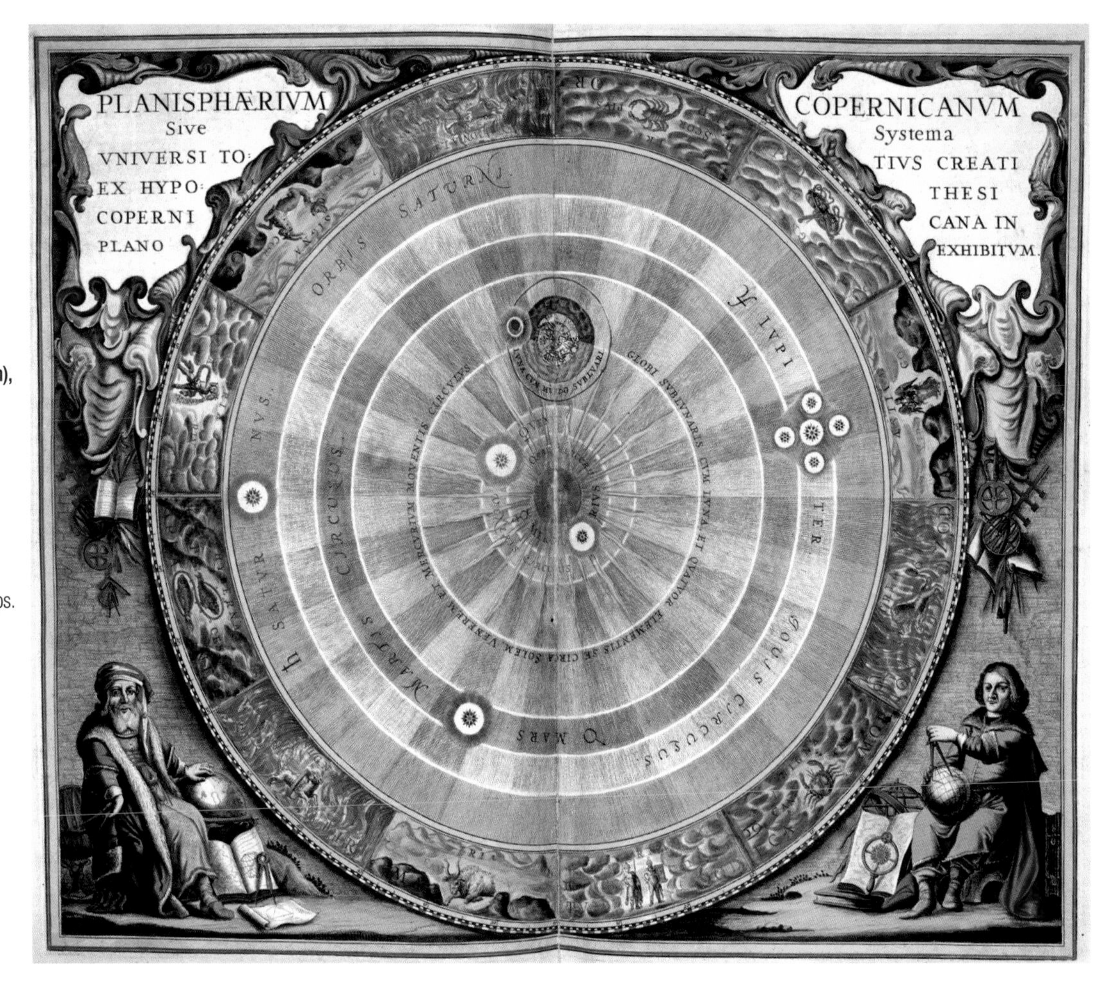

A 1708 chart depicting the Copernican system: with the Sun at the centre of the Universe surrounded by the orbits of Mercury, Venus, Earth (with Moon), Mars, Jupiter and Saturn. Copernicus is seen, bottom right. (The British Library Board Maps. C.6.c.2, after 22)

craftsmen had known for a couple of centuries that lenses made objects appear closer, it was Lippershey's combining them that drew items closer still … and, in the hands of the 45-year-old Galileo Galilei (1564–1642) at the end of 1609 and start of 1610, 'closer' would encompass the Solar System. Although an English explorer, mathematician, and astronomer, Thomas Harriott (1560–1621), turned his 'Dutch Trunke' to the Moon in July 1609, before the Italian, it was Galileo who understood and recorded what he saw: Jupiter and its moons – a miniature solar system analogous with Earth and its planets – rotating solar sunspots, the phases of Venus, Saturn's rings, and the 'innumerable stars' of the Milky Way. His observations accorded with Copernican theory and he was outspoken in its support, as was the philosopher and cosmologist Giordano Bruno (1548–1660). It was an endorsement that would force Galileo before the 1633 Inquisition, induce his contrition and sentence him to house arrest. Bruno, who never renounced his opinions and held other heretical views, had no such lingering demise – he was burned to death at the stake.

But heliocentric astronomy prevailed and in the 18th century science triumphed. In his rooms at Trinity College, Cambridge, Sir Isaac Newton (1642–1727) experimented with prisms, overturning classical optics by using a 'pinhole' to split and reconstitute Aristotle's 'pure' heavenly white light into its constituent 'impure' colours. After pondering the famous 'apple falling to the ground' he went on to astound with his theories of gravity in the three-volume *Principia*, proclaiming that all 'out there' answered to this force. New telescopes were invented, including his now famous revolutionary reflector that reflected, rather than refracted, light and at last removed obscuring chromatic aberration.

With King Charles II's need for improved naval navigation came the Royal Observatory, Greenwich Mean Time, and the Meridian. Science was being tackled by the greats: Reverend John Flamsteed (1646–1719), the first Astronomer Royal of the Royal Observatory; Robert Hooke (1635–1703); Edmond Halley (1656–1742), who predicted the return of his now eponymous comet; Sir Christopher Wren (1632–1723), and others. Consequently, Earth became safely navigable, the Sun was deemed 93 million miles away, and the Solar System, to its farthest known planet, Saturn, was safely understood … until the later discovery of more: Sir William Herschel's (1738–1822) gas giant Uranus in 1781; John Couch Adams (1812–92) and Urbain J.J. Le Verrier's (1811–77) predictions that led the Berlin astronomer Johann Gottfried Galle (1812–1910) to identify Neptune in 1846; the now demoted dwarf planet, Pluto, discovered by Clyde Tombaugh (1906–97) in 1930.

The cosmos beckoned. Astronomy was an unfettered movable feast. Monster telescopes enabled more and more objects to be found … and to confound. In the 19th century, questions long unasked – the true distance and nature of stars – were not only asked, but answered. The conclusion: stars were mind-cripplingly distant. Friedrich W. Bessel's (1784–1846) 61 Cygni was 60 million million miles away. Friedrich Struve's (1793–1864) star, Vega, was almost three times farther.

The journey had started: Sir John (son of Sir William) Herschel's photographic developments would capture progress: in 1802, English chemist William Wollaston (1766–1828) repeated Newton's experiment but substituted his 'pinhole' light source with a fine slit and found inexplicable black lines in sunlight: a self-taught Munich physicist, Joseph Von Fraunhofer (1787–1826) devised the modern spectroscope; replacing a prism with a diffraction grating in 1821 he systematically mapped 574 of these dark, but still mysterious, 'Fraunhofer' lines. In 1849, Leon Foucault (1819–68) discovered they were sodium, but it was two Germans, Gustav Kirchoff (1824–87) and Robert Bunsen (1811–99), who realised each element produced its own set of bright emission or dark absorption lines – unique fingerprints – in their gases, and that sodium itself was produced in our Sun.

We soon started to identify many elements in our Star, not least of which was helium, discovered by England's self-taught astronomer Sir Norman Lockyer (1836–1920). Stellar composition was dissected, understood and, through the efforts of Annie Jump Cannon (1863–1941) and Cecilia Payne-Gaposchkin (1900–79), later classified and catalogued. Astrophysics had arrived. We learned that, like us, stars have lives. They were born in giant molecular clouds and lived by fusing hydrogen atoms into helium to produce prolific heat and light; a journey in itself lasting half a million years. They died, sometimes catastrophically, like the supernova witnessed by the ancients, or they lingered, decaying slowly over millennia, their elements dispersing for future nuclear reactions ... future stars ... future everything.

Technology has taken us 'out there'. In the 20th century it was telescopes, specifically ever-expanding giant telescopes mirroring their visionaries: the Third Earl of Rosse, William Parsons' 72-inch 'Leviathan of Parsonstown' at Birr Castle, Ireland (built 1845); George Ellery Hale's (1868–1938) 40-inch Yerkes refractor (still the largest ever made) and 200-inch Palomar reflector (built 1948); and John D. Hooker's 100-inch (2.5m) aluminised mirror at the Mount Wilson Observatory (built 1917). Suddenly fuzzy nebulae (Latin for 'cloud') fused into spirals, could be sketched, photographed, and studied. What were they? Stars with equal luminosity and period brightness – Henrietta Leavitt's Cepheids – were used as 'standard candles' to quantify their distances over the grandest scale of all. Together with the work in the 1930s of the ex-lawyer, Edwin Hubble (1899–1953), and the once mule-driver Milton Humason (1891–1972), using the great Hooker scope, these 'extra galactic nebulae' or 'island universes' were resolved into ... galaxies ... many, many galaxies. Now known to number over 250 billion, each galaxy is phenomenally distant and home to billions of stars. Studies of their spectra revealed they were moving, rushing away in all directions, the most distant rushing faster still. Mankind was truly seeing 'out there'. Like the shell surrounding an unborn chick, the sky had cracked open and in flooded light, learning, and scientific liberation.

The 100-inch mirror Hooker Telescope, Mount Wilson, California: the world's largest telescope until 1948 and, having been upgraded, still in use today.

We asked what could it be, how could this 'new' Universe have evolved? Theories abounded, among them the astrophysicist and cosmologist Fred Hoyle's (1915–2001) Steady State Theory suggesting constancy; Georges Lemaître's (1894–1966) Big Bang, suggesting creation of all from nothing; and Einstein's great General Theory of Relativity, likening space to a tensile fabric spanning the Universe. But where was the evidence? It would come with the invention of radio telescopes and their ability to 'see' deeper than optical 'eyes'. If the Universe was cooling down, variations in a sea of background radiation had to be detectable. In 1965, these legacy radio waves were inadvertently 'found' by Arno Penzias (b.1933) and Robert Wilson (b.1936). The 1990s saw the sophisticated COBE and WMAP satellites, and in 2009 ESA's Planck Infrared Space Observatory, map this 'Background Radiation' – the evidence of the primordial 'soup' whose ingredients were fluctuations and irregularities, mysterious 'dark matter' and elusive 'dark energy', mind-melting distances and increasing expansion ... a menu of 'all' and from which all had been created. Could there be any more, any further? Where did it all 'go'?

Who can know? The harnessing of other energies across the electro-magnetic spectrum continues to open up the 'out there'. But, like a glass ceiling, with increasing technology the observable threshold slips tantalisingly from our grasp. Yet ... once, the night sky was completely beyond us. This is now no longer. The Milky Way galaxy was once our Universe. No longer. Is the Universe our Universe?

Our minds imagine and our ingenuity triumphs. The mighty Hubble Space Telescope transformed vision. Its successor, NASA/ESA's James Webb Space Telescope, will do more. Unmanned spacecraft have orbited and mapped, sniffed and sampled, analysed and blasted surfaces of other worlds. Robotic rovers, *Pathfinder*, *Spirit*, *Opportunity* and *Curiosity* have scrabbled along Martian crevices and crawled into craters, while the stationary *Phoenix* scratched at the very possibility of life. Emissaries bearing pioneering names of the past – Gian-Domenico Cassini (1625–1712), Christiaan Huygens (1629–95) and Galileo Galilei – have graced the giant outer planets, hurtling into their atmospheres or immersing in their moons. NASA's *Pioneer* and *Voyager* probes head now for the heliopause – the boundary of the Sun's magnetic influence – and intergalactic space. More missions to the gas giant planets, and their satellites, are planned. Sophisticated and probing robotic rovers, such as NASA's *Curiosity*, have left the drawing board. A new Solar Dynamics Observatory already offers instantaneous and startling images of our ever-changing Sun.

Our Star, once a revered unquestioned god, is now intimately understood; it lives and breathes. Exoplanets – planets around other stars – are now sought and swiftly found, with over 2000 on record at time of going to press. Innovative telescope technology, such as NASA's *Solar Dynamics Observatory*, offers instantaneous and startling images of our ever-changing Sun. Mercury has been mapped by NASA's *Messenger* spacecraft. New missions now speed to the gas giant planets, and their satellites, with more planned. ESA's *New Horizons* spacecraft has flown past Pluto, the dwarf planet sentinel of the outer solar system, transmitting stunning previously unseen images, and now heads for a distant Kuiper Belt object. *Rosetta* has orbited Comet 67P Churyumov-Gerasimenko and despatched a lander, *Philae*, to its surface. NASA's *Voyager 1* probe has left the solar system and crossed into interstellar space. *Voyager 2* will follow.

The Karl G Jansky Very Large Array, New Mexico: 27 linked radio telescopes listen for signals in outer space.

Should we keep our eyes averted? No! We must look up. We owe it to these pioneers, to Armstrong's footsteps in the lunar dust, to ourselves and to this wondrous creation. As tiny and inconsequential as Dr Carl Sagan's 'pale blue dot' of an Earth compared to the cosmos, this book is just a beginning … an introduction and practical journey … for all … for ALL. Open those gallery doors, step inside, and *see*.

Hubble Space Telescope: after its final upgrade, and with its aperture door open, the HST views the Universe from above Earth's restricting atmosphere and across a wider range of the optical spectrum to reveal objects in stunning detail at staggering distances.

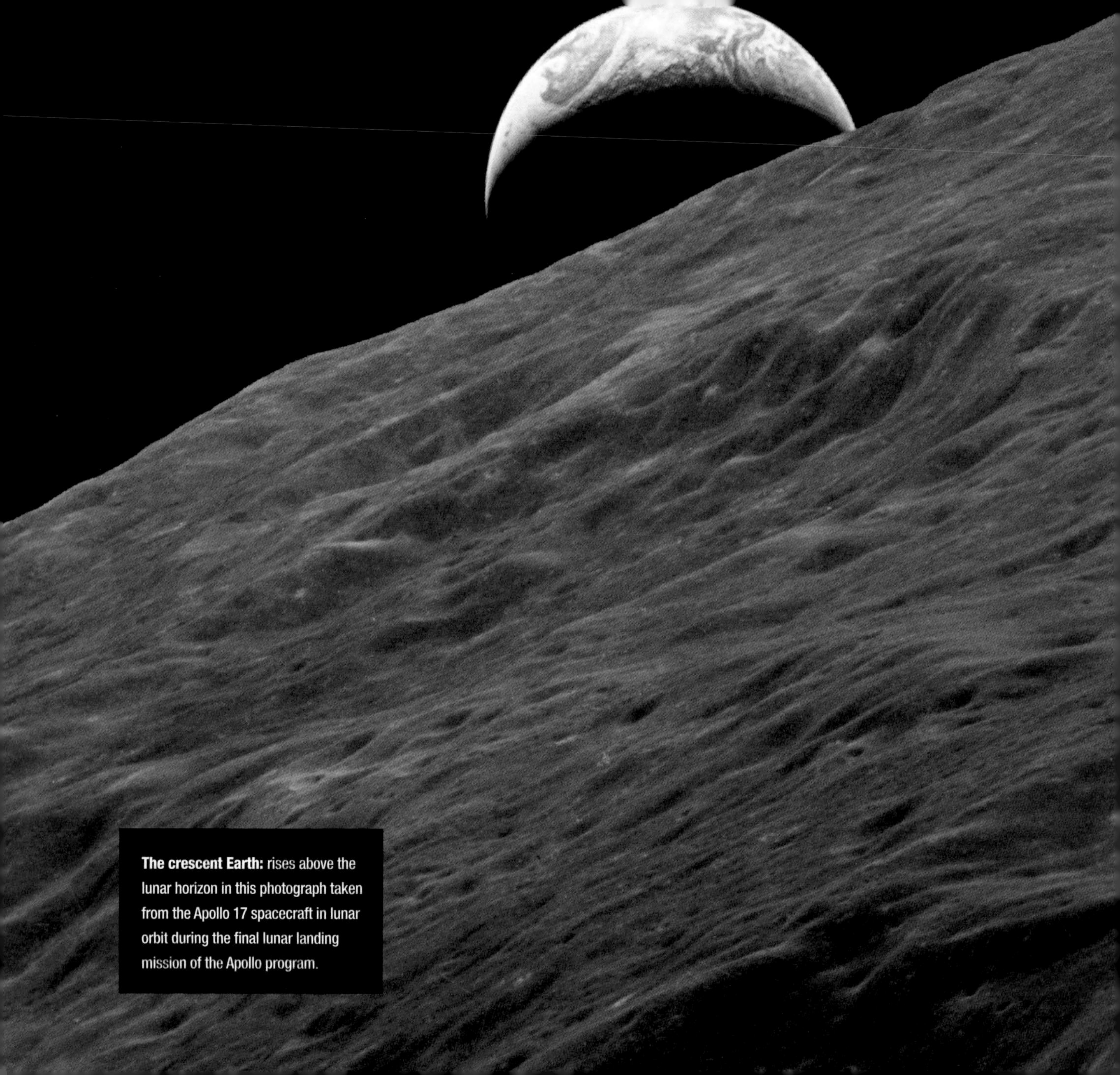

The crescent Earth: rises above the lunar horizon in this photograph taken from the Apollo 17 spacecraft in lunar orbit during the final lunar landing mission of the Apollo program.

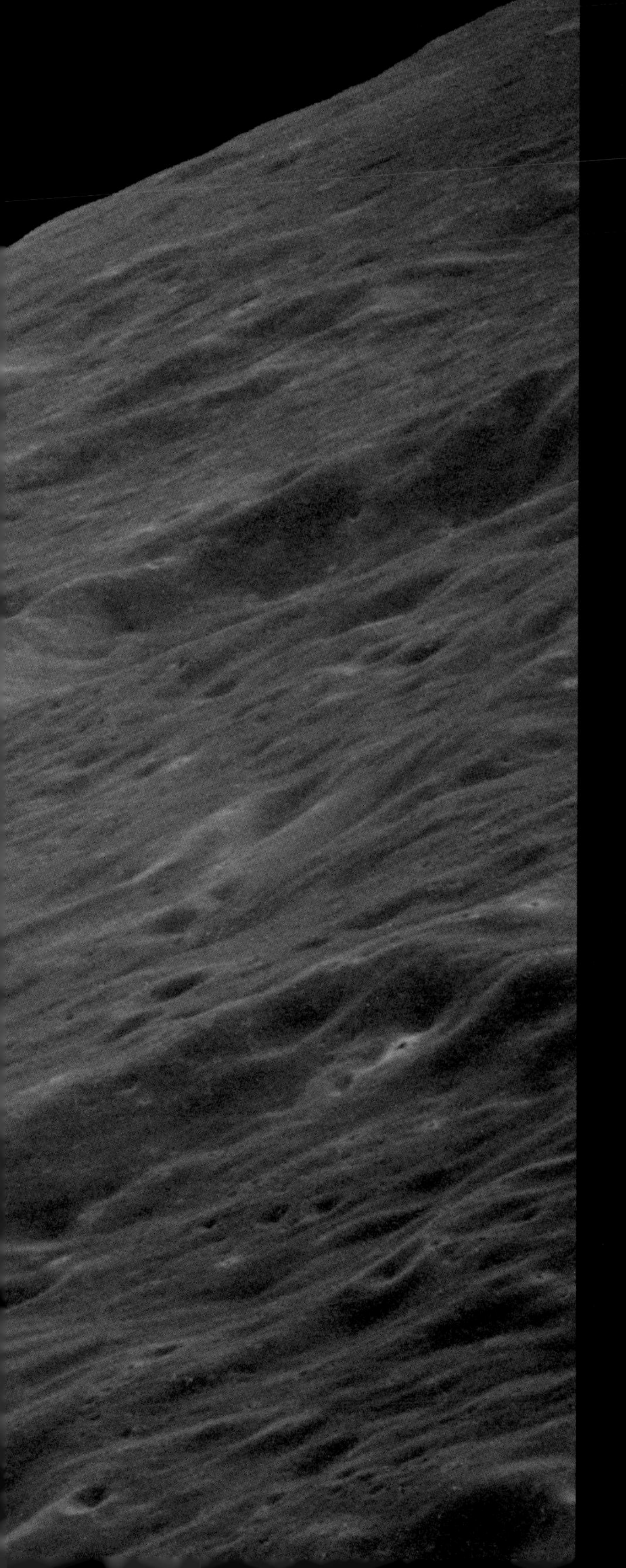

The Solar System

1

Our Sun is just another star located in the backwater of our Milky Way Galaxy. Our Galaxy contains over 400 billion suns but is itself a speck of froth on a cosmic ocean. The cosmos is home to more than 250 billion known galaxies, each home to billions of stars. In fact, there are, in aggregate, more stars 'out there' than all the grains of sand on all the beaches on 10,000 Earth-like planets. A paralysing thought. Yet, place just three of those grains inside a vast cathedral, and the cathedral is still more tightly packed with sand than space is with stars. Although enormous, they appear tiny when compared to the space between them: draw a line from our planet right through the observable Universe and it will probably never pass through a star! They appear 'close' but they are very, very far away...
and astronomers are still exploring...

'Who, my friend, can scale Heaven?'

The Epic of Gilgamesh,
Sumer, third millennium BC

The Sun

Let's start with our Sun – our insignificant star. Remember, all stars are suns. Ours appears immense because … it is. It is the most massive object in our Solar System, 330,000 times more massive than our 'pale blue dot' of a planet; 109 Earths (diameter 7,926 miles/12,756km) would fit comfortably across its diameter and, in terms of volume, a million would fit inside. It also appears enormous because we are 'close' to it in astronomical terms: an average of 93,000,000 miles (150,000,000km) away, about 400 times more distant than the Moon – one Astronomical Unit (AU). But it is vital to keep perspective: if the Sun was scaled down to the size of a basketball then our planet would appear no larger than a grain of rice, and we would have to place that grain some 30 metres away to truly conceptualise the scale of Earth's distance and smallness.

■ BIRTH

But what is the Sun? How and when was it born? How does it live? How and when will it die? When we gaze up at night, we see what appears to be gaps or dark holes amidst the stars. These are giant clouds of cold gas and dust. We see them because they 'block' the light from more distant stars. Comprised of tiny grains of ices, carbon-rich molecules, and minerals, these festoon the Universe and provide the raw material for star formation. How?

Just as Earth takes 24 hours to rotate once on its axis, so everything else 'out there' is moving. Space is a chaotic maelstrom. Turbulence swirls and eddies like cream in coffee, but on a truly epic scale over staggering distances. This tumult, caused by heat, shock waves from star explosions, magnetic fields or the movement of the Galaxy itself, initially prevents a cloud's collapse, but over tens of thousands of years the clouds cool, become less turbulent, and magnetic fields fade. Dense clumps form in this new quiescent climate and gravity takes over to induce collapse. A proto-stellar 'seed' forms from the tiny dust grains – a process reliant on their individual gravity and taking millions of years – and the centre of the clumps heat up as pressure bounces the particles against each other. To conserve momentum, the entire cloud – or solar nebula – flattens into a rotating disc. Like a spinning ice-skater with outstretched arms, the disc spins rapidly but, as it condenses, like the skater pulling her arms to her sides, it spins faster and faster. Slower spinning material discharges on to the central baby proto-star whilst fast-spinning material is thrown off at the poles in giant perpendicular jets – ionised bipolar outflows. These dissipate momentum and reach velocities of over 220,000mph (354,000kph), extending vast distances into space. This in turn allows more material to feed the proto-star, a process taking tens of millions of years. This accretion disc, as it is known, can be 100 to 1,000AU in diameter.

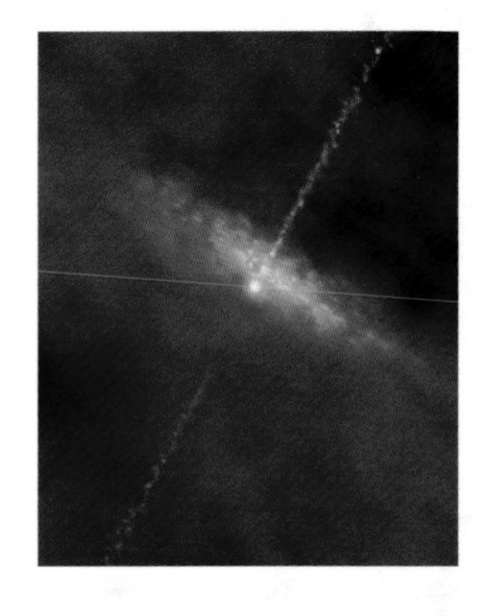

The proto-star: gas jets emanating from a proto-star in the solar nebula.

The 'Black Cloud' Barnard 68: a dark molecular cloud 'blocking' light from distant stars.

Eventually, the jets and radiation clear away the surrounding cloud until little gas is left. The nascent sun slowly contracts over the next 100 million years, increasing its core pressure and temperature to millions of degrees and it is at these high temperatures that hydrogen atoms fuse to create helium – nuclear fusion. A massive nuclear explosion ensues, producing heat and light: a star is born. This is the birth of our Sun, starting some 4.6 billion years ago. It is a nuclear powerhouse. Having settled into its stable hydrogen-burning existence – known as the main sequence – it converts 600 million tonnes of hydrogen into 596 million tonnes of Helium every second, radiating the missing 4 million tonnes in the form of photons (particles of light energy) into the Universe.

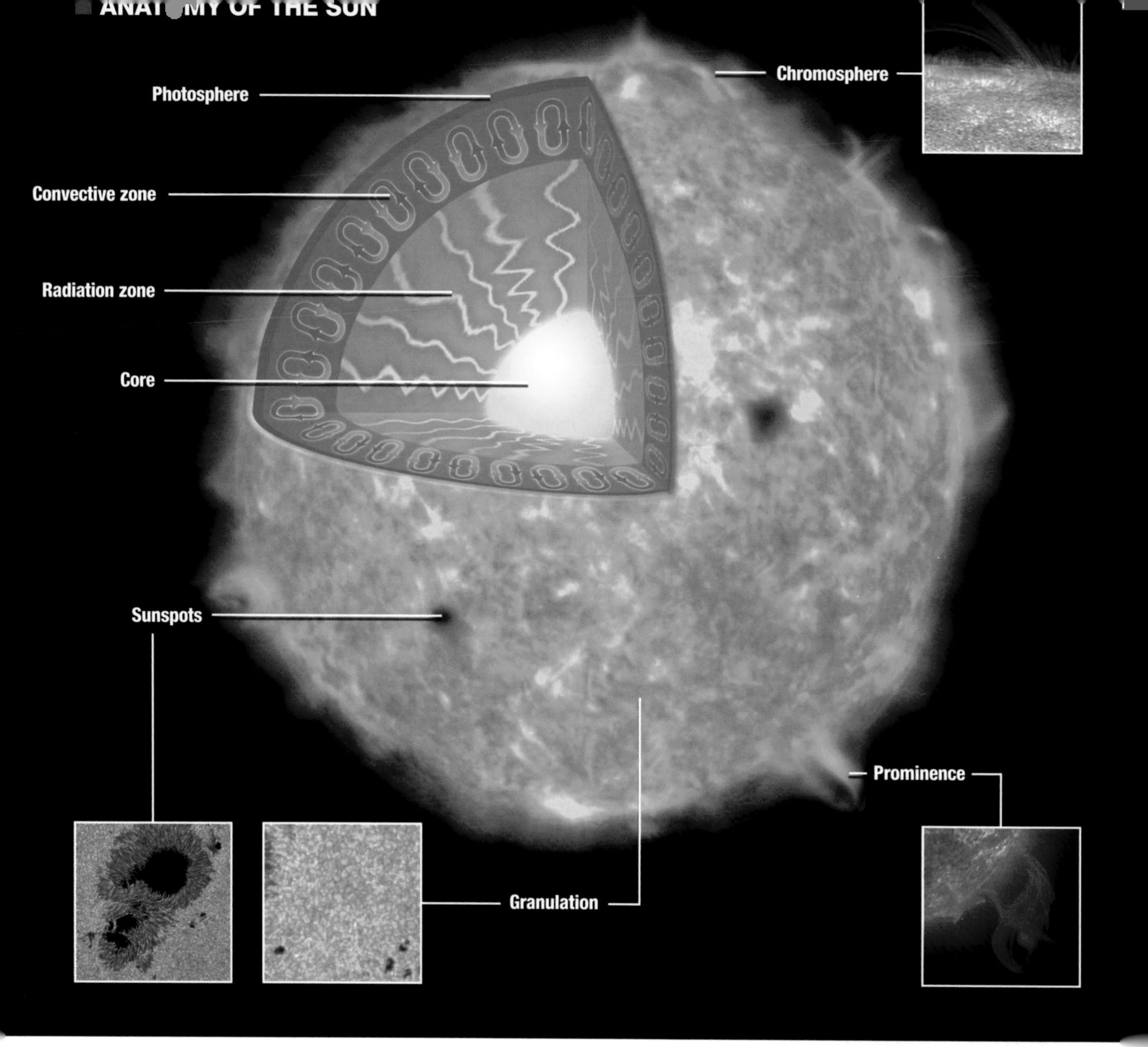

COMPOSITION AND ATMOSPHERE

Energy is generated in its core (250,000 miles wide), at temperatures of around 15 million degrees, and radiates from atom to atom to within 60,000 miles (96,500km) of the 'surface'. This is known as a 'random walk' and takes around a million years. It then convects this energy to the surface, the photosphere, via enormous columns of gas. Looking down into these gas columns – granulation – is like looking through billions of straws, each some 800 miles (1,287km) wide and contributing to a larger supergranulation pattern, forming a broiling cauldron some 250 miles (400km) deep.

Above the photosphere is the pink-tinted tenuous chromosphere through which tongues of gas – spicules – lick to a height of 6,000 miles (9,650km) and beyond that the corona; a rarefied halo of gas at some 1,000,000°C extending millions of miles into space. But the Sun's influence extends way

ELEMENTS IN THE SUN

Element	% by number of atoms	% by mass
Hydrogen	92.0	73.4
Helium	7.8	25.0
Carbon	0.03	0.3
Nitrogen	0.008	0.1
Oxygen	0.06	0.8
Neon	0.008	0.05
Magnesium	0.002	0.06
Silicon	0.003	0.07
Sulphur	0.002	0.04
Iron	0.004	0.2

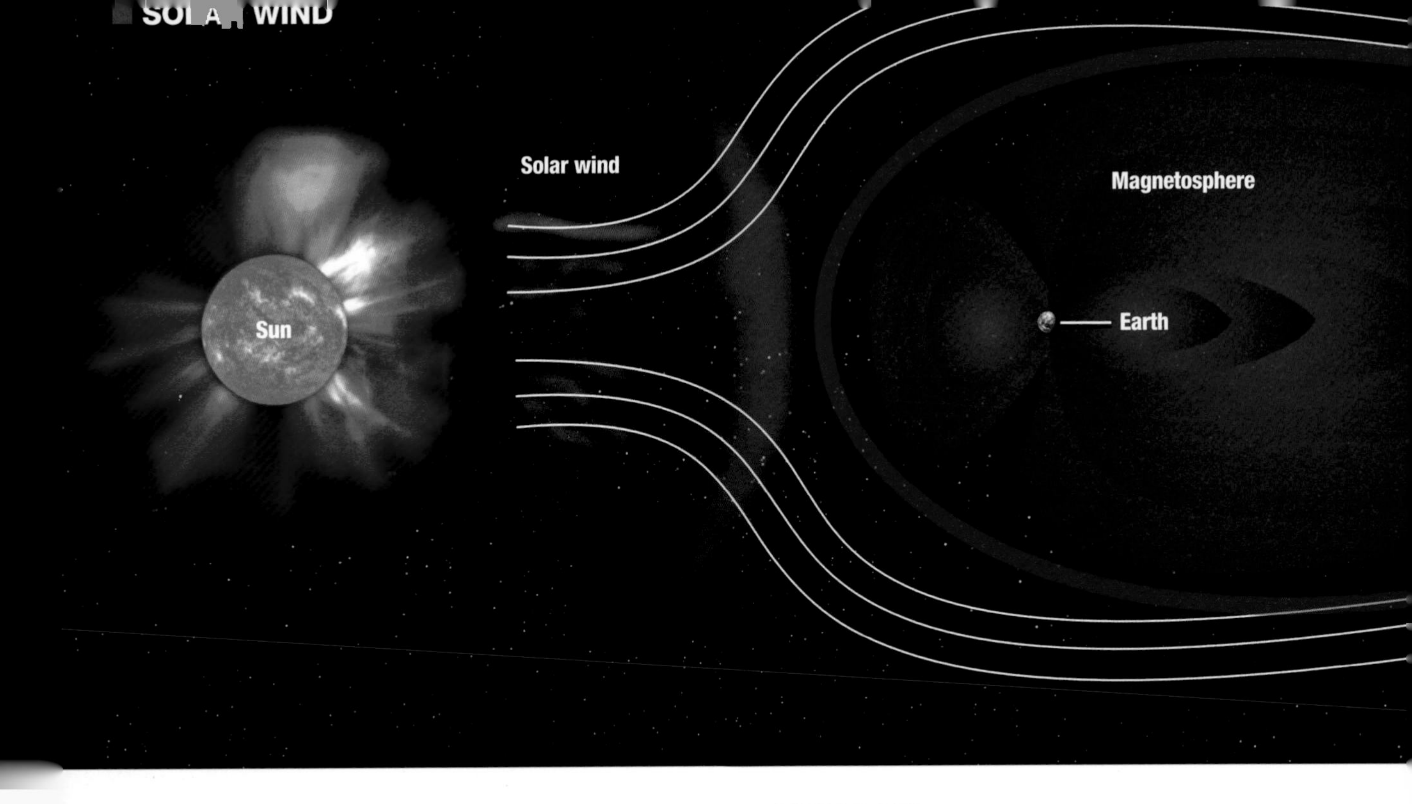

Solar wind: buffets objects in the Solar System. Earth is shielded by its magnetosphere; a large cavity within which the planet's magnetic field dominates over interplanetary magnetic fields.

beyond this atmosphere in the form of charged particles and plasma (mostly electrons, protons and alpha particles). Known as the solar wind, these gases, too hot to be confined by solar gravity, escape from the upper atmosphere. There are two components: the slow wind moving at approximately 400 km/s, and the fast – originating in coronal holes – at around 750 km/s.

The Sun, too, has looping magnetic field lines that need to reconnect. If Earth is in the way, these lines will break and use our planet, like a magnet, to reconnect and continue the charge. The Sun's charged particles interact with our upper atmosphere, particularly at the poles, and are drawn downward. They lose energy but become sufficiently 'relaxed' to regain it and re-emit the energy in the form of light, creating magnificent auroral displays – the Aurora Borealis and Aurora Australis – the northern and southern lights.

Aurora Australis: red and green colours predominate in this view of the Aurora Australis photographed from the Space Shuttle Discovery in May 1991.

SUNSPOTS

The Sun maintains its shape through the phenomenon of hydrostatic equilibrium: gravity pulls material inward whilst radiation pressure from internal thermonuclear reactions pushes outward – a balance at every point is achieved. But, since it is a gaseous body, it experiences differential rotation, taking 25 days to spin once at the equator and 36 days at the pole.

When the solar wind reaches Earth it interacts with the planet's magnetic field. Particles from the Sun enter the Earth's magnetotail and, under specific conditions, are accelerated back towards the planet where they enter the atmosphere at the polar cusps. Here, they excite molecules of atmospheric gases which, depending on their composition, emit light of specific colours to produce the magnificent auroral displays of the Aurora Borealis and Aurora Australis – the northern and southern lights.

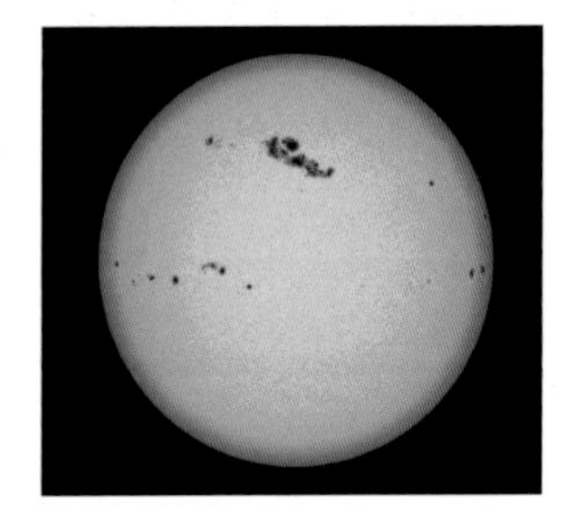

Sunspots: 'dark, cooler blemishes' couriered by the Sun's rotation.

Just as all 'out there' has a life cycle so sunspots follow an approximate 11-year period – solar cycle – alternating between a minimum, with a few spots some 30° above the equator, to a maximum, with well over a hundred spots drifting to around 5° latitude. There is a 22-year magnetic cycle too: leading spots in the northern hemisphere of the Sun often have a uniform north-seeking polarity whilst the opposite occurs for those below the equator. After 11 years, the polarity is reversed. Those in the

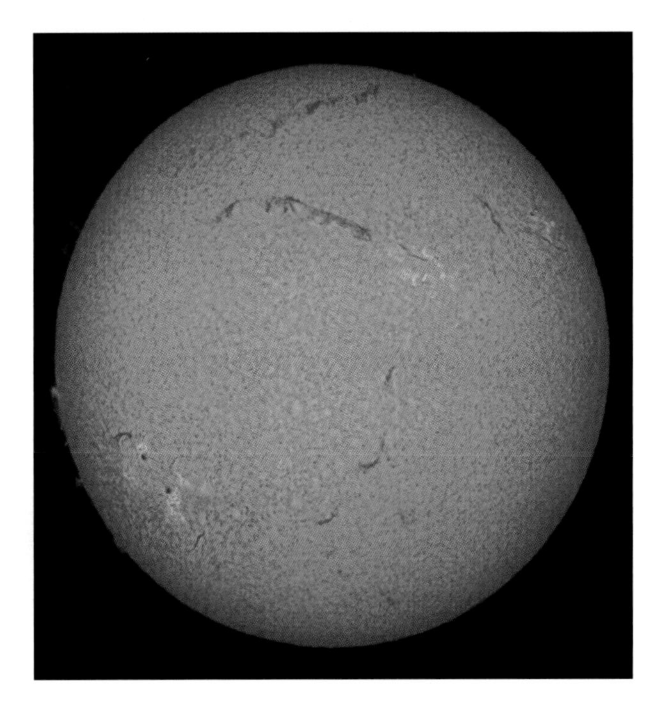

A Hydrogen-Alpha view in this image capturing the red light of hydrogen: dark ribbons, known as filaments, become filaproms when viewed straddling the limb, and prominences when viewed beyond the Sun's 'edge'.

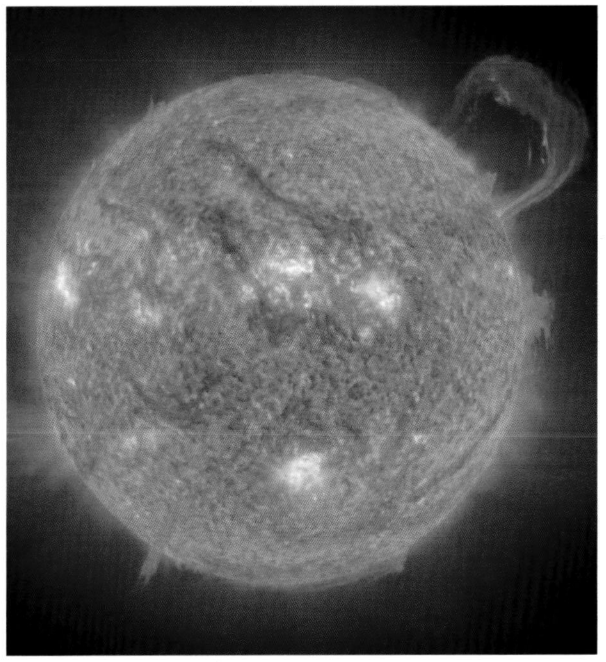

An enormous prominence: appearing as a giant loop of cooler gas, it will either drain down onto the surface or erupt into space, typically releasing as much energy as a few million hydrogen bombs.

north become south-seeking and vice versa.

PROMINENCES

Linked with sunspots are many other activities. Most notable are the spectacular coronal prominences that appear as red, flame-like protuberances rising high above the Sun's limb out to distances of over a million miles. Rising in graceful arcs and following magnetic field lines, they can last from hours to days, blasting matter away from the Sun at speeds of over 1,000mph before cooling and falling back to the photosphere.

FLARES

The most awesome events on the surface of the Sun are solar flare explosions where ions and electrons race away from the surface, crash into the chromosphere, and transform moving energy into heat. This heat then evaporates into the corona, increasing gas density and radiating it out into space. A typical flare lasts for five to ten minutes and releases a total amount of energy equivalent to a few million hydrogen bombs. Larger flares last for hours and emit enough energy to electrically power the USA for 100,000 years. One such flare, on 6 July 1996, released a total amount of energy 40,000 times greater than the 1906 San Francisco earthquake:
an everyday hiccough in the life of an average star (see page 19)!

Flares arise when magnetic fields pointing in opposite directions interact and destroy each other, releasing energy. If interactions are sufficiently immense, then enormous quantities of coronal material can be ejected at speeds of over two million mph into interplanetary space – coronal mass ejections. Like the solar wind, when this material reaches Earth's upper atmosphere there is increased auroral activity and sometimes major disruption: electric currents change, power stations burn out, cities are without power
for hours and radio transmissions are interrupted. Increased heat causes Earth's upper atmosphere to expand, creating friction for spacecraft and inducing drag that slowly brings down low-orbiting satellites and rocket debris.

SUNQUAKES

Like earthquakes, the after-effects of such flares on the Sun are sunquakes; high-energy electrons race down the magnetic field lines back on to the Star's surface and create compression waves. These radiate outward like ripples on a pond, from 35,000mph (56,327kph) to a massive 400,000mph (643,700kph). They even travel through the Sun to recombine on the far side.
Such seismic quakes are 'audible' to astronomers (see *Solar and Heliospheric Observatory*, page 158). Consequently, it is known that the entire Star 'breathes'in and out every six minutes.

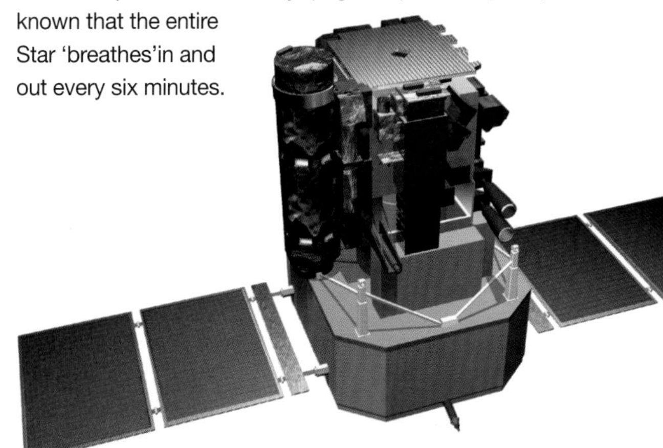

SOLAR ECLIPSES

A solar eclipse occurs whenever a New Moon passes directly between the Sun and the Earth. Since the Moon's orbit is inclined around 5° to the ecliptic, we do not witness an eclipse at every New Moon. Mostly, the Moon passes above or below our Star. Eclipses of the Sun can be total, partial or annular depending on the area covered by Earth's natural satellite.

A total eclipse of the Sun, whether lasting just a few seconds or as long as seven and a half minutes, is an unforgettable event seen only by travelling to a place where the narrow shadow, or umbra, of the Moon strikes Earth's surface. As our planet rotates, this shadow traces a dark path up to around 200 miles (300km) wide and thousands of miles long. Anyone observing within this narrow pathway will witness the total eclipse. During an eclipse daylight darkens, temperatures perceptibly lower, a breeze will ease, wildlife quietens, and planets, such as Venus and Mercury, may become visible. As the diminishing sliver of sunlight heralds the final few seconds, the phenomenon of Baily's beads (after English astronomer, Francis Baily, who first described them, in 1836) blazes forth; sunlight blasts between the mountains on the lunar edge resembling a glistening necklace of pearls. Occasionally, one 'pearl' outshines the others, giving rise to the diamond ring effect.

With the Sun completely masked by the Moon, the stunning corona – the outer atmosphere of tenuous gas extending millions of miles into space, normally lost in the intense glare – reveals its blinding splendour. Giant pinkish prominences of hydrogen gas lick away from the Sun's limb. Then, with the flash of the diamond ring, Baily's beads reappear and the Moon slowly heaves onward in its orbit. The eclipse is at an end.

Due to a remarkable coincidence – the Moon is about 400 times smaller than the Sun, but it also happens to be about 400 times closer – the Moon and Sun appear virtually the same size in the sky, but the Moon's distance and, therefore, angular size will change as a result of its elliptical orbit. When the Moon is farthest away – at apogee – it is too small to cover the Sun. An annular eclipse results, so-called because a ring, or annulus, of sunlight is left around the Moon's disc. For guidance on how to safely observe a solar eclipse see Chapter 3.

The shadow of the Moon: as seen from the International Space Station, 230 miles above the planet (29 March 2006).

FUTURE TOTAL SOLAR ECLIPSES

Date	Duration of totality	Where visible
9 March 2016	4.5 mins	Indonesia, Pacific Ocean
21 August 2017	2.7 mins	Pacific Ocean, USA, Atlantic Ocean
2 July 2019	4.5 mins	South Pacific, South America
14 December 2020	2.2 mins	South Pacific, South America, South Atlantic Ocean
4 December 2021	1.9 mins	Antarctica
20 April 2023	1.3 mins	Indian Ocean, Indonesia
8 April 2024	4.5 mins	South Pacific, Mexico, Eastern USA
12 August 2026	2.3 mins	Arctic, Greenland, North Atlantic, Spain
2 August 2027	6.4 mins	North Africa, Arabia, Indian Ocean
22 July 2028	5.1 mins	Indian Ocean, Australia, New Zealand
25 November 2030	3.7 mins	South Africa, Indian Ocean, Australia
30 March 2033	2.4 mins	East Russia, Alaska
20 March 2034	4.1 mins	Nigeria, Cameroon, Chad, Sudan, Egypt, Saudi Arabia, Iran, Afghanistan, Pakistan, India, China
2 September 2035	2.5 mins	China, Korea, Japan, Pacific

TOTALITY SEQUENCE: FOUR CONTACTS

A total solar eclipse has four stages: first contact, second contact (or start of totality), third contact (or end of totality), and fourth contact.

Partial Phase: after first contact, when the lunar disk first 'bites' the solar disk, the Moon takes roughly an hour to slowly slide across the face of the Sun.

Diamond ring: sunlight blasts through the mountains on the lunar limb.

Baily's beads: the last rays of sunlight slip through the valleys on the lunar edge.

Second contact: with a flash of pinkish light from the chromosphere, totality begins.

TOTAL ECLIPSE

When the Moon's shadow hits the Earth's surface, a total eclipse results for observers within the area of the umbra. For those observing within the transition region area of the shadow, the penumbra, a partial eclipse results.

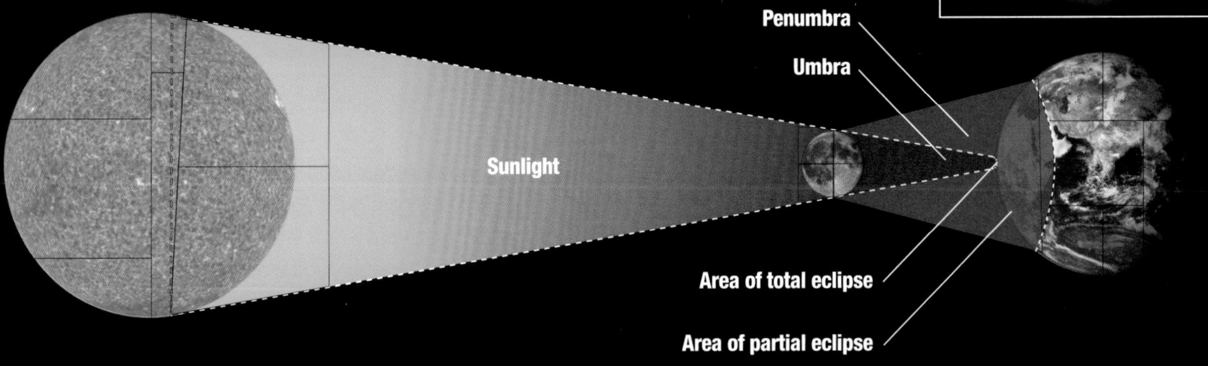

ANNULAR ECLIPSE

If the New Moon is at its farthest point in its orbit around Earth and passes in front of the Sun, an annular eclipse results. The Sun's disk will not be completely covered.

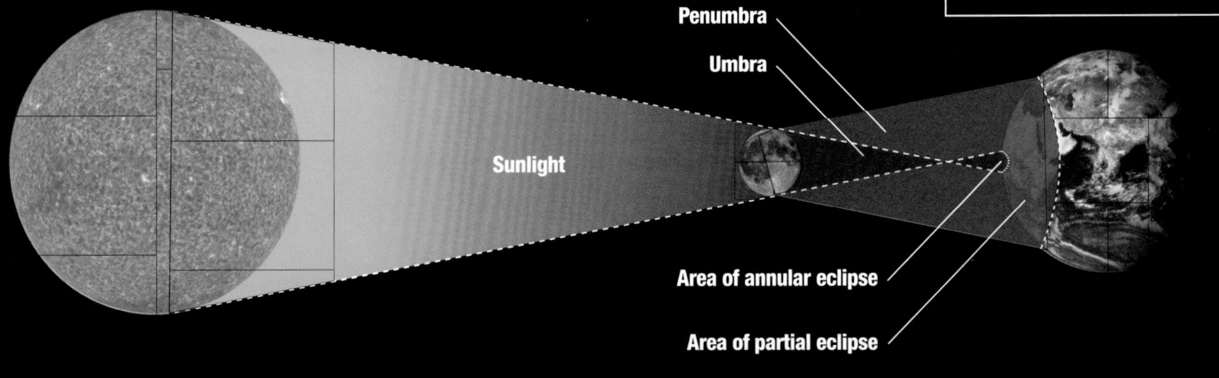

Totality: the only time when the Sun's corona, or outer atmosphere, is visible. Prominences – enormous gaseous eruptions – can be seen.

Third contact: a burst of light from the Moon's eastern edge signals the end of totality. The corona disappears as daylight slowly returns.

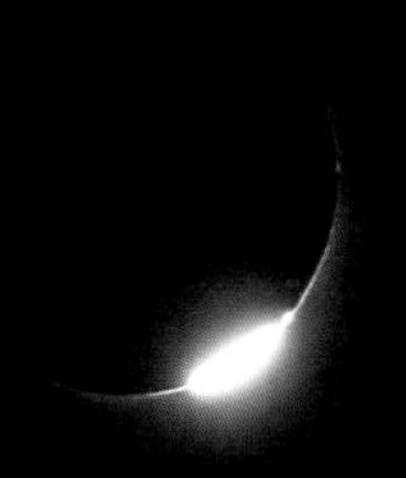

Baily's beads: sunlight again pushes through the lunar highlands.

Diamond ring: increased sunlight signifies the Moon's eastward transit but it will be around an hour before our satellite finally exits the Sun's disk – known as fourth contact.

The planets: fire to ice

The Sun's effects are felt not just by our planet, 93,000,000 miles away, but by all the other disparate siblings in our Star's neighbourhood: the eight planets, dwarf planets, moons, comets, asteroids, and other debris. How were they adopted into this fragile architecture? The answer is, they were not: each formed from the 0.2% leftover material in the molecular cloud that spawned our Sun.

■ BIRTH

Residual dust grains in the spinning disc surrounding the infant Sun collided and coagulated into larger bodies, like dust piling up beneath your bed. In the hotter area nearest the newborn Sun, metals and silicate minerals with high melting points became solid, forming bodies up to a certain size: the four small rocky planets: Mercury, Venus, Earth, and Mars. Beyond the 'ice line', where methane and water are solid, growing planets became sufficiently large to gather more gas, mainly hydrogen. This was how the gas giants Jupiter and Saturn evolved and the more distant giants of Uranus and Neptune. It is known as the accretion theory. But now conjecture and computer simulations question this 'neat' formation. Perhaps formation was not as straightforward in the 100 million years it took for the planets to form, but was instead an arena of frenzied activity, like a game of cosmic billiards. How could metre-sized objects coalesce if buffeting each other? They would ricochet like cosmic dodgem cars and crash into the Sun.

Rather than a gradual build-up, there may have been a sudden cloud collapse: giant Jupiter formed, but then gobbled the gas and dust to leave only leftovers for the inner rocky and outer ice planets. Perhaps Jupiter and Saturn formed closer to the Sun but migrated further out. Perhaps pockets of low turbulence created low-pressure vortices in the cloud, allowing material to amass. Perhaps Jupiter and Saturn experienced a conjunction, their interaction whipping Uranus and Neptune into the distant orbits they inhabit today. Other smaller bodies may have been grabbed by Jupiter's enormous gravity and hurled from the Solar System altogether, ending up in the hypothetical Oort Cloud (see *Comets*, page 47). Perhaps there were other planets the size of Earth batted way beyond the cosmic boundary; too distant to reflect sunlight and in wildly elliptical orbits 1,000 or even 10,000AU from the Sun. In this 'oligarchic' scenario they lurk in the Solar System's shadows, yo-yoing around the ecliptic crease. Some may have been whacked from the Solar System altogether

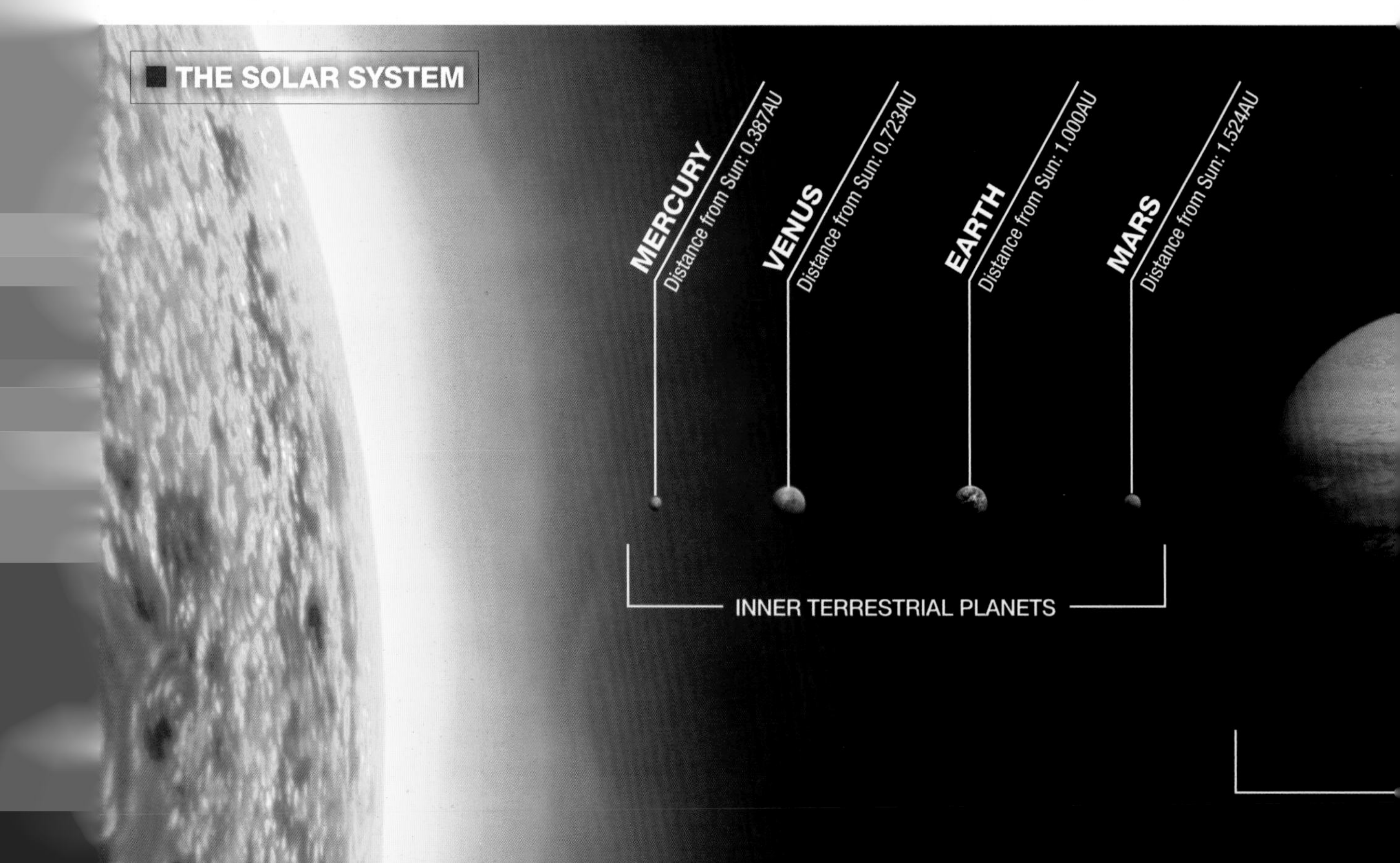

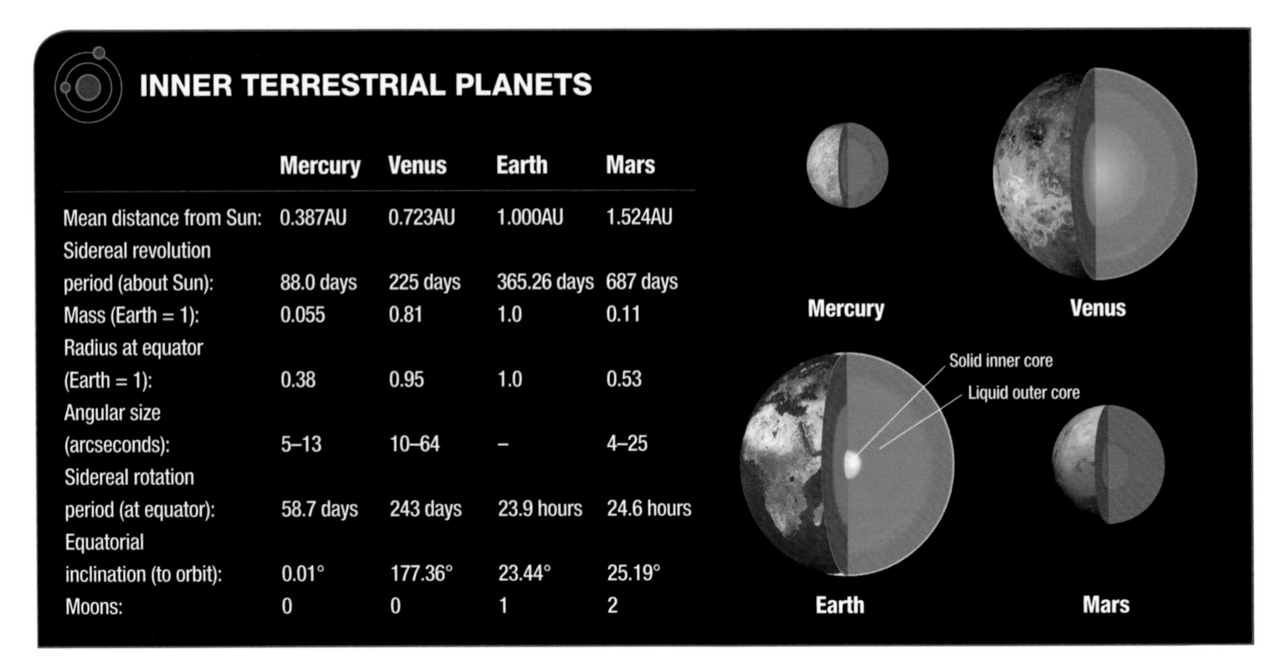

INNER TERRESTRIAL PLANETS

	Mercury	Venus	Earth	Mars
Mean distance from Sun:	0.387AU	0.723AU	1.000AU	1.524AU
Sidereal revolution period (about Sun):	88.0 days	225 days	365.26 days	687 days
Mass (Earth = 1):	0.055	0.81	1.0	0.11
Radius at equator (Earth = 1):	0.38	0.95	1.0	0.53
Angular size (arcseconds):	5–13	10–64	–	4–25
Sidereal rotation period (at equator):	58.7 days	243 days	23.9 hours	24.6 hours
Equatorial inclination (to orbit):	0.01°	177.36°	23.44°	25.19°
Moons:	0	0	1	2

– the planetars – to loop around the Galaxy in orbits taking tens, to thousands, to millions of years to complete.

Nowhere was spared; like a giant food mixer, ingredients smashed everywhere, especially in a period known as the Late Heavy Bombardment when Earth was pounded by meteorites, some four billion years ago, around 500 million years after the Sun formed. Yet … it is an incredible thought that everything within the gravitational grasp of the Sun formed from one inconsequential molecular cloud, from a grain of dust a micron (0.001mm) in size; something no larger than a particle found in smoke.

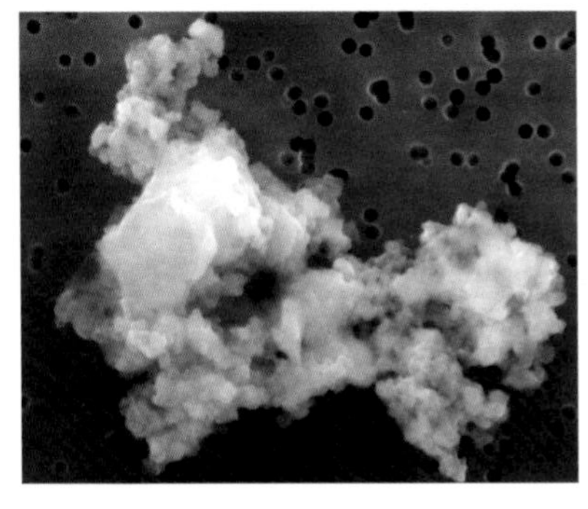

A grain of dust: measuring a tenth the width of a human hair this is the material from which our entire Solar System was formed.

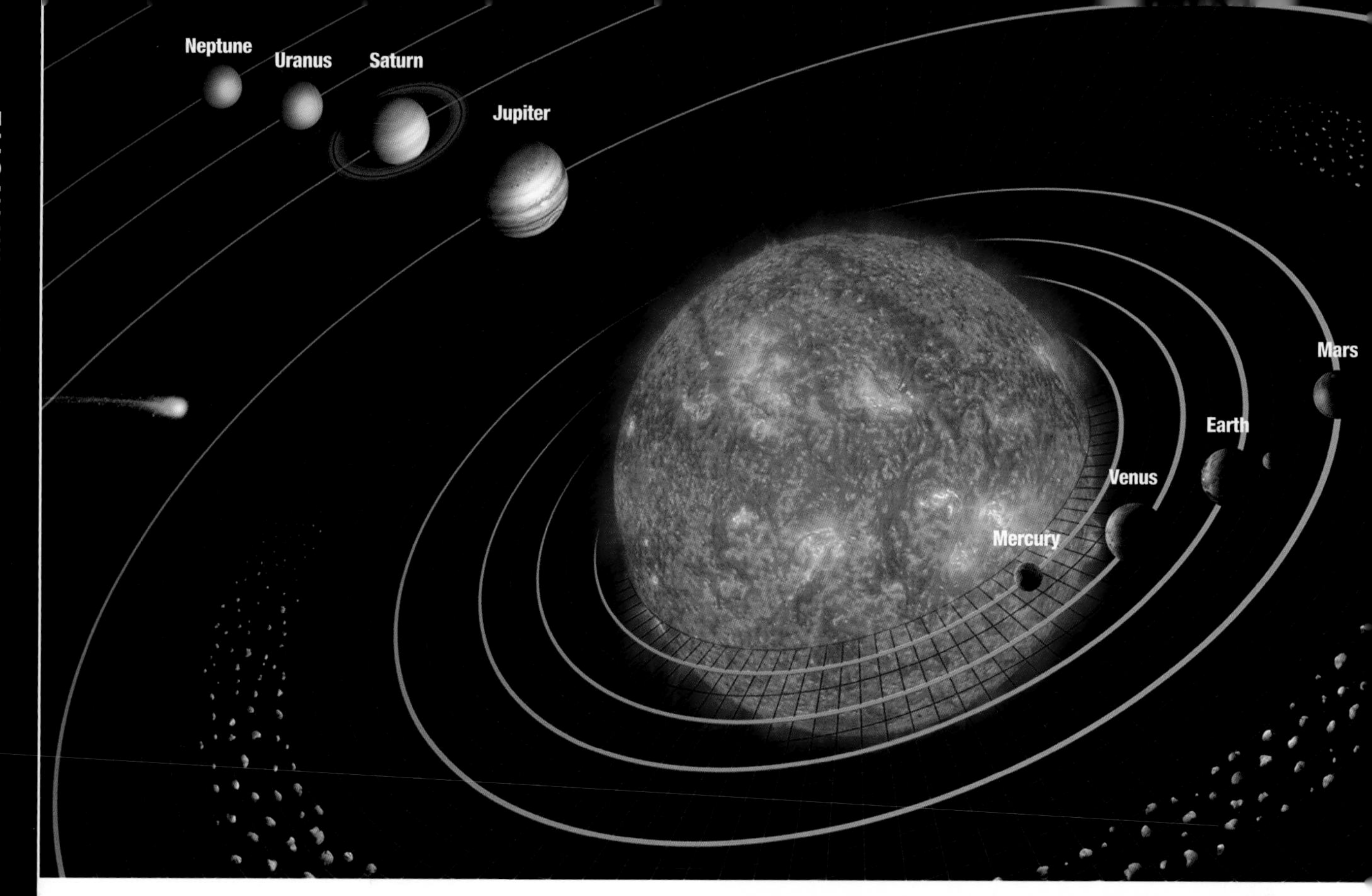

PLANETARY MOVEMENT

The word 'planet' derives from Greek, meaning 'wanderer' because they appear to wander independently across the night sky: the pale yellow 'light' of a planet viewed one evening will have noticeably moved against the backdrop of stars when viewed a week later. This distinguishes them from stars that appear 'fixed', which, of course, they are not; they are also moving within their neighbourhood, just as everything 'out there' is moving, but they do not appear to move in relation to each other in our brief human lifetimes. However, as a result of Earth's rotation about its axis, the entire canopy of stars does visually move from east to west throughout the night – at a rate of around 15° an hour.

Here are a few helpful terms that describe the positions of the planets in their orbits:

SUPERIOR PLANET

Any planet whose orbit is larger than that of Earth, ie Mars, Jupiter, Saturn, Uranus, and Neptune.

INFERIOR PLANET

A planet closer to the Sun than the Earth, ie Mercury and Venus.

OPPOSITION

Periodically, Earth passes between a superior planet and the Sun. That planet then appears in the opposite direction in the sky from the Sun. At this time, the planet rises at sunset, is above the horizon all night, and sets at sunrise. We look one way to see the Sun and in the opposite direction to see the planet. The planet is in opposition (see Fig a).

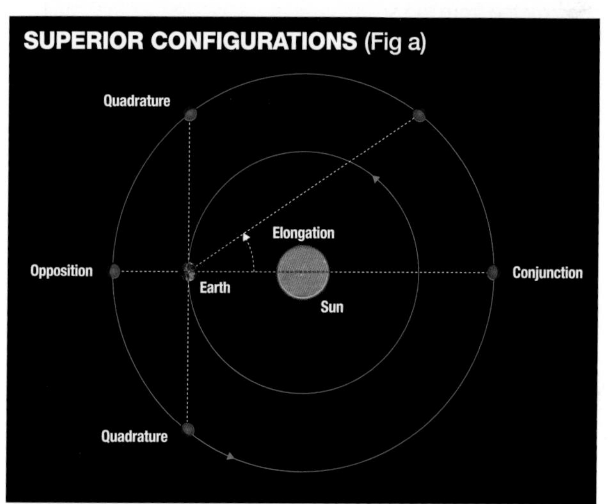

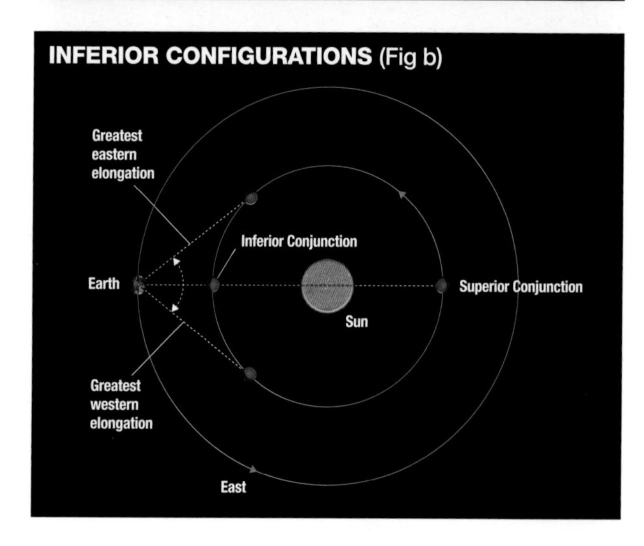

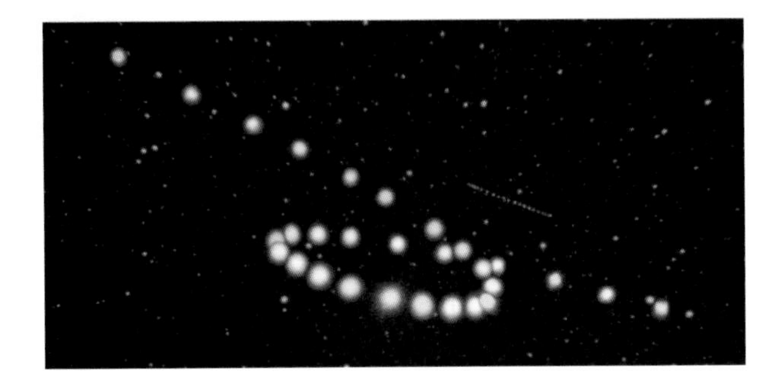

Why would Mars appear to move backwards? About every two years the Earth passes Mars as they orbit around the Sun. As it does so, the red planet appears to temporarily have a westward motion against the backdrop of the stars.

RETROGRADE MOTION

When a planet appears to move backwards across the sky. Like vehicles on a motorway, planets 'motor' at varying speeds, overtaking and being overtaken. When overtaking, the car being overtaken appears to momentarily stop and reverse. Similarly, when Earth passes another planet, the planet being overtaken appears to reverse against the backdrop of stars.

Of course, neither is reversing; they travel along in the orbital plane – ecliptic – which is the apparent path of the Sun and all the planets as seen from Earth. It is the 'planetary highway' constructed from the spinning solar nebula disc of their birth (see also *Celestial sphere and constellations*, page 88).

CONJUNCTION

Sometimes, a superior planet is on the other side of the Sun from Earth. It is in the same direction as the Sun and is, therefore, not visible. It is in conjunction (see Fig a).

INFERIOR CONJUNCTION

When an inferior planet passes between Earth and the Sun it is in the same direction from Earth as the Sun, and is in inferior conjunction (see Fig b).

SUPERIOR CONJUNCTION

When an inferior planet passes on the far side of the Sun from Earth, and is in the same direction as the Sun, it is at superior conjunction (see Fig b).

QUADRATURE

When a superior planet appears 90° away from the Sun, so that a line from the Earth to the Sun makes a right angle with the line from the Earth to the planet, it is at quadrature – the planet rises or sets at either noon or midnight (see Fig a).

ELONGATION

The angle formed at the Earth between the Earth–planet direction and the Earth–Sun direction is called the planet's elongation, so the elongation of a planet is its angular distance from the Sun as seen from Earth. At conjunction, a planet has an elongation of 0°, at quadrature 90°, and at opposition 180° (see Fig a).

GREATEST EASTERN/WESTERN ELONGATION

An inferior planet can never be at opposition because its orbit lies completely within that of Earth. Its largest angular distance from the Sun, whether on the east or west side, is known as its greatest eastern or western elongation (see Fig b).

PLANETARY DATA

The planets orbit at varying distances from the Sun. Mercury, Venus, Earth, and Mars are known as the inner terrestrial or rocky planets, composed primarily of rock and metal, typically having high densities, slow rotation, solid surfaces, no rings, and few satellites. Jupiter, Saturn, Uranus, and Neptune are known as the outer Jovian, gas giant, or gas planets. The asteroid belt between Mars and Jupiter forms the boundary between the inner and outer Solar System.

OUTER GAS GIANT (JOVIAN) PLANETS

	Jupiter	Saturn	Uranus	Neptune
Distance from Sun:	5.204AU	9.582AU	19.201AU	30.047AU
Sidereal revolution period (about Sun):	11.9 years	29.7 years	84.3 years	165 years
Mass (Earth = 1):	318	95.2	14.6	17.1
Radius at equator (Earth = 1):	11.2	9.5	4.0	3.88
Angular size (arcseconds):	31–48	15–21	3–4	2.5
Sidereal rotation period (at equator):	9.84 hours	10.2 hours	17.9 hours	19.2 hours
Equatorial Inclination (to orbit):	3.13°	26.73°	97.77°	28.32°
Moons:	67	62	27	14

MERCURY
FIRST PLANET FROM THE SUN

Average distance from Sun:	58 million km
Rotation period:	59 Earth days
Orbital period:	88 Earth days
Mean orbital velocity:	172,341km/h
Diameter:	4,900km
Mass:	0.05 Earths
Mean surface temperature:	340 degrees Kelvin

Although the Chinese and Egyptians knew of this planet, the appropriately named Mercury was named after an ancient Roman deity and was commonly identified with the Greek god Hermes, the fleet-footed messenger. Being closest to the Sun – an average distance of 36,000,000 miles (57,936,300km) – and with an orbit of only 88 days, 40% the size of Earth's, it whizzes along at 30mps whilst taking only 59 days to rotate on its axis. It appears to have an iron-dominated core that takes up 75% of its radius and nearly half its volume, and it is denser than Earth. Surrounded by a rocky crust – home to myriad craters, giant impact basins, and broad lava plains – it resembles a cannonball the size of the Moon with a diameter less than three-hundredths the width of the Sun.

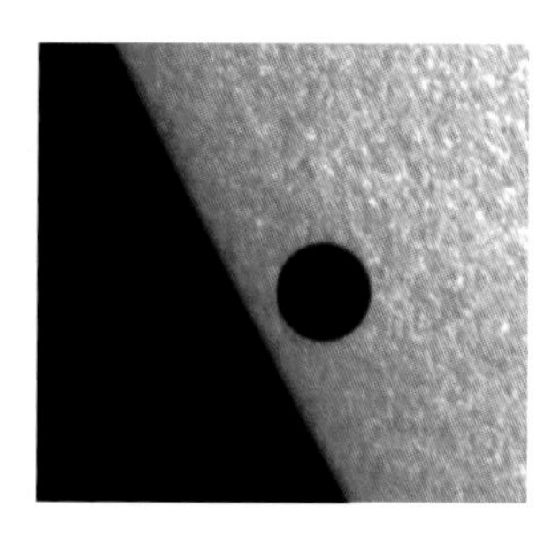

Transit of Mercury: the planet appears tiny as it crosses the Sun's disk.

Its escape velocity is too low and surface temperature too high to retain any substantial atmosphere, but sodium and potassium do ooze from its surface, possibly as a result of interaction with the solar wind. It also has a small magnetic field, albeit less than 1% as strong as Earth's, indicating a molten core. Temperature varies from a massive Sun-facing 800°F (430°C) at noon, almost hot enough to melt lead, to a nose-diving 150°K after sunset, the widest range in the Solar System. But due to its axis of rotation being almost perpendicular to the orbital plane, low-lying areas of its polar regions are in continuous shadow with temperatures never above -180°C. Even though it is so close to our Star, radar probing of highly reflective areas hints at signs of trapped ice!

Mercury: its copious craters captured in a recent close flyby of NASA's *Messenger* spacecraft.

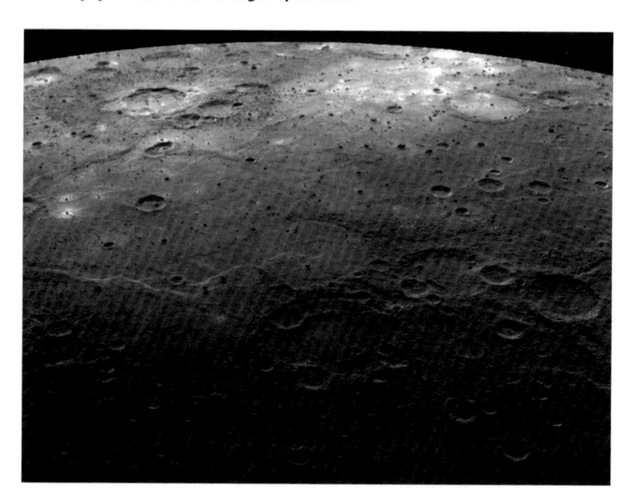

NASA's recent *Messenger* mission has mapped Mercury in all its glory, revealing small bowl-shaped cavities to multi-ringed impact basins hundreds of kilometres across. Between these craters are gently rolling hilly plains as well as smooth plains filling depressions of varying sizes. The smoother plains are criss-crossed by numerous compression folds; as Mercury's interior cooled after its formation, it may have contracted and its surface deformed. These folds can also be seen covering craters, indicating that they are recent. The largest known crater is Caloris Basin with a diameter of 1,550 km, the impact suspected of creating it being so powerful that it caused lava eruptions and left a concentric ring over 2 km tall around the impact site. Diametrically opposite this immense basin is a large, unusually hilly terrain known as 'Weird Terrain', its theorised origin being that shock waves generated during the Caloris impact travelled around the planet, converging at this point 180° away. Other recent images obtained by *Messenger* have also found evidence for pyroclastic flows from low-profile shield volcanoes.

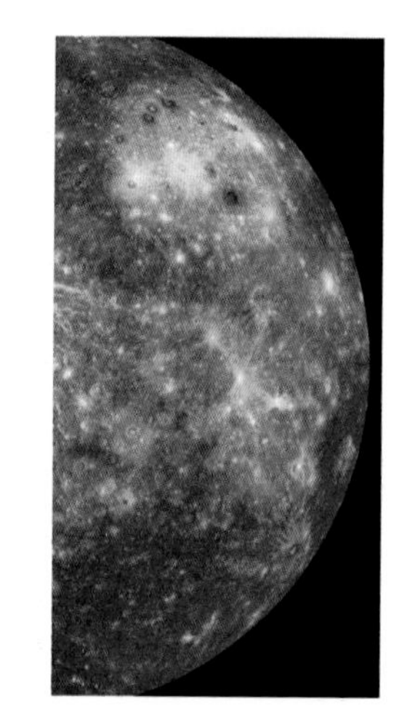

Caloris Basin: the eastern limb of Mercury as seen by the *Messenger* spacecraft from a distance of 13,000km (8,000 miles). The Caloris Basin is about 1,550km (960 miles) in diameter.

VENUS
SECOND PLANET FROM THE SUN

Average distance from Sun:	108 million km
Rotation period:	243 Earth days (retrograde)
Orbital period:	225 Earth days
Mean orbital velocity:	126,077km/h
Diameter:	12,100km
Mass:	0.8 Earths
Mean surface temperature:	735 degrees Kelvin

Venus transit - Hinode: Venus captured by Hinode's Solar Optical Telescope. Venus has just begun its journey across the face of the Sun. Its atmosphere is visible as a thin, glowing border on the upper left of the planet.

Named after the Roman goddess of love and beauty, and sometimes seen as Earth's twin, Venus is the planet most like Earth in mass and size. The planet's magnitude, or brightness, is exceeded only by the Sun and Moon. At night it casts a shadow and can even be seen in daylight if you know exactly where to look. This is due to its closeness to Earth and a blanket of dense clouds of water vapour, sulphuric acid, and sulphur dioxide 40 miles (65km) above its surface; these are highly reflective to sunlight and penetrable only by radar. Not bad for a planet a mere 25,000,000 miles (40,233,600km) away at its closest reach! Temperatures approach a furnace of 460°C, the hottest in the Solar System, and a barometer reading would be a hundred times higher than on Earth.

The planet rotates from east to west – opposite to the other planets – and its day (lasting as long as 243 days) is longer than its year (225 days). Since, like Mercury, it is located inside Earth's orbit, it has phases like the Moon that are apparent in binoculars and telescopes. Lacking water or ice, and with low surface wind speeds, its complex geological features, produced by widespread forces within the crust and volcanic eruptions, reveal a history preserved for hundreds or millions of years. Knowledge gained from spacecraft, especially the Soviet *Venera* series, the US *Magellan* radar orbiter and today's *Venus Express*, reveals that around 80% of the surface consists of lightly cratered, relatively flat rolling volcanic plains less than half a billion years old. Above these are individual mountains and mountain ranges as well as two full-scale continents: the largest, Aphrodite, is the size of Africa and stretches one-third of the way around the equator and the other, Ishtar, the size of Australia, is home to the highest region on the planet, the 7-mile (11km) high and 540-mile (869km) long Maxwell Mountains.

Every 500 million years, episodic volcanic eruptions repeatedly regenerate the surface, with enormous flows of highly fluid lava cutting through craters and carving channels thousands of miles long. The volcanoes range in size from the largest, Sif Mons (400 miles/643km wide, 2 miles/3.2km high), to many thousands of smaller ones in the form of cones or circular, flat-topped 'pancake' domes – where thick, viscous lava has exuded slowly from below. The massive atmosphere is 98% carbon dioxide and creates a runaway greenhouse effect whereby infrared radiation from the heated rocky surface is unable to escape and simply cocoons and fuels the furnace within. Even though its rotation is slow, atmospheric winds whiz around the entire planet in only 96 hours. At the equator, some clouds reach hurricane force but reduce to a leisurely 2–4mph (3.2–6.4kph) at the poles. Enormous vortices are subsequently created. Hardly a heavenly place!

Surface lava domes: an area 160km (96 miles) by 250km (150 miles) in the Eistla region of Venus showing prominent circular volcanic 'pancake' domes measuring 65km (39 miles) in diameter and less than 1km (0.6 miles) in height.

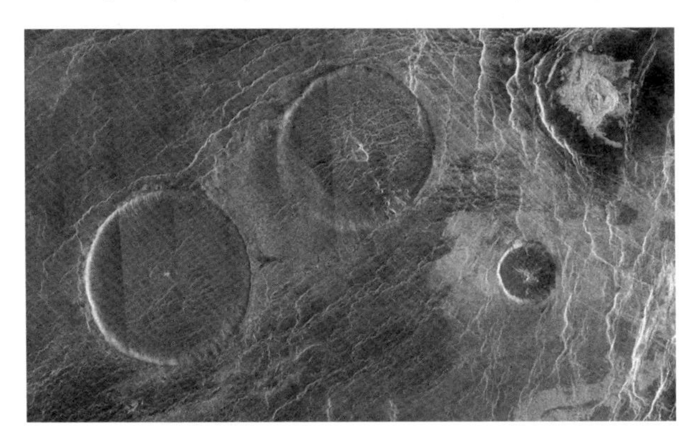

EARTH
THIRD PLANET FROM THE SUN

Average distance from Sun:	150 million km
Rotation period:	24 hours
Orbital period:	365 days
Mean orbital velocity:	107,229km/h
Diameter:	12,700km
Mass:	1 Earth
Mean surface temperature:	288 degrees Kelvin

When NASA's *Voyager* 1 completed its reconnaissance of the giant outer planets in the 1970s–'80s, it headed for interstellar space. At a distance of 3.7 billion miles it took a parting 'family portrait' of the Solar System. Each signal took five-and-a-half hours to reach Earth. The resultant mosaic of pictures was 6m long (see below). The Sun was so bright it was erased for contrast. Letters denote dots that are the planets; all are indiscernible but digitally enhanced in the inset images. Mars and Pluto were too small to be recorded. Mercury was lost in the glare of the Sun. Earth was the size of a pixel. It eloquently conveyed our minuscule planet, encircling its tiny parent Star in an inconsequential Solar System. Professor Carolyn Porco, NASA Imaging Specialist at the time, inspected thousands of such pictures. Trying to brush aside what seemed a speck of dust, she discovered it was Earth, our vulnerable 'Goldilocks' home so stunningly abundant with life, reduced to an almost invisible speck in the cosmic dark.

Highly reflective as a result of its surface being uniquely covered in 70% water, whether in liquid or solid form, Earth also has a thin atmosphere of 78% nitrogen, 21% oxygen, and traces of argon, water vapour, carbon dioxide and other gases. Beyond is the magnetosphere extending some 37,000 miles (60,000km) in the direction of the Sun and reaching downstream as far as the orbit of the Moon.

Its interior, studied by the transmission of seismic waves, comprises the uppermost layer, or crust. Beneath the oceans this crust covers 55% of the surface and is typically five miles (8km) thick. The continental crust covers the other 45% and is from 12 to 43 miles (20 to 70km) thick. Both comprise igneous rocks buried by sedimentary and metamorphic rocks produced by weathering and erosion. Beneath, the mantle of igneous silicate rocks extends to a depth of 1,800 miles (2,900km). Under this mantle is the highly dense core, 4,350 miles (7,000km) wide – substantially larger than the planet

SATURN AND EARTH

Imaged on July 19, 2013 with the wide-angle camera on NASA's Cassini spacecraft at a distance of approximately 753,000 miles (1.212 million kilometers) from Saturn, Earth is 898 million miles (1.44 billion kilometers) away, appearing as a blue dot at centre right - only the third time ever that Earth has been imaged from the outer solar system. The dark side of Saturn, its bright limb, the main rings, the F ring, and the G and E rings are clearly seen. NASA/JPL-Caltech/Space Science Institute

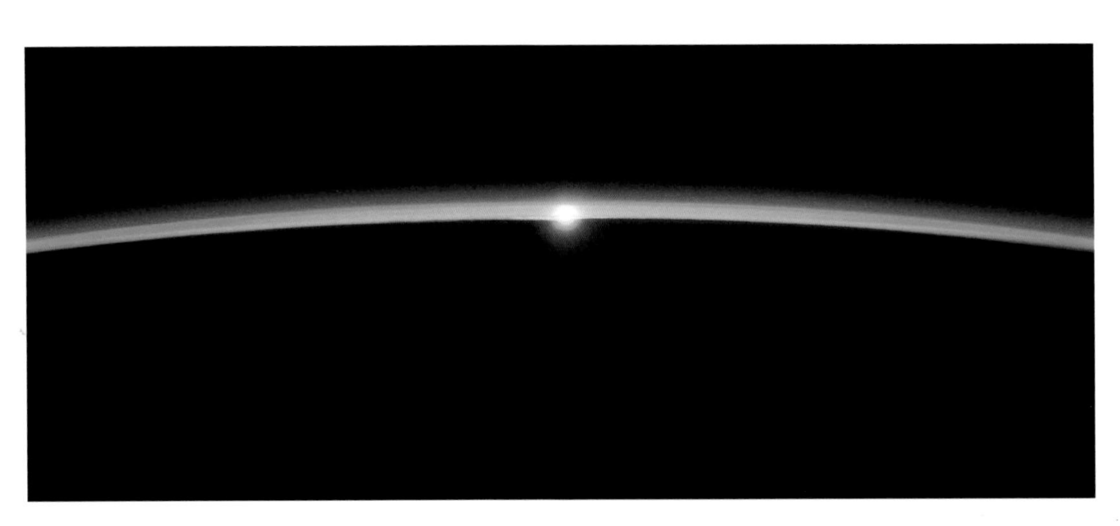

The thin blue line: the thin line of Earth's atmosphere and the setting Sun are featured in this image photographed by the crew of the International Space Station.

Mercury. Here, heat is carried upwards by convection to the crust and then escapes via conduction through solid rock or by the release of molten lava in volcanic eruptions.

Our planet is an ever-changing home. Geologic activity reshapes its surfaces. The study of plate tectonics reveals that the crust and upper mantle (lithosphere) are divided into a dozen areas whose boundaries fit like a giant jigsaw. These can pull apart along rift zones, slip underneath each other at subduction zones, or slide parallel to each other at fault zones. Volcanic activity occurs at rift and subduction zones and over mantle hotspots. Earthquakes are frequent at both subduction and fault zones and mountain ranges result where continental plates collide. But while most terrestrial landforms can be traced to these internal processes, water, wind, and ice also modify the surface of our planet, as does chemical interaction with the atmosphere and oceans. The most pronounced sculptures, from mountain peaks to the deepest canyons, are the work of water and ice erosion.

The Grand Canyon in northern Arizona near Point Sublime: a chasm some 280 miles long and a mile deep, sculpted over the millennia by wind, water and ice. Natural colour.

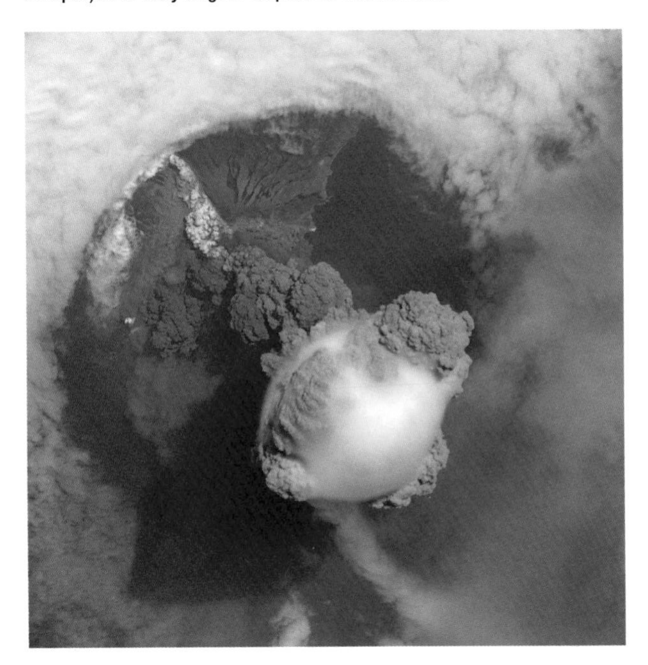

Our volatile planet: captured by the International Space Station, Sarychev volcano (Russia's Kuril Islands, north-east of Japan) in an early stage of eruption on 12 June 2009.

We enjoy seasonal weather because Earth is tilted at an angle of 23.5° to the horizontal plane of the ecliptic – not because of its distance from the Sun throughout its 365-day orbit. Consequently, the northern and southern hemispheres experience opposing seasons. In June the southern endures winter because it is leaning away from the Sun while the northern enjoys midsummer because it is turned toward the Sun. The South Pole experiences six months of continuous darkness. Six months later in December, when Earth is on the opposite side of the Sun, the northern hemisphere leans away from the Sun: it will be midwinter, but midsummer for the southern. The South Pole meanwhile basks in six months of continuous sunlight – the midnight Sun. The two occasions of the year when the Sun simultaneously reaches its most northerly or southerly declination (23.5° latitude north or south) are known as the solstices: summer solstice occurs around 21 June and winter solstice around 22

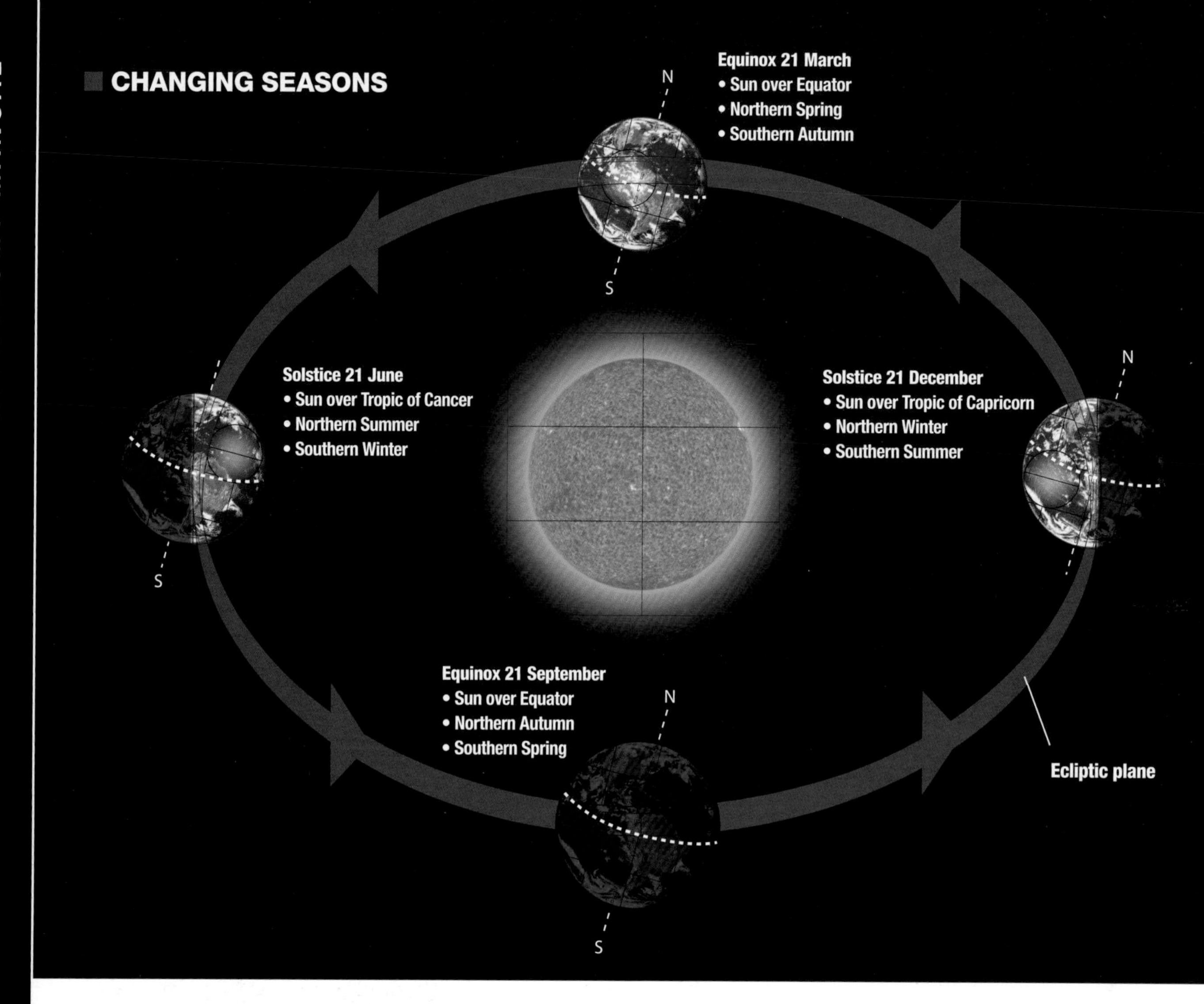

December. Midway between – around 21 March and 21 September – both hemispheres experience equal hours of daylight and night and these are known as the spring (vernal) and autumnal equinoxes (see also *Celestial sphere and constellations*, page 88).

During winter, the Sun appears low in the sky so the days are shorter and the nights longer. The Sun's rays are resultantly more spread out and have to travel through a greater density of atmosphere: some heat is absorbed before it reaches the planet. Summer days are long and hot because the Sun is high and the rays closer together, yielding more intense heat.

As the planet rotates, the atmosphere and oceans redistribute heat to cooler areas. This gives us the phenomenon of weather. On a slow-moving planet like Venus, the circulatory pattern is simple: warm air rises at the equator and migrates to the poles. On Jupiter and Neptune, faster rotation creates only east–west flows. But Earth's rotation is intermediate and the result is large cyclonic weather systems.

Our planet serves a movable feast, a sedimentary recycling not dissimilar to the recycling in the Universe itself. If Earth's internal activity had been as quiescent as the Moon, we would read a map of impact craters. Whether scarred by comets, asteroids, or other space debris, its surface would be a geologist's dream. But these have been lost to geologic activity or concealed by vast vegetation. The 4.6-billion year life history of Earth keeps many of its secrets still.

The Earth straddling the limb of the Moon, as seen from above Compton crater. Captured by NASA's *Lunar Reconnaissance Orbiter.*

THE MOON
EARTH'S NATURAL SATELLITE

Average distance from Earth:	385.5 thousand km
Rotation period:	27.3 Earth days
Orbital period:	27.3 Earth days
Mean orbital velocity:	3,400km/h with respect to Sun
Diameter:	3,500km
Mass:	0.01 Earths
Mean surface temperature:	220 degrees Kelvin

The Earth is unique among the inner planets in that its single natural satellite is unusually large. The current popular hypothetical 'Giant Impactor Theory' suggests the Moon's very existence was a result of the chaos in the solar nebula disc during the Solar System's formation 4.6 billion years ago. Earth may well have been struck by a body the size of Mars, known as Theia. This would have been the largest impact our nascent planet could have withstood without being totally shattered and may well have been responsible for its 23° axial tilt. It would have penetrated Earth's mantle, released enormous amounts of energy, and ejected vast amounts of material into space. Over millions of years, these shards would have condensed into a ring of orbiting material similar to Saturn's rings that ultimately accreted to form the Moon.

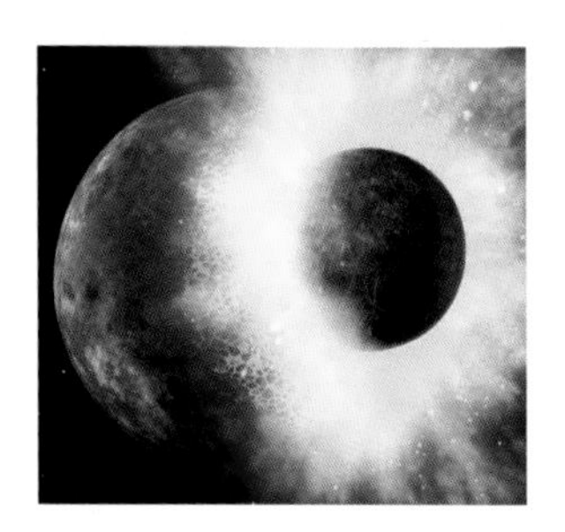

Planetary smash: an artist's impression of a planetary collision perhaps responsible for the creation of our Moon.

The Moon was once much closer to Earth, appearing 15 times larger and revolving more rapidly around our planet. Indeed, it is thought that Earth once had a five-hour day. But the Moon is now naturally slowing Earth's rotation due to a transfer of our planet's rotational momentum to the Moon's orbital momentum as tidal friction slows the Earth's rotation. The subsquent increase in the Moon's speed is causing it to slowly recede from Earth, increasing its orbital period and the length of its month. Laser beams targeting retro-reflectors left by the Apollo astronauts confirm our satellite is receding at a rate of 4 cm a year. One hundred years from now, Earth's day will be 2 milliseconds longer than it is now.

For at least five million years after its formation, as with the rest of the Solar System, the Moon lacked a protective atmosphere and was battered by planetary debris. Its surface, unchanged by wind or water erosion, tells a story, with bright highlands smothered in impact craters ('crater' is the Greek word for 'cup' or 'bowl') or large dark impact basins known as 'maria' (singular, mare) or 'seas'. Maria flooded with lava some 3.9 billion years ago, and their crater walls have clearly eroded and the lava long since solidified. They too have been subject to further bombardment and are pitted with smaller craters.

Craters range in age and size, from small, shallow indentations, to 57-mile (92km) wide, two-mile (3.2km) deep expanses, like Copernicus, or the 140-mile (225km) wide Clavius, each possessing high terraced walls, central mountain peaks, and each millions of years old. Bright rays of ejected material, blasted out during impacts, spike hundreds of miles across the dark grey terrain. Aim a pair of binoculars or a small telescope at its surface, cast your eye along the terminator (the line of demarcation between lunar day and night), and observe these stunning mountains and valleys, shadowed or illuminated by sunlight or Earthshine.

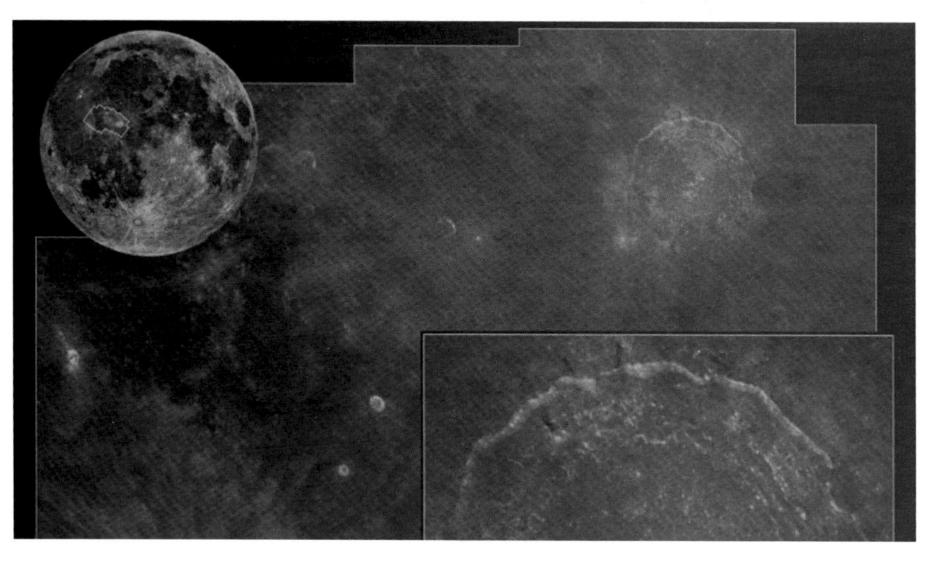

Copernicus Crater: created by giant meteor impact. Note the terraced walls and blanket of blasted material at lower right.

LUNAR PHASES

The Moon is reined in its monthly orbit by Earth's gravity and passes through a cycle of phases. It also takes an equal amount of time to turn once on its axis, so the same face, or side, is always turned towards us – something known as captured rotation – but the face emerges 50 minutes later each night. Consequently, any area of the Moon experiences two weeks of daylight and two weeks of night whereby the surface temperature ranges from 110°C to -170°C. Until the advent of space probes, the Moon's 'far side' – with its bright, cratered uplands, few dark maria, and thicker crust – eluded us.

As the satellite orbits, we see varying amounts of its sunlit hemisphere, from new, when it lies between the Earth and the Sun and is invisible, through crescent, half (first quarter since the Moon is quarter of the way in its orbit), gibbous, full (opposite side of the sky to the Sun) through third quarter, shrinking crescent, and back to new again. The whole cycle takes 29.5 days to complete and is the origin of our calendar month, or 'moonth'.

LUNAR TIDES

The Moon's gravity also affects Earth, as does the gravity of the Sun. Both generate a rhythmic rise and fall of the ocean tides, although the Moon doubly so since it is much closer. When the Moon and Sun are in a line the pull is strongest so the tides are greatest – the spring tides – and these occur at full Moon and new Moon. However, when the Moon and Sun pull at right angles, at first quarter and last quarter, the pull is less so the tides are smaller – known as neap tides. Two bulges arise in the oceans, one facing the Moon where the gravitational attraction is greatest, and one on the opposite side where Earth itself is effectively pulled towards its satellite, leaving the water to 'pile up' behind. Earth's rotation, its restricting landmass, the wind, friction within variable-depth oceans and against ocean floors all play a part. Many places on the planet have two tides a day as Earth rotates beneath these bulges (one every 12 hours and 25 minutes) and these become apparent 50 minutes later each day as the Moon progresses in its orbit. They are gradually slowing Earth's rotation too; our planet's day is getting longer. Although minuscule – less than two-thousandths of a second per century – in over 200 million years from now there will be a 25-hour day.

LUNAR PHASES AND THE TIDES

As we view the Moon orbiting Earth its appearance changes (outer images) from new to full and back again, whilst its gravitational influence creates the tides.

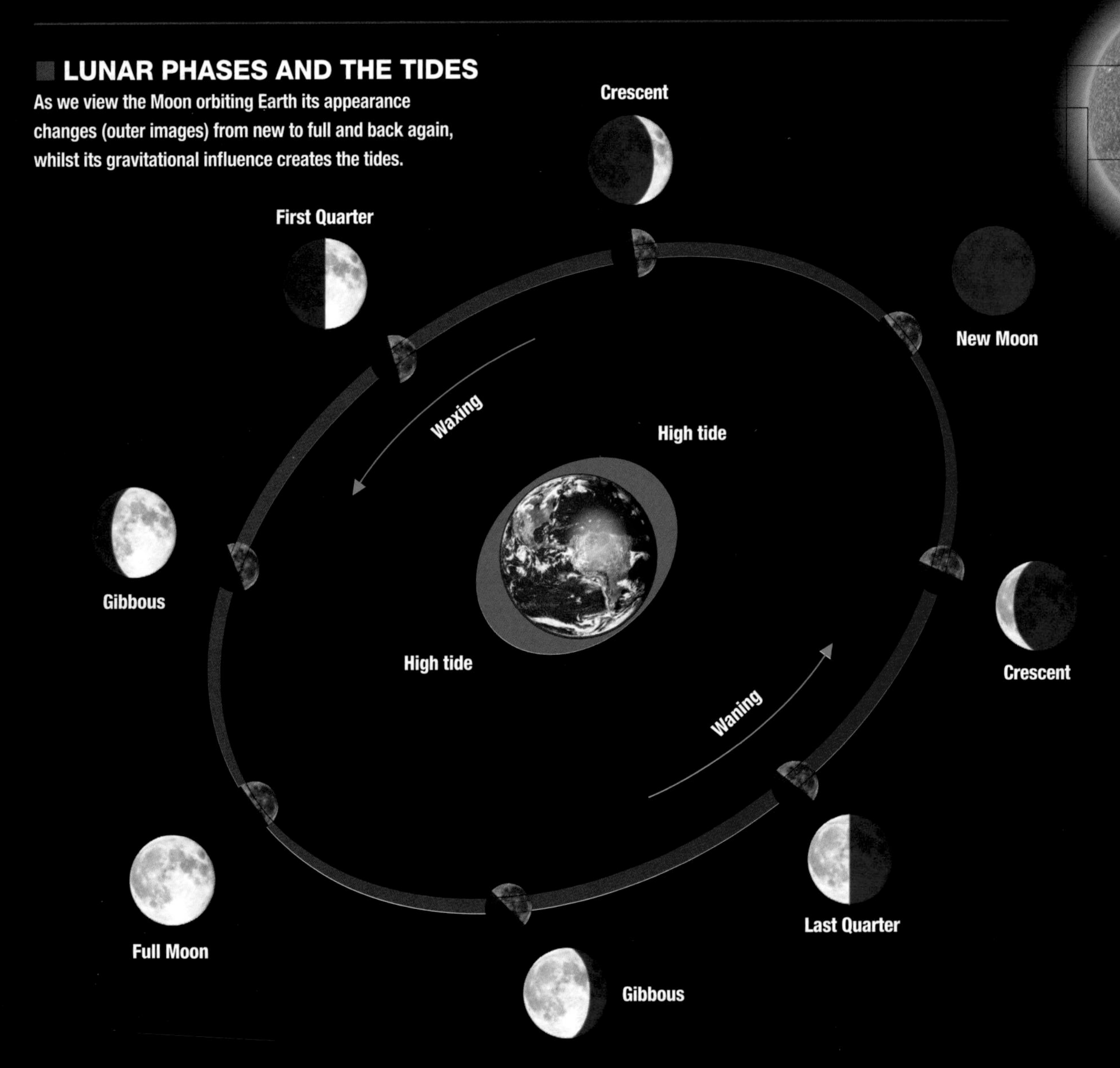

LUNAR ECLIPSE

n a total lunar eclipse the Moon initially moves into Earth's penumbra, the outer shadow. Its appearance changes little until our planet has almost reached the umbra when the Moon begins to darken for approximately an hour.

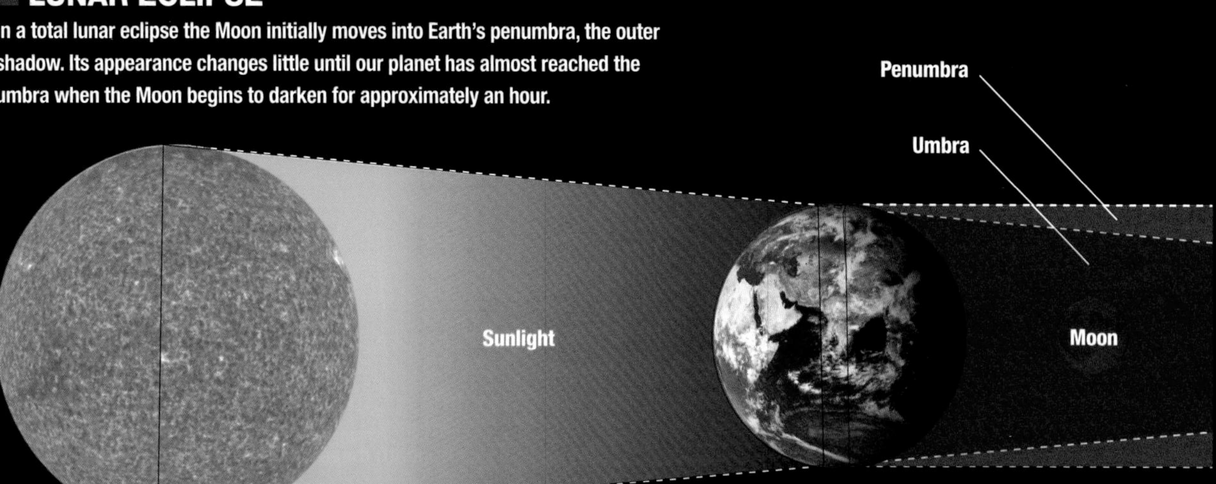

Periodically, at new Moon, our satellite appears to glide across the face of the Sun, giving rise to a spectacular total solar eclipse. If the full Moon passes into the shadow of the Earth there is a total lunar eclipse. A partial eclipse occurs when only some of the Moon's disk is in shadow. This does not happen at every new or full Moon since the orbit of the Moon is tilted with respect to the Earth's; only when their orbits intersect can the Sun, Earth, and Moon line up exactly.

A total eclipse of the full Moon starts with a limb darkening as it moves into Earth's shadow until gradually, over a period of an hour, the entire face is immersed. The Moon does not vanish; residual sunlight is refracted through Earth's atmosphere, yielding a dark red or coppery colour. Only rarely does it become entirely dark; an opaque atmosphere of thick cloud, or high-altitude volcanic dust from recent eruptions, can block the passage of sunlight into the Earth's shadow.

FUTURE TOTAL LUNAR ECLIPSES

Date	Location
31 January 2018	Asia, Australia
27 July 2018	Asia, Africa
21 January 2019	North and South America
26 May 2021	Eastern Asia, Australia, Pacific, Americas
16 May 2022	Americas, Europe, Africa
8 November 2022	Asia, Australia, Pacific, Americas
14 March 2025	Pacific, Americas, Western Europe, Western Africa
7 September 2025	Europe, Africa, Asia, Australia
3 March 2026	Eastern Asia, Australia, Pacific, Americas

LUNAR ECLIPSE SEQUENCE

As the Moon progresses into Earth's shadow, our planet's atmosphere acts like a lens or prism, bending sunlight into the shadow to create a dark coppery colour, essentially capturing every sunrise and sunset occurring around the globe.

MARS
FOURTH PLANET FROM THE SUN

Average distance from Sun:	228 million km
Rotation period:	24.6 hours
Orbital period:	687 Earth days
Mean orbital velocity:	86,871km/h
Diameter:	6,700km
Mass:	0.1 Earths
Mean surface temperature:	227 degrees Kelvin

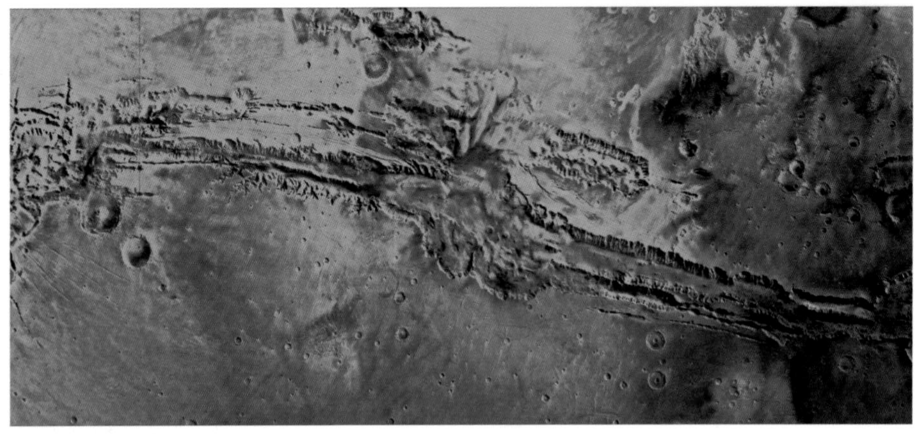

Valles Marineris: a vast canyon system that runs along the Martian equator. At 4,000km long, 200km wide and 11km deep.

Professor Percival Lowell: at his Flagstaff Observatory.

Myths surround Mars, so named after the Roman God of War. The 19th-century Italian astronomer Giovanni Schiaparelli recorded long straight lines – canali – criss-crossing its surface. A French astronomer, Camille Flammarion, endorsed them in 1892 in a 600-page compendium. Later, in 1894, they were interpreted as channels, or canals, by Percival Lowell, a wealthy Bostonian, used for transporting water from the poles to the arid equatorial regions. They implied artificial origin. In 1938 this fascination was fuelled by Orson Welles' radio adaptation of H.G. Wells' book *War of the Worlds*, a tale of Martian invasion that terrified New Jersey listeners. Mars was further glamorised with Edgar Rice Burroughs' tales of gentleman adventurer John Carter, and an ethereal 'Barsoom'. But, to date, no signals of life have been found. Mars remains an alluring orange globe in the night sky, its mass only 11%, and atmospheric surface pressure only 1%, that of Earth. Yet, in many ways, Mars does resemble Earth: with its 25° axial tilt it has seasons, is only a little colder, and has thin polar caps of frozen water and carbon dioxide ices that spread or shrink as the seasons progress.

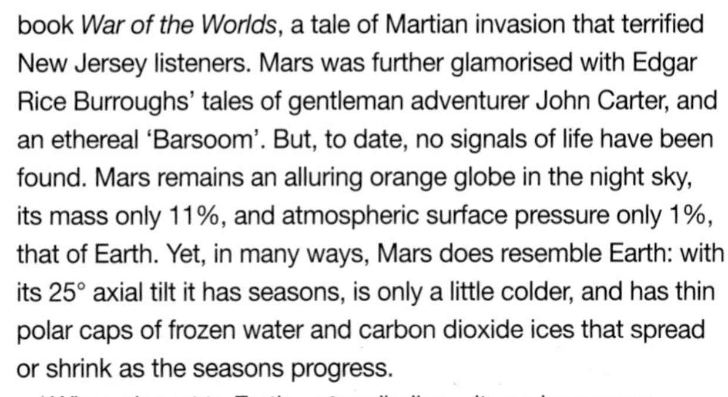

When closest to Earth – at perihelion – it can be a mere 35,000,000 miles (56,327,000km) away, although eccentricity can place it some 250,000,000 miles (402,336,000km) distant. But good 'seeing' reveals the surface as being divided into ancient cratered uplands in the southern hemisphere and younger volcanic plains in the north, with the youngest areas existing in the Tharsis Bulge – home to large shield volcanoes topped by huge calderas, an area where the absence of plate tectonics has denied heat escape and instead led to a giant bulge. Weathering, including fierce global dust storms that engulf the planet for weeks, erode the features but some, like Solis Lacus (the 'Eye of Mars') and Syrtis Major (originally known as the 'Hourglass Sea', but now classical Roman for Libya's 'Gulf of Sidra', and also the first documented surface feature of another planet), a shield volcano, are readily visible in amateur telescopes. With temperatures almost always below freezing point, Mars appears a barren place, rich in iron oxide – hence its epithet, the Red Planet.

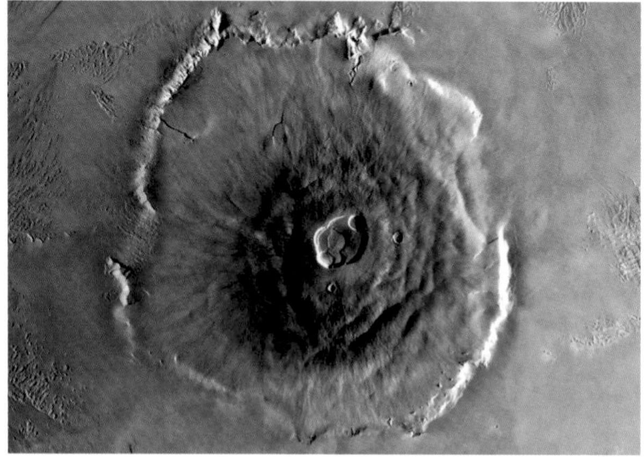

Olympus Mons: a mosaic from NASA's *Viking 1* orbiter.

***Curiosity* rover at Mount Sharp:** NASA's *Curiosity* rover captures a self-portrait at the 'Mojave' site on Mount Sharp.

Perhaps its two most spectacular features are the gargantuan volcano Olympus Mons, and the stunning Valles Marineris – Mariner Valley, a system of canyons some four miles deep and 2,500 miles (4,000km) long, probably created when four nearby volcanoes pushed up and stretched the crust. This 'Grander Canyon' dwarfs Earth's 280-mile (450km) Grand Canyon. Indeed, Earth's number-one tourist attraction would be lost in one of its many tributaries since the entire Martian chasm would stretch from the USA's California to Virginia! The largest known volcano in the Solar System is equally mesmerising. With its base measuring 600 km across (comparable in area to the state of Arizona), Olympus Mons is a hundred times larger than any volcano on Earth and looms a staggering 15 miles higher than the surrounding plain.

The climate may once have been warmer and wetter; water may have existed. Run-off channels in the old highlands have been interpreted as valleys of ancient rain-fed rivers that emptied into basins perhaps containing shallow seas. Larger outflow channels hint at sudden catastrophic floods, their source perhaps regions where abundant water has frozen as subsurface permafrost.

Flanked by its two appropriately named moons, Phobos (fear) and Deimos (panic), themselves thought to be captured asteroids (of which more later), the God of War continues to arouse interest. Following on from the many Mars missions, the planet is now monitored by NASA's orbiting Mars *Odyssey* spacecraft, *Mars Reconnaissance Orbiter*, MAVEN – Mars Atmosphere and Volatile Evolution space probe, the NASA/ESA *Mars Express*, one of the two ageing but enduring surface rovers, *Opportunity*, the other, *Spirit*, currently mired in Martian sand and the recent NASA rover, *Curiosity*. More missions are planned for the second decade of the 21st century to characterise and understand Mars as a dynamic system. By analysing its past and present environment, climate cycles, geology and biological potential, conditions may be discovered that are conducive to life and future manned exploration. The story continues…

MOONS OF MARS

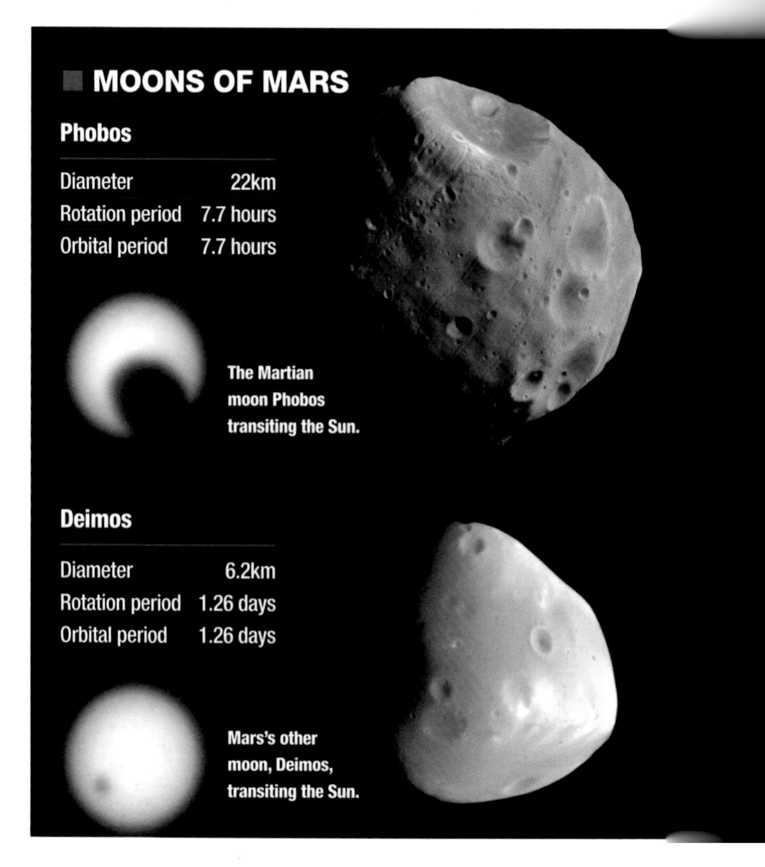

Phobos

Diameter	22km
Rotation period	7.7 hours
Orbital period	7.7 hours

The Martian moon Phobos transiting the Sun.

Deimos

Diameter	6.2km
Rotation period	1.26 days
Orbital period	1.26 days

Mars's other moon, Deimos, transiting the Sun.

JUPITER
FIFTH PLANET FROM THE SUN

Average distance from Sun:	779 million km
Rotation period:	9.9 hours
Orbital period:	11.9 Earth years
Mean orbital velocity:	47,051km/h
Diameter:	143,000km
Mass:	318 Earths
Mean cloud top temperature:	165 degrees Kelvin

With two and a half times the mass of all the other planets put together, and with over 300 times the mass of Earth, Jupiter, as its ancient Roman name suggests, is king. Together with Saturn, Uranus, and Neptune, it is a gas giant, far more immense and significantly less dense than the smaller, predominantly rocky planets of the inner Solar System. Eleven Earths could sit comfortably across its equator and 1,400 would fit inside. Its enormous magnetosphere trapping hot ionised gas, or plasma, extends 1.9 to 4.3 million miles (3 to 7 million kilometres) away from the Sun and stretches in a windsock shape at least as far as Saturn's orbit – 466,000,000 miles (750,000,000km) away. Situated 5AU from the Sun, it takes just under 12 years to complete one orbit but only a staggering ten hours to rotate once on its axis: the result is flattened poles and dynamic weather patterns.

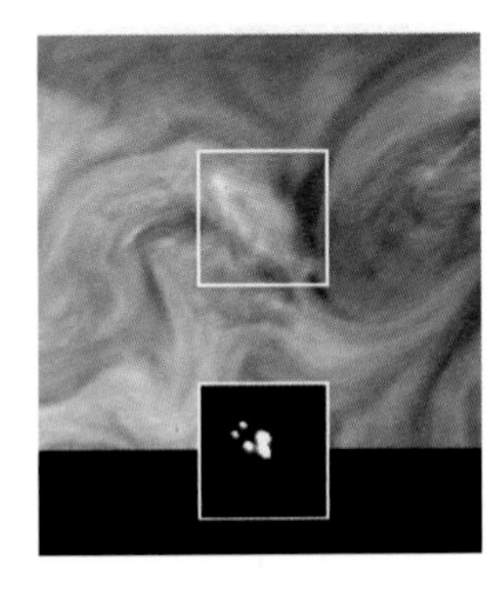

Lightning storm on Jupiter: the inset shows areas of lightning at the same location as the storm.

Its atmosphere of hydrogen, helium, methane, and ammonia creates stunningly coloured cloud formations, the wind speed smearing these into bands above and below the equator; light-coloured zones indicate higher, cooler clouds in areas of updraft; dark zones mark warmer downdrafts. There are many oval-shaped, anti-cyclonic, high-pressure storm systems swirling around the planet, the largest being the famous Great Red Spot – a seething whirlpool in the southern hemisphere about four times larger than Earth and rising five miles (8km) above the surrounding clouds. Apparent for over 300 years, it is the largest known storm in the Solar System. In recent years, as a result of a merger between three white oval storms, there has been a smaller companion – Red Spot Junior – that turned from brown to red with the ultraviolet light of the Sun. And the entire planet, like Saturn and Neptune, has a major internal heat source, obtaining as much (or more) energy by convection in its interior as by radiation from the Sun.

Like our Solar System in miniature, Jupiter has 67 known moons. Even with binoculars, Jupiter's four main Galilean satellites (discovered by Galileo Galilei in 1610) are easily visible. Io, Europa, Ganymede, and Callisto are strikingly different from the other moons. Io, the closest, is 217,500 miles (350,000km) above Jupiter's clouds and about the size of Earth's Moon. Since Jupiter is 1,400 times larger than our planet, however, this has a huge impact on the satellite. Moreover, the bigger Ganymede and Europa pull Io into an eccentric orbit so that it travels nearer and farther from the planet. This twists and flexes its surface, making it rise by over 10m (32ft) a day. The Jupiter-facing side bulges by up to six miles! Since it is inside Jupiter's magnetosphere, it builds up an electrical charge which discharges on to the giant planet's surface – a continuous flow of three million amps activating tremendous storms. Such flexing is like bending a paper clip. Enormous heat generates in Io's interior, which fuels the power for immense volcanic activity. The surface is covered in spectacular plumes of sulphur and sulphur dioxide. Lava blasts from these volcanoes, some the size of Texas, at a rate of ten tons a second. Consequently, its pizza-like appearance changes on an almost daily basis.

Jupiter with Great Red Spot (GRS) and storms: the GRS is a giant storm at least 300 years old with counterclockwise winds in excess of 250 mph.

When this material falls on to the planet's surface, or is captured within its giant magnetosphere, the particles interact with the upper atmosphere and induce spectacular aurora displays.

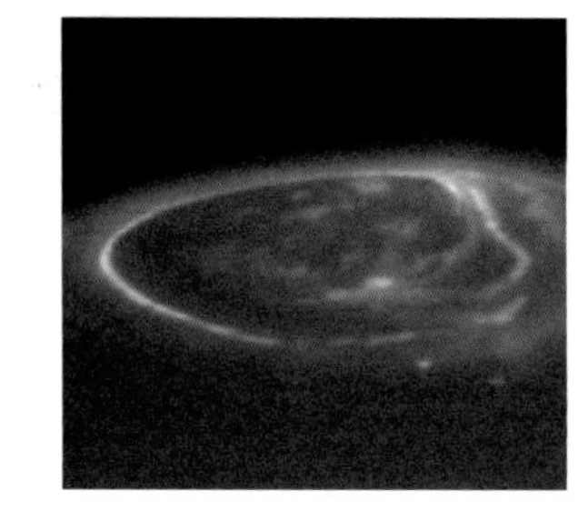

Jupiter Aurora: Jupiter Aurora imaged by the Space Telescope Imaging Spectrograph (STIS) on NASA's Hubble Space Telescope.

Contrastingly, Europa's scarred appearance resembles the Arctic Ocean. Known as a capped ocean because of its 30-mile (48km) thick layer of surface ice, the water or slush beneath could be as much as a hundred miles deep, warmed by internal heat similar to that driving Io's volcanoes. Fractures and fissures, again caused by tidal tugging from Io and Ganymede, criss-cross its icy shell; upwellings suggest soft ice and evidence of surface salt. Together with its paucity of craters inferring its surface is constantly recycled and, therefore, young, it is a Mecca for astro-biologists seeking subterranean volcanoes where, like the 'black smokers' in extreme environments on Earth, there could be life.

Ganymede is the largest moon in the Solar System and the only one known to have an internally generated magnetic field. Callisto's surface is ancient and heavily cratered. Like Earth's Moon, it provides a record of bombardment from the Solar System's creation.

Trapped along with the moons inside Jupiter's magnetosphere is Jupiter's ring system. Comprised of fine dust particles, it has four components: a thick inner 'halo ring', a bright, thin 'main ring', and two wide, thick outer 'gossamer rings'. They extend to an outer edge of around 80,161 miles (129,000km) from the centre of the planet and inward to around 18,642 miles (30,000km), and have a depth of 30km to 12,500km. All probably originated from ejected dust as a result of high-velocity micrometeorite bombardment of the tiny moons, or other small bodies.

With the recently announced new joint NASA/ESA mission, aiming to arrive at Jupiter in 2026, and orbit Ganymede and Europa, many more secrets will undoubtedly be revealed.

But before we head for Saturn, let's refresh perspective and return to our basketball/Sun analogy. Even though Jupiter is clearly immense, it would still only be the size of a grape when compared to our basketball Sun ... and would have to be placed 150m (46ft) away.

Europa: Thera and Thrace are two dark, reddish regions of enigmatic terrain that disrupt the older icy ridged plains on Jupiter's moon Europa.

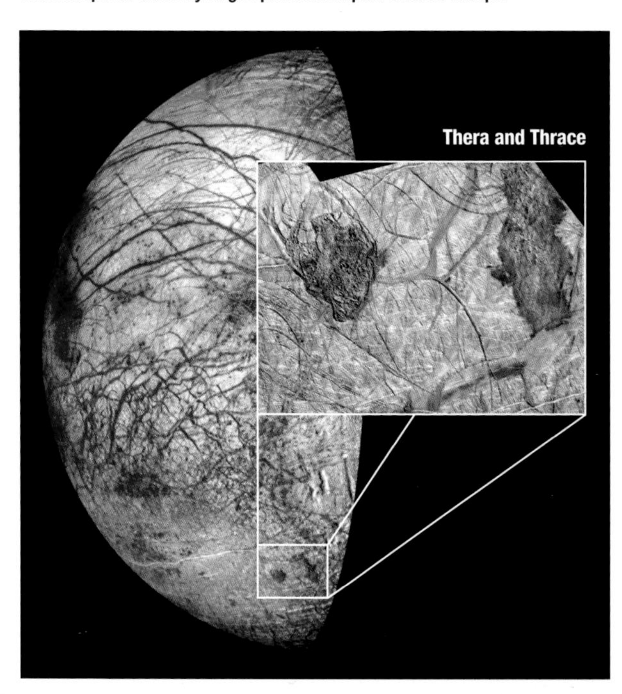

GALILEAN SATELLITE DATA

The four largest moons of Jupiter, known as the Galilean satellites.

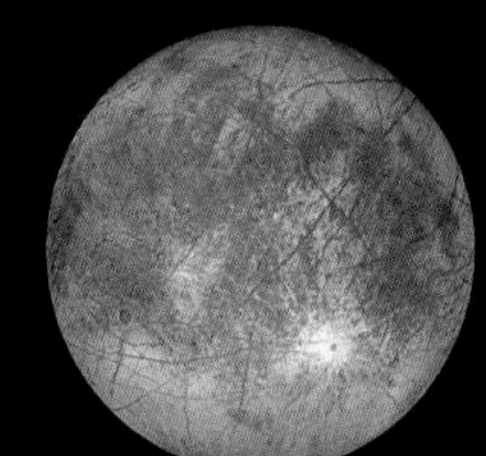

Ganymede

Diameter (km)	5,260
Visual mag.	4.6
Orbital period (days)	7.16

Callisto

Diameter (km)	4,800
Visual mag.	5.6
Orbital period (days)	16.69

Io

Diameter (km)	3,630
Visual mag.	5.0
Orbital period (days)	1.77

Europa

Diameter (km)	3,140
Visual mag.	5.3
Orbital period (days)	3.55

SATURN
SIXTH PLANET FROM THE SUN

Average distance from Sun:	1,427 million km
Rotation period:	10.6 hours
Orbital period:	29.7 years
Mean orbital velocity:	34,821km/h
Diameter:	121,000km
Mass:	95 Earths
Mean surface temperature:	134 degrees Kelvin

Saturn is named after the ancient Roman god of the harvest. Its graceful ring system is surely one of the sky's most breathtaking sights in any size of telescope. Following NASA's *Pioneer* and *Voyager* missions of the 1970s–'80s, its mysteries have been wondrously unveiled, with the Cassini Orbiter currently returning gorgeous close-up views of the globe and its diaphanous rings.

Saturn is a large gaseous ball with a mass 95 times that of Earth, but with an average density less than water. Its rocky core is perhaps the size of 15 Earths and the entire planet could hold 700 of ours. Yet, if we could place its 75,000-mile diameter on one of our oceans, it is so 'light' it would float! And remember, we can see it because it reflects the sunlight it receives from 1.4 million km away (9AU). Since it is so distant, it takes 29 years to orbit our Star but, like Jupiter, only ten hours and 47 minutes to rotate once on its axis. Again, this induces tempestuous weather with winds in excess of 1,000mph (1,600kph), and an equator that bulges 7,300 miles (11,750km) wider than at the poles!

Composed mainly of hydrogen and helium, but with poisonous ammonia, the atmosphere's true colour ranges from shadowed blues and green to creamy pastels, and shields a layer of ammonium hydrosulphide ice particles with a lower layer of water droplets containing ammonia. Cassini's infrared camera has probed a mere 20 miles down to reveal the deepest clouds ever seen; at pressures twice that of Earth at sea level they can race around at supersonic speeds. It is a climate of numerous doughnut-shaped clouds, cyclones, and severe electrical storms with lightning bolts triggering radio waves a thousand times more powerful than those on Earth. Its overall appearance, like Jupiter, is a result of these winds smearing the clouds into narrow abundant belts. Saturn, too, has spectacular auroral displays, and for the same reason as Earth, but here the solar wind is pushing charged particles past the planet's atmosphere at speeds in excess of 900,000mph (1,448,000kph).

Saturn: a full colour view of Saturn's northern hemisphere captured by the *Cassini* spacecraft from a distance of 1.1 million km (680,000 miles).

The stunning bright ring system, starting 4,000 miles (6,437km) above its surface, is comprised mainly of water ice, from porous micron-sized fluffy dust particles to those the size of a house. It is a dynamic, ever-changing environment and spans a distance almost equal to that between Earth and its Moon (around 240,000 miles/386,000km). The three main rings are lettered in order of their discovery: the outer A (9,000 miles/14,480km wide), B (16,000 miles/25,750km) and C (10,500 miles/16,700km). These alternate with a bewildering array of other more faint rings starting with the inner D, C, B, A, F, G, and E, and all are shaped and sculpted by gravitational interactions with Saturn's myriad smaller embedded satellites or neighbouring 'shepherd' moons. The outer A ring features two small 'gaps' known as Encke and Keeler, regions seemingly swept out by the tiny moons Pan and Daphnis, and between A and B is the famous 3,000-mile (4,800km) wide Cassini Division. Yet these are not gaps, since they are filled with dark 'dirty' ice. The broad, tenuous outer E ring is home to several icy satellites, including Mimas, Tethys, Dione, and Enceladus (310 miles/500km), the last moon – with its folds and criss-crossing fractures – 'feeding' the dazzling, and seemingly uncontaminated, ring via icy geysers from its south polar regions.

A more recent revelation, using NASA's Spitzer Space Telescope, has been the discovery of an even more distant and nearly invisible ring. This starts some 3.7 million miles (6 million kilometres) from the planet and extends 7.4 million miles (12 million kilometres); its diameter is equivalent to lining up 300 Saturns side by side. Its vertical height is around 20 times the diameter of the planet, and about one billion Earths stacked together would neatly fit inside. (See *Spitzer*, page 156.)

Interestingly, despite their enormous width, all the rings are mainly only 10–50 miles deep – the equivalent of the thickness of a sheet of paper compared to the size of a football pitch. However, NASA's Cassini orbiter has recently revealed numerous 'ruffles and dust clouds' and 'bumps' as high as the Rocky Mountains. One ridge, whipped up by the gravitational pull of the moon Daphnis, looms as high as 2.5 miles.

Moon shadows from Janus (middle) and Mimas (bottom), and a series of spokes, adorn this image taken about a month after Saturn's August 2009 equinox.

Amidst Saturn's many moons is Titan; 3,200 miles wide (5,150km) and first discovered by Christiaan Huygens in 1655, it is the second largest moon in the Solar System and the only one yet known to have an atmosphere – a photochemical orange-coloured haze of 95% nitrogen with traces of methane. While Earth's atmosphere extends 37 miles (60km) into space, Titan's extends ten times that distance and is ten ti thicker. In January 2005, piggybacked on the Cassin ESA's Huygens probe was released and landed in wh thought to be a floodplain of once liquid methane but w actually 'muddy' and strewn with small icy pebbles. Late radar images from Cassini reveal lakebeds and desert-like undulating sand dunes at heights of 100–150m (330–490ft). As a result of its interior being heated by tidal interaction with Saturn, there are hints of possible cryovolcanism too, where icy ammonia and methane, stored as cryomagma, explode into the atmosphere like a fizzy drink released from a shaken bottle. These explosions and sand dunes, like the Moon affecting our planet's ocean tides, are subject to Saturn's gravity, although the gas giant's pull is 400,000 times greater!

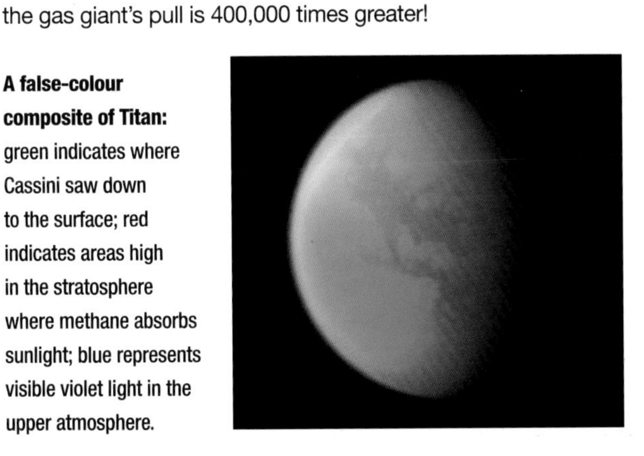

A false-colour composite of Titan: green indicates where Cassini saw down to the surface; red indicates areas high in the stratosphere where methane absorbs sunlight; blue represents visible violet light in the upper atmosphere.

SATURN'S SATELLITES AND RING STRUCTURE

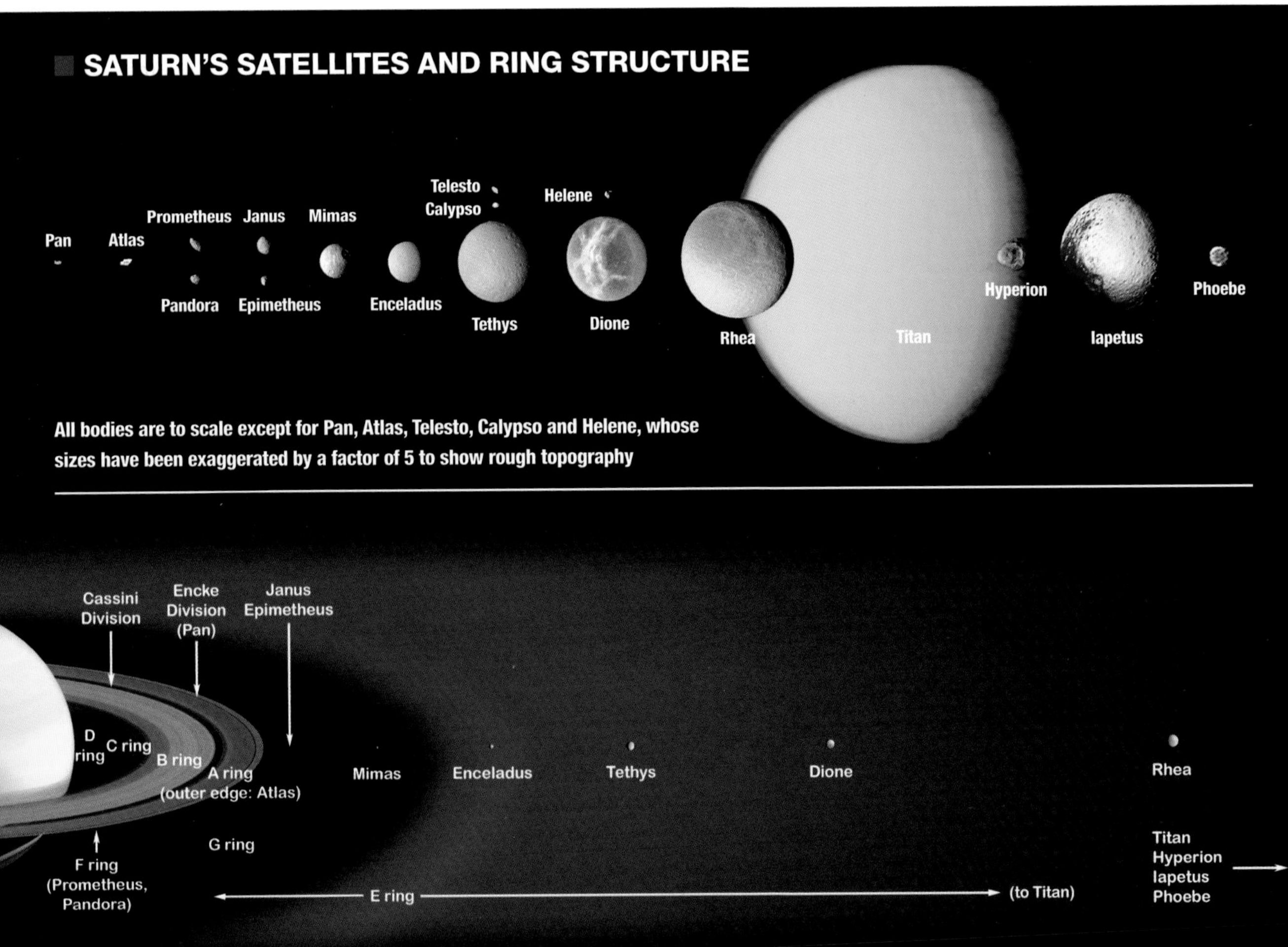

URANUS
SEVENTH PLANET FROM THE SUN

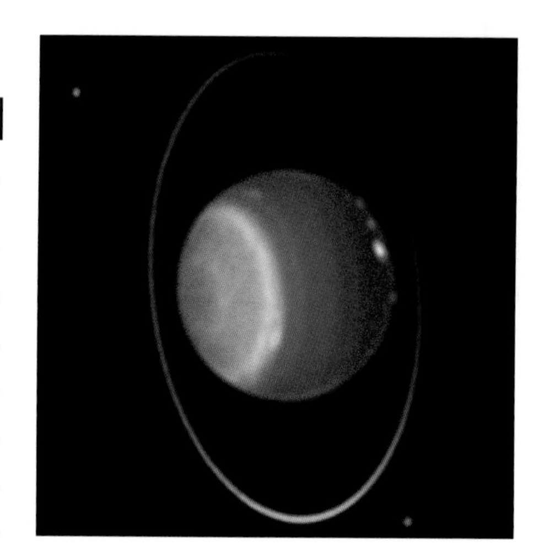

Average distance from Sun:	2.7 billion km
Rotation period:	17 hours
Orbital period:	84.3 Earth years
Mean orbital velocity:	24,607km/h
Diameter:	51,000km
Mass:	14.5 Earths
Mean cloud top temperature:	53 degrees Kelvin

Considered somewhat bland after NASA's *Voyager* 2 visit in 1986, Uranus, named after a Greek sky god, has shown itself to be a dynamic gas giant planet with some of the brightest clouds, numerous cloud bands, circulatory spots, and 11 rings. Tilted at an angle of 98° to its plane of orbit, and elliptical in shape due to its 17.9-hour rotation, this blue-green apparently featureless globe rolls along on its side. Consequently, during its 84-year orbit around the Sun, each of its poles in turn faces its parent Star to enjoy a day that is 42 years long. With a seemingly impenetrable atmosphere of hydrogen, helium, methane, water, and ammonia, 80% of this is contained within an extended liquid core, its density increasing with depth; but unlike Jupiter, Saturn, and Neptune, there is no measurable internal heat source. At 19.2AU from the Sun it is still able to reflect the radiation it receives, but temperatures in its upper clouds are around -330°F (-200°C); ammonia exists in the form of ice crystals. And initial appearances can be deceiving – recent Hubble Space Telescope images have revealed a dark cloud about 27° north of the equator: a giant vortex two-thirds the size of the USA.

Of greater interest are the 27 eclectic moons, most uniquely named after Shakespearean or Alexander Pope characters. Miranda, the innermost and smallest of the five major satellites, has a surface unlike any other, with giant fault canyons 12 times deeper than those of Earth's Grand Canyon; this dramatic landscape indicates it was once shattered and reassembled after a collision in its early history. Ariel has the brightest and possibly youngest surface; a palimpsest of small craters overlapping larger ones indicating fairly recent low-impact collisions. Umbriel, the darkest and oldest of the five larger moons, sports large craters and a mysterious bright ring on one side. Oberon, the outermost, is likewise old, heavily cratered and shows little sign of any internal activity.

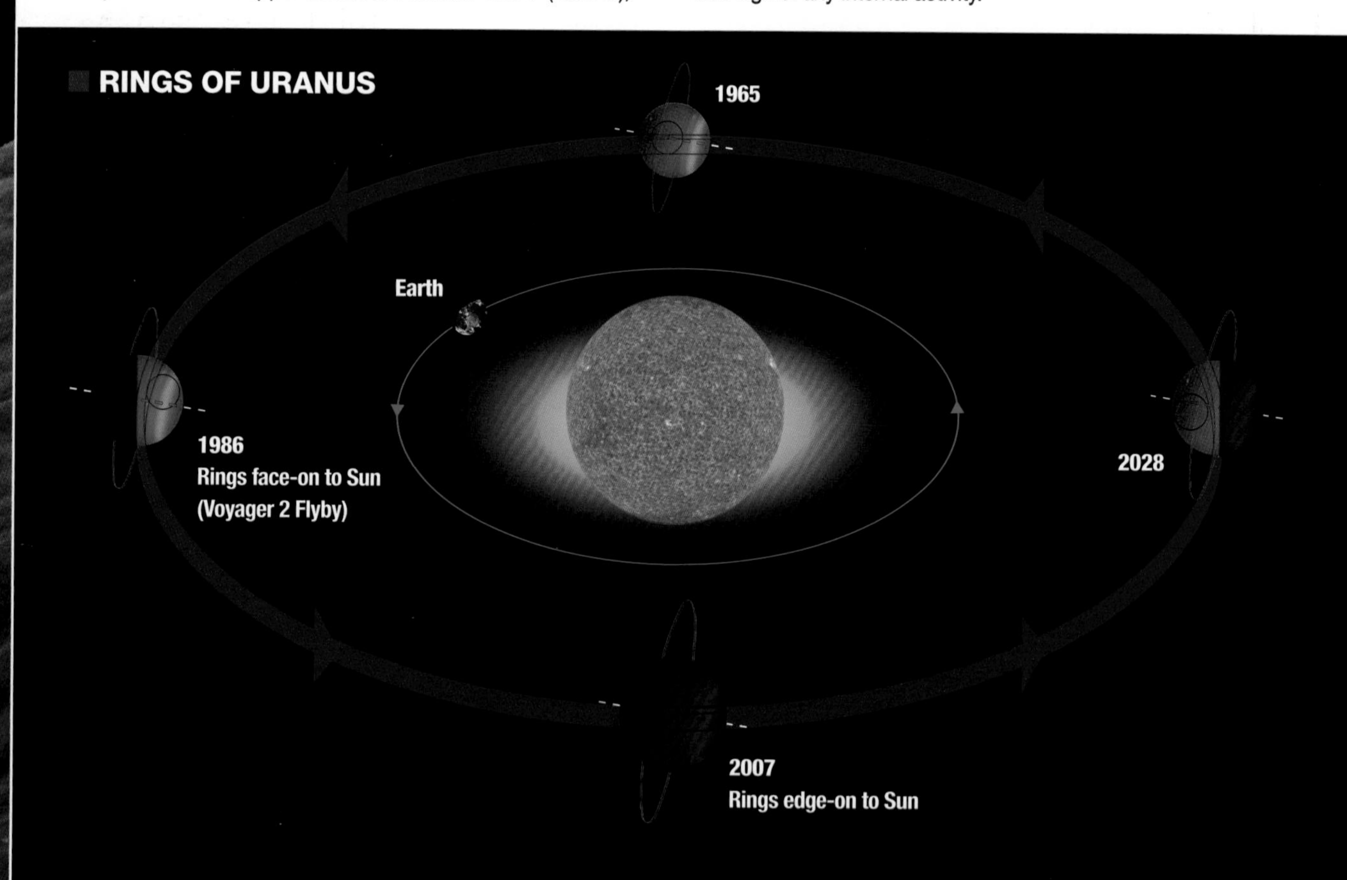

NEPTUNE
EIGHTH PLANET FROM THE SUN

Average distance from Sun:	4.5 billion km
Rotation period:	16 hours
Orbital period:	165 Earth years
Mean orbital velocity:	18,720km/h
Diameter:	49,000km
Mass:	17.1 Earths
Mean cloud top temperature:	55 degrees Kelvin

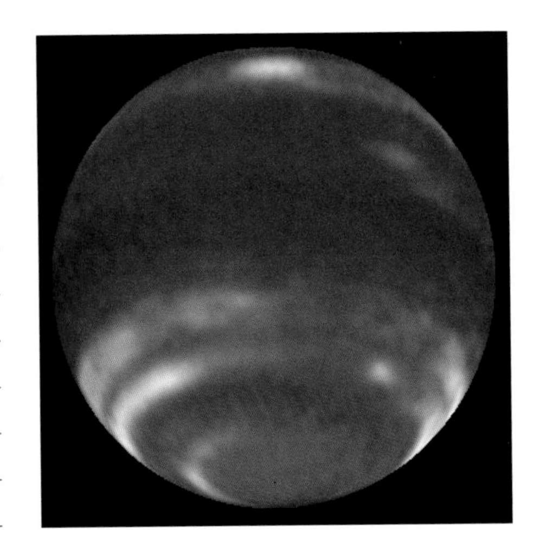

Named after the Roman god of the sea, Neptune is the azure sentinel of our Solar System since Pluto's demotion to dwarf planet status. Nearly 2.8 billion miles (4.5 billion kilometres) from the Sun, sunlight is nevertheless scattered by traces of methane in its atmosphere, hence its beautiful deep blue colour. Bright cirrus clouds of methane ice crystals cast shadows on the cloud deck 30 miles (50km) below. Although not visible to the naked eye, it is a jewel to behold. Its deep atmosphere has no solid lower boundary so speeds unchecked from east to west, and, like Jupiter and Saturn, generates convection currents from within. Despite its greater distance from the Sun, the gas giant's winds are three times stronger than Jupiter's, nine times stronger than those on Earth, and reach gusts of 600 metres per second (2,100kph) – the fastest in the Solar System. Around the equator, they approach supersonic speeds. With its temperatures averaging -214°C (-353°F) it is a dazzling icy orb. As with the other gas giants, hydrogen and helium predominate in its atmosphere and it possesses a magnetosphere similar in size to Uranus but smaller than Saturn and Jupiter: the density and, therefore, pressure of the solar wind decreases with distance.

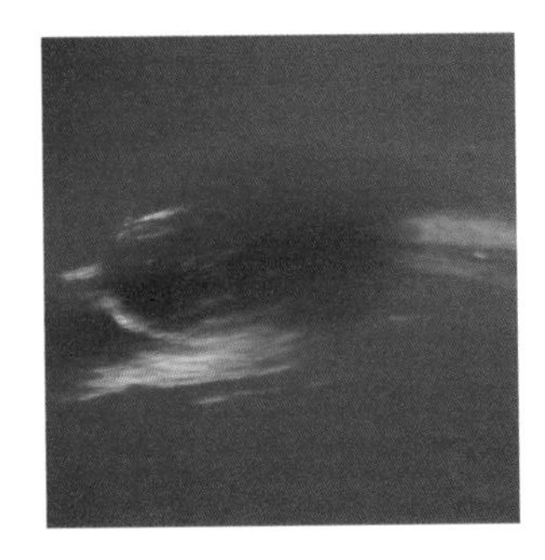

Neptune storms: the Great Dark Spot, the bright feature, 'Scooter', the Dark Spot 2 and accompanying clouds have since migrated to the equator and disappeared.

Like Jupiter, Saturn, and Uranus, it also has rings, six in total, in the form of incomplete arcs confined by one of the 14 moons, Galatea.

The largest of the moons is Triton. Three-quarters the size of Earth's Moon, it orbits in a direction opposite to the planet's rotation. Furthermore, it is spiralling inward; eventually Neptune's gravity will tear the moon apart, resulting in a ring system resembling that of Saturn. With a surface temperature around -235°C (-391°F), this moon is the coldest body yet visited in the Solar System. Yet, despite these freezing conditions, *Voyager 2* discovered geysers spewing icy material upwards of five miles (8km) high into its tenuous atmosphere. The fallout from the plumes can be seen as diffuse dark streaks on its frozen surface. And the atmosphere, even though it is around 30AU from its star, appears to be warming up.

Triton: in reality, no part of Triton would appear blue. The bright southern hemisphere is generally pink, as is the brighter equatorial band.

LOOKING BACK

Before leaving the Solar System, revise perspective: Mercury, Venus, Earth, and Mars are all within 155 million miles (250 million kilometres) of the Sun. Jupiter orbits the Sun over three times farther than Mars, and Saturn is nearly twice the distance of Jupiter. It took NASA's Cassini-Huygens spacecraft eight years to arrive … travelling at 68,000mph (109,000kph)! If we had hitched a ride on *Voyager*, travelling at 50,000mph (80,000kph), it would have taken us five years to reach Uranus from Saturn. If we return to our basketball analogy, then the reclassified dwarf planet Pluto, once the outermost denizen of the family, would be the size of a grain of sand … but would have to be placed almost a mile from our basketball Sun. To think of it another way: at an average vehicle speed of 70–90mph (112–145kph), it would still take 12,000 years to drive from one side of Pluto's orbit to the other. By passenger jet, non-stop, 2,000 years. And this is local … very, very local.

For observing the planets with an amateur telescope, see Telescope observing suggestions in Chapter 3.

Cosmic billiards

The outer limit of the Solar System is littered with icy detritus, surrounded by a hypothetical sphere of frozen debris extending from 10,000 to 100,000AU from the Sun. Within are newly designated dwarf planets, tumbling rocky asteroids and careening ghostly comets. Everywhere dust fuels majestic meteors and their showers of 'shooting stars'. Objects have been corralled into a 'belt' or exiled to a distant scattered disc. It is the legacy of Solar System birth... ceaseless cosmic billiards... let's take a closer look...

DWARF PLANETS

The International Astronomical Union (IAU) convened in Prague, Czech Republic, on 24 August 2006 reclassified 'planets' into three distinct categories:

- **Planets** – celestial bodies that (a) are in orbit around the Sun; (b) have sufficient mass for their self-gravity to overcome rigid-body forces so that they assume a nearly round shape; and (c) have cleared the neighbourhood around their orbit.
- **Dwarf planets** – celestial bodies that (a) are in orbit around the Sun; (b) have sufficient mass for their self-gravity to overcome rigid-body forces so that they assume a nearly round shape; (c) have not cleared the neighbourhood around their orbit; and (d) are not satellites.
- All other objects, except satellites, orbiting the Sun, collectively referred to as 'small solar system bodies'.

Eris, Pluto, Ceres, Makemake, and Haumea are now recognised as dwarf planets and, with the exception of Ceres, comprise five of the over 1,000 known icy trans-Neptunian objects (TNOs) located in the region of space known as the Kuiper Belt or scattered disc. Ceres, located in the asteroid belt (see below) between Mars and Jupiter, is not a TNO. Eris, Pluto, Makemake, and Haumea are further classified as plutoids, since they are dwarf planets located beyond the orbit of Neptune or, as is the case with Makemake and Haumea, they are exceptionally bright. Satellites of plutoids are not plutoids themselves. Needless to say, the number of dwarf planets will undoubtedly rise! There are estimated to be at least 70,000 100km-plus diameter TNOs, with Eris and its moon, Dysnomia, currently the most distant known objects in the Solar System, apart from long-period comets and space probes.

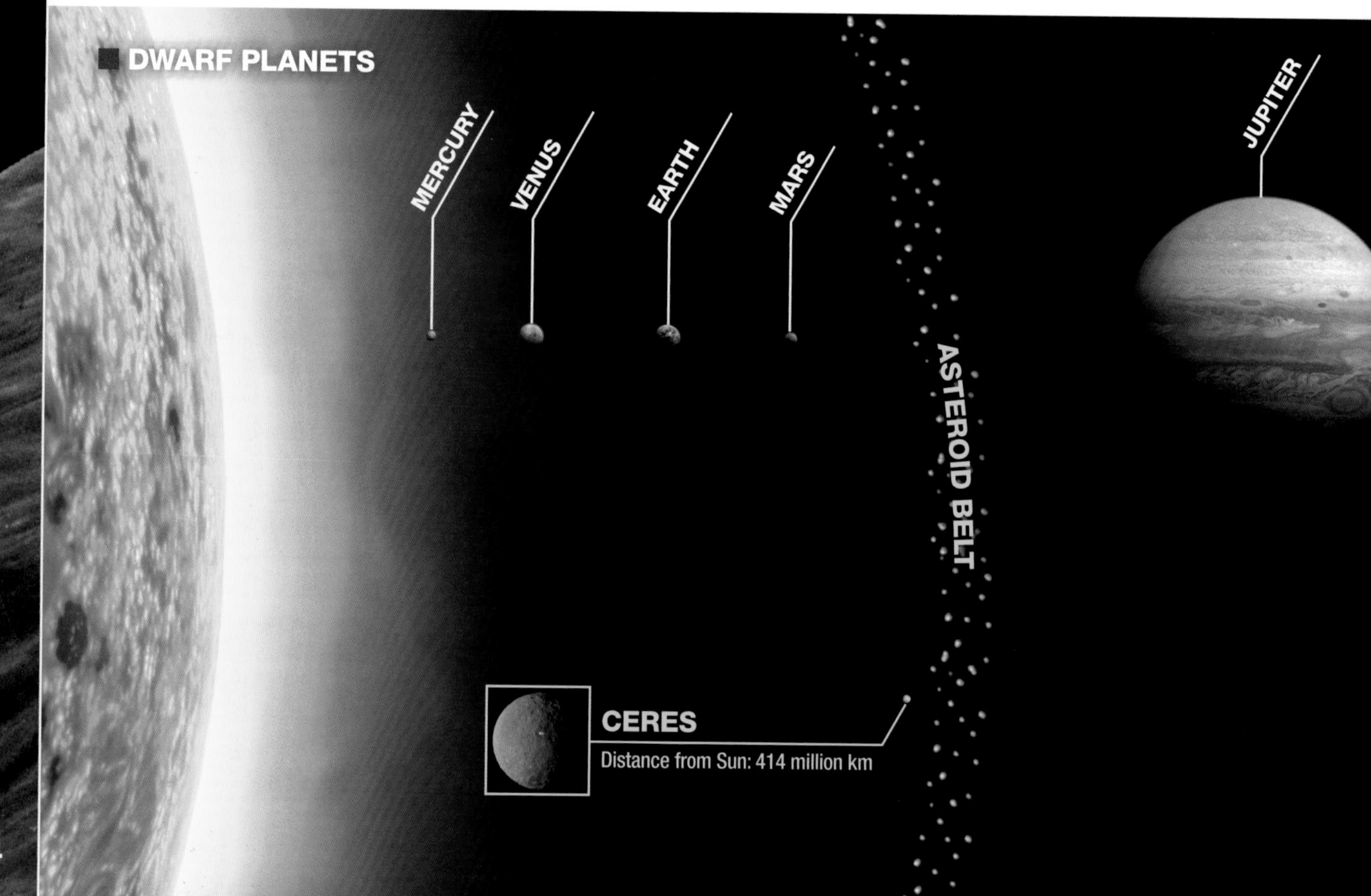

DWARF PLANETS – ORBITAL DATA

Name	Average distance from Sun (km)	Diameter	Moons	Orbital period (years)	Inclination	Average speed	Minor planet category
Eris	9.87 billion	2,398km	1	560	44°	3.4km/s	Plutoid/TNO/SDO
Pluto	5.9 billion	2,370km	5	249	17°	4.7km/s	Plutoid/TNO
Ceres	414 million	938km	0	4.6	10°	17.8km/s	Dwarf planet, main belt
Makemake	8 billion	1,500km	0	310	29°	4.4km/s	Plutoid/TNO/KBO
Haumea	7.5 billion	2,000km	2	283	28°	4.4km/s	Plutoid/TNO

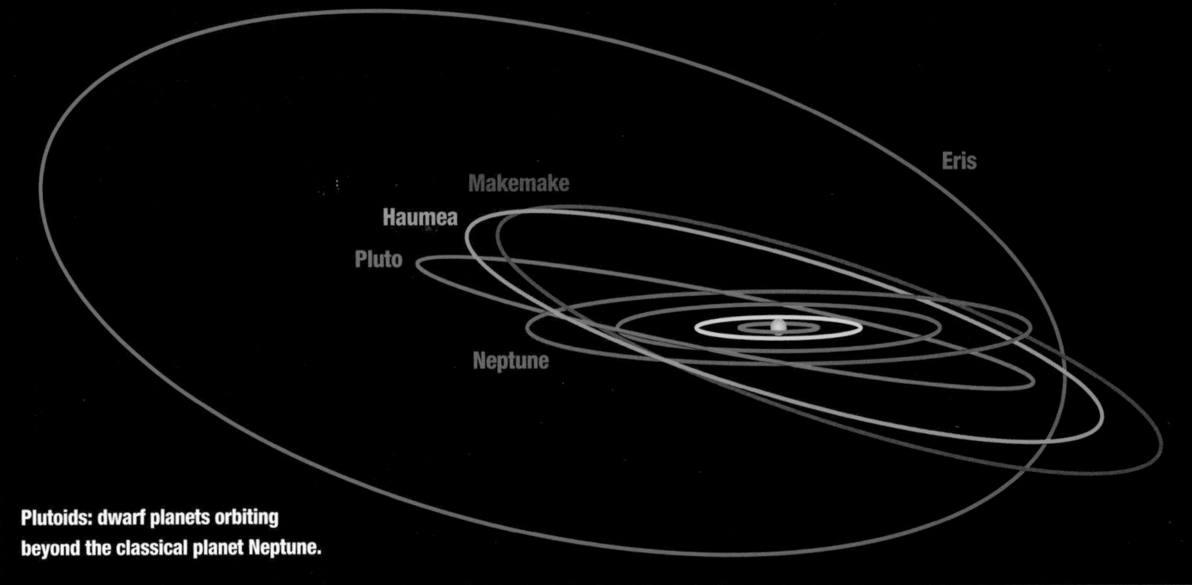

Plutoids: dwarf planets orbiting beyond the classical planet Neptune.

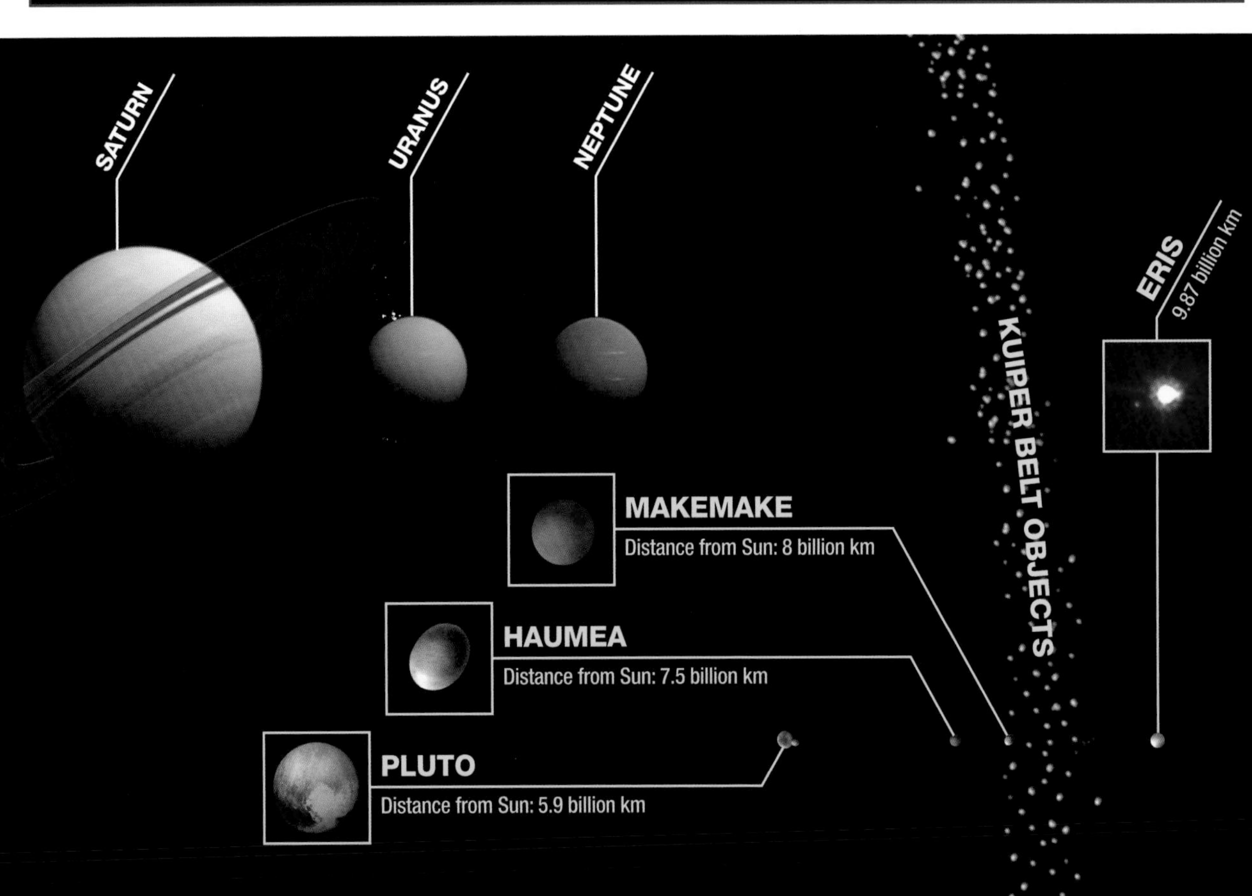

ASTEROIDS

The Solar System is peppered with smaller bodies. They range in size from tiny specks of dust – meteoroids – to asteroids, flying mountains measuring up to 600 miles across. They were so-named by Sir William Herschel since they appeared 'star-like'. Along with comets and other small bodies, these are the near-pristine remnants of the solar nebula that spawned the Solar System 4.6 billion years ago.

Also referred to as minor planets or planetoids, they are irregularly shaped, cratered piles of rubble of rock and metal. Most orbit in the asteroid belt between the orbits of Mars and Jupiter. This belt holds more than 200 asteroids larger than 60 miles in diameter, although scientists estimate it contains far more than 750,000 asteroids larger than three-fifths of a mile in diameter, along with millions of smaller ones. The asteroid belt is itself an interplanetary ghost town, since in the early days of the Solar System's formation there were perhaps 1,000 times that number; many were ejected during the first 100 million years due to gravitational interactions with the planets or other much larger asteroids. A few, Vesta (diameter 360 miles/580km) and Pallas (diameter 335 miles/540km), for example, orbit farther out. There are also some asteroids whose orbits carry them closer to the Sun (Aten, Icarus, Hephaistos).

Asteroid Ida and its moon Dactyl: scientists found the moon – the first discovered orbiting an asteroid – when NASA's *Galileo* spacecraft flew past Ida in 1994. Ida is around 32 miles (52 km) in length, irregularly shaped and believed to be like stony or stony iron meteorites.

Asteroids may be the remains of a failed planet, a large chunk of rocky material too small to have sufficient gravity to accrete into a larger object. It could be that a planet did exist but was blasted apart by the impact of another planetesimal during the Solar System's formation. Perhaps Jupiter's enormous

ASTEROID BELT

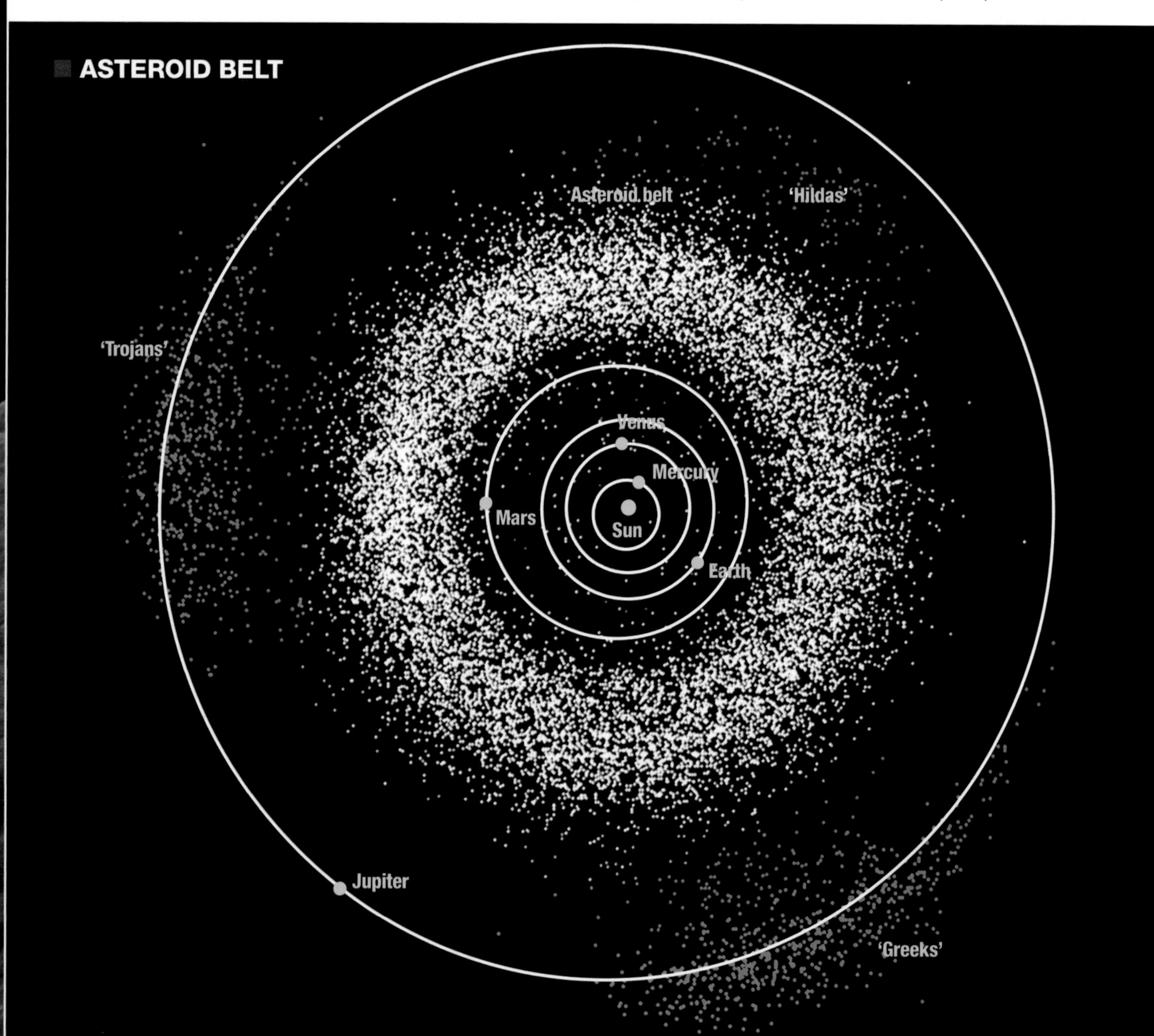

Asteroid Itokawa: has the appearance of a 'contact binary': two smaller asteroids loosely joined together. Images from Japan's *Hayabusa* spacecraft reveal a fragile, boulder-strewn landscape and a paucity of craters.

gravitational pull acted like a giant vacuum cleaner, preventing accretion. Matching the chaos of the entire Solar System, the asteroids themselves are not smoothly distributed throughout the belt. At certain distances from the Sun there are gaps – Kirkwood gaps. These gaps correspond with the locations of 'orbital resonance' with Jupiter. Resonance occurs when two orbiting bodies exert a regular, periodic gravitational influence on each other, usually due to their orbital periods being related by a ratio of two small integers. This regular interaction, in most cases, results in orbits ultimately becoming unstable; bodies exchange momentum and shift orbits until the resonance no longer exists, thereby creating gaps.

Outside the main belt, a group of asteroids, the Trojans – after Homer's legendary Trojan War heroes – orbit at the same distance as Jupiter but 60° ahead of (the Greeks, with 624 Hektor, a Trojan spy) or behind (the Trojans, with Greek spy 617 Patroclus) the planet. These areas are known as Lagrangian Points. Between the outer planets beyond Jupiter there is a small group of icy asteroids called Centaurs that resemble the nuclei of comets rather more than the solid core of asteroids; the first, Chiron, was discovered in 1977.

Periodically, as a result of the gravitational influences of Mars and Jupiter, collisions in the asteroid belt cause fragmentation. These fragments are then hurled into the Solar System in Earth-crossing orbits. There are three groups of these near-Earth asteroids (NEAs): the Amor asteroids that cross the orbit of Mars but not Earth; and the Apollo and Aten groups, both of which cross Earth's orbit. The Atens' average orbital distance from the Sun is less than that of Earth and ground-based telescopes are regularly used to monitor the skies for possible threats. There are currently believed to be roughly 1,000 of these NEAs greater than 0.6 miles/1km in diameter, ie large enough to cause global problems should they strike our planet. Such strikes are predicted to occur every 100,000 years or so.

■ COMETS

Also swooping grandly through the Solar System are the stunning comets: dirty snowballs of ice with a dusty crust measuring just a few kilometres across. These cosmic apparitions, historically inspiring awe or fear, spend most of their lives unseen, lurking in the shadows of the farthest reaches of the Solar System. But, as we have seen, everything 'out there' is moving. Disturbances from passing planets or, more distantly, the tug of the Milky Way galaxy or an occasional passing star, can easily perturb a chunk of ice and send it on a highly elliptical orbit towards the inner Solar System.

As a comet approaches the Sun, it warms up and releases gas and dust to form a tenuous coma (head), sometimes ten times the diameter of Earth. In some cases two spectacular tails, one of gas and the other dust, both millions of miles long, stream away from the coma in the headwind of charged particles from the Sun – the solar wind. The gas tail of ionised molecules is bluish in colour and almost straight whereas the dust tail is curved; its particles lag behind the comet's motion and appear yellowish since they reflect sunlight.

At least one comet collides with the Sun every year. A comet passing close to a planet will either impact, be ejected onto a hyperbolic trajectory, permanently exit the Solar System, or be perturbed into a smaller orbit. The cometary dust dissipates in space where it is then swept up by planets or falls into the Sun. It can be seen when it reflects sunlight and is known as the zodiacal light, best observed in the west in the few hours after sunset or in the east before sunrise.

Comet 67P: Single frame enhanced NAVCAM image of Comet 67P/C-G taken on 21 September 2015. The comet is oriented with the small lobe to the left and the large lobe to the right. Jets of gas and dust are seen all around the sunlit portion of the nucleus and are particularly clear around the central neck region with the ejected material seen extending towards the edge of the image frame.

Cometary perihelion: solar wind blasts a cometary tail away from our Star during its closest approach. When viewed from Earth, the comet appears to zip tail first through space as it exits the solar neighbourhood.

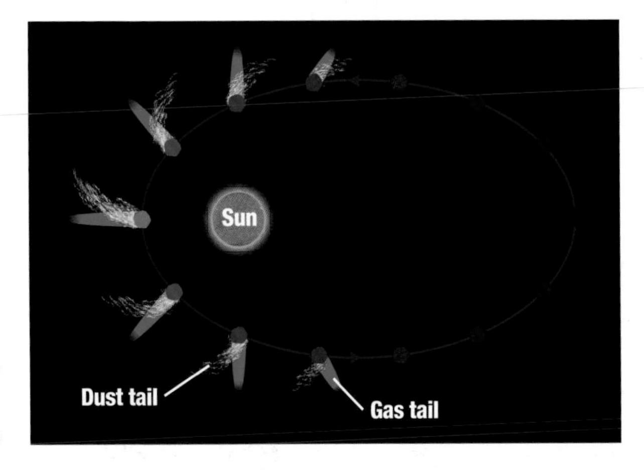

SHORT-PERIOD COMETS

Comets originate from two areas and are divided into two broad categories according to their orbital periods. Short-period comets are defined as those with orbital periods of less than 200 years. With an orbital period of 76 years, and last seen in 1986, the short-period Halley's Comet was regarded as a portent of doom, as before the Battle of Hastings in 1066. They hail from the 'local' outer Solar System in the realms of the planet Neptune some 30AU to 60AU from the Sun. This area is known as the Kuiper Belt, after the Dutch-American planetary astronomer Gerard P. Kuiper (1905–73), who suggested it as another source of comets, in addition to the Oort Cloud (see below).

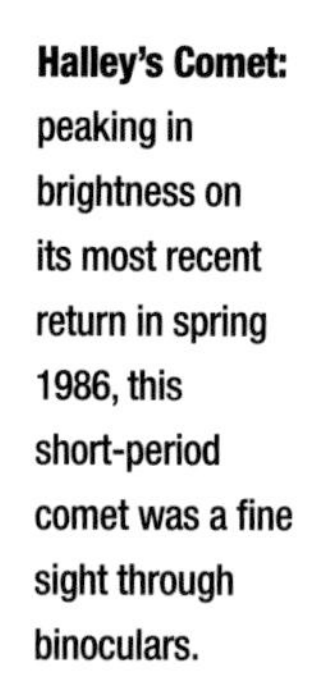

Halley's Comet: peaking in brightness on its most recent return in spring 1986, this short-period comet was a fine sight through binoculars.

Comet Hale-Bopp: the coma, or head, of a comet can be thousands of miles wide and tens of millions of miles long.

LONG-PERIOD COMETS

Long-period, infrequent comets, such as Hale-Bopp that last appeared in 1997 but was discovered by Alan Hale and Thomas Bopp on 23 July 1995, have orbital periods of thousands or even millions of years and are thought to originate much farther away in a hypothetical region known as the Oort Cloud, after the Dutch astronomer, Jan Hendrik Oort (1900–92). This halo, at a distance of 10,000AU to perhaps 100,000AU from the Sun – roughly halfway to the nearest star, Proxima Centauri – is believed to consist of a swarm of trillions of cometary nuclei surrounding the entire Solar System. Each is perhaps a kilometre across with a total mass amounting to several times that of Earth. They are icy outcasts tossed billions of years ago from the Sun's immediate vicinity by the gravity of the giant planets. Indeed, Hale-Bopp's next closest passage by the Sun – its perihelion – is not until the year 4380!

THE OORT CLOUD

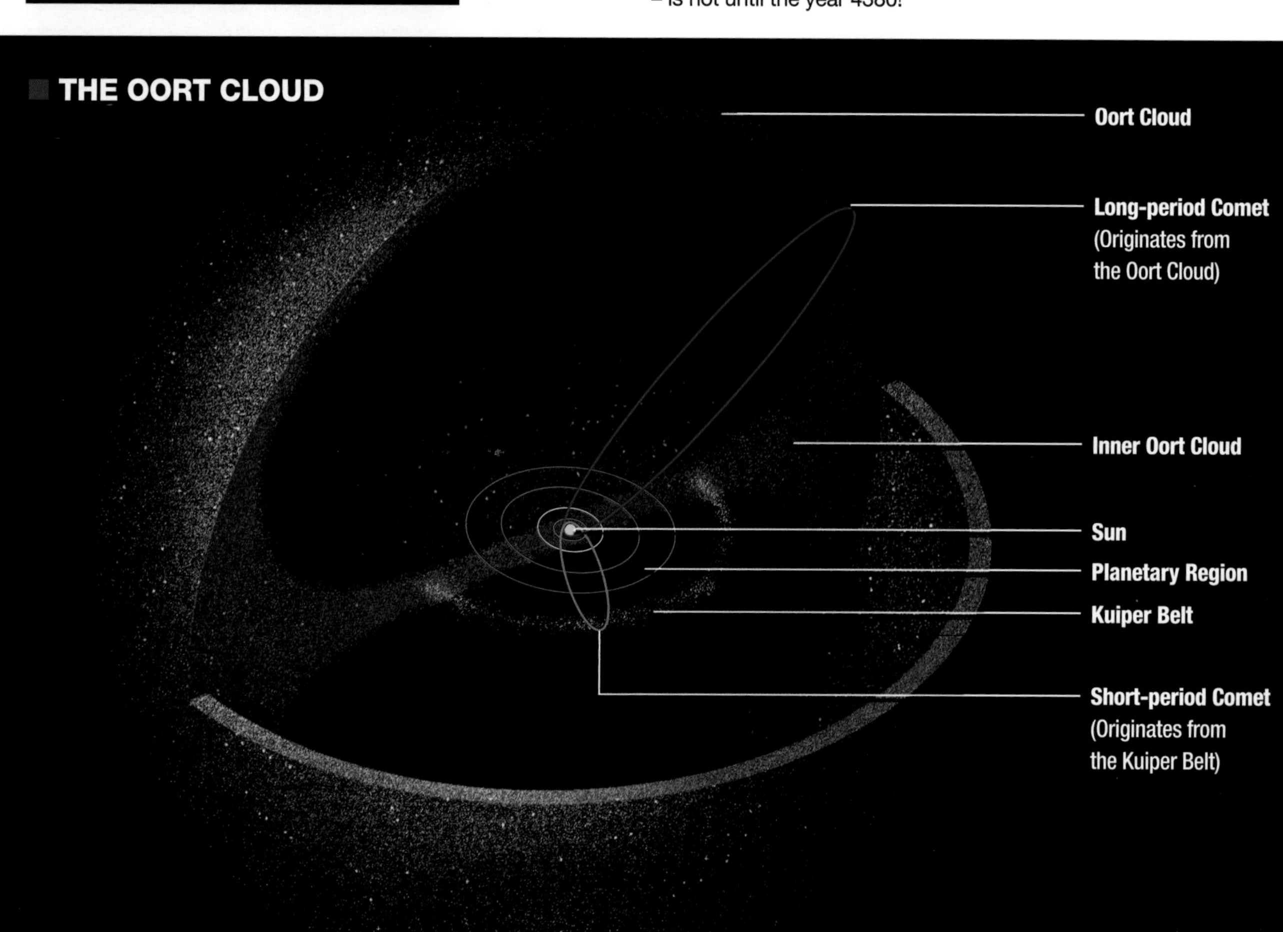

METEORS

Interplanetary dust particles (meteoroids), whether from asteroid fragments or comets, are everywhere in the Solar System. Earth, as it travels around the Sun, vacuums them up at a rate of around 40,000 tonnes a year. Most debris is microscopic and vaporises in the upper atmosphere, at around 60 miles (100km) altitude, to produce meteors, otherwise known as 'shooting stars'.

Beneath a dark country sky, roughly a dozen meteors can be seen per hour any night of the year. This figure can double after midnight when an observation point on Earth's surface has rotated from being in the shadow of the planet's forward motion to becoming the leading edge: debris impacts head-on.

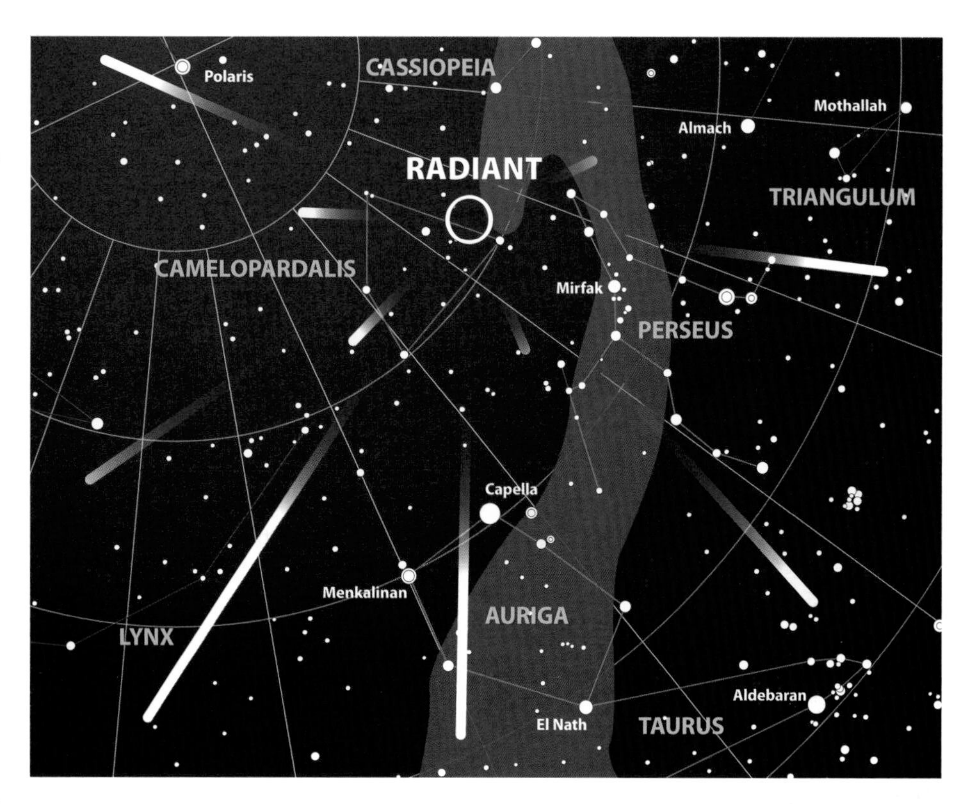

The radiant of the Perseids Meteor Shower: the point in the sky from which the 'shooting stars' appear to radiate.

METEOR SHOWERS

Dependent on Earth's orbital position around the Sun, sometimes many meteors can be seen, as in 1966 during the Leonid meteor storm when observers witnessed 40 meteors a second. This phenomenon is known as a meteor shower and is caused by Earth passing through leftover dense debris from a comet that has transited the Solar System. The showers return annually because Earth passes through the same meteoroid swarm every year. Each shower has a peak Zenith Hourly Rate (ZHR), the maximum number of meteors per hour that can be seen under perfect conditions. Because of the effects of perspective, meteors appear to radiate from a particular point in the sky – the radiant – in a manner similar to train tracks converging in the distance. Most showers are named for the constellation (star pattern) where this convergence appears to originate.

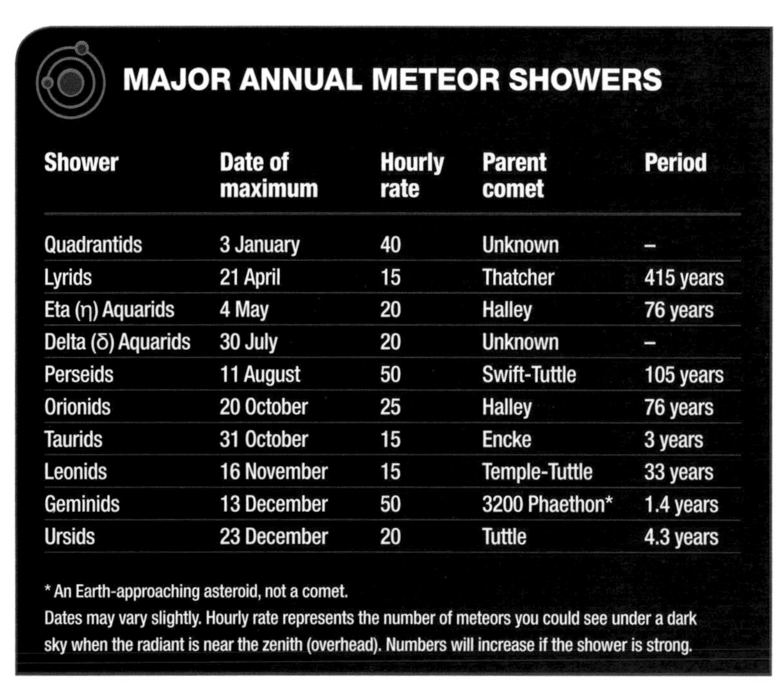

MAJOR ANNUAL METEOR SHOWERS

Shower	Date of maximum	Hourly rate	Parent comet	Period
Quadrantids	3 January	40	Unknown	–
Lyrids	21 April	15	Thatcher	415 years
Eta (η) Aquarids	4 May	20	Halley	76 years
Delta (δ) Aquarids	30 July	20	Unknown	–
Perseids	11 August	50	Swift-Tuttle	105 years
Orionids	20 October	25	Halley	76 years
Taurids	31 October	15	Encke	3 years
Leonids	16 November	15	Temple-Tuttle	33 years
Geminids	13 December	50	3200 Phaethon*	1.4 years
Ursids	23 December	20	Tuttle	4.3 years

* An Earth-approaching asteroid, not a comet.

Dates may vary slightly. Hourly rate represents the number of meteors you could see under a dark sky when the radiant is near the zenith (overhead). Numbers will increase if the shower is strong.

A Leonid meteor explodes: producing a fireball amidst a spectacular shower in 1998.

■ METEORITES

Microscopic particles simply drift down through the atmosphere and settle softly on Earth's surface. Larger meteoroids ignite dazzling fireballs. But an object heavier than around 1kg (0.04oz) can survive its fiery passage and impact the planet. When this happens it is known as a meteorite, and these 'immigrants' may well have come from the asteroid belt, the Moon, or even Mars – a distance of some 35 million miles.

Many of the earliest meteorites have been lost in a fog of mythology. Religious texts speak of 'stones from the heavens'. Pliny the Elder, the Roman encyclopaedist from the 1st century AD, described in his *Natural History* a brown rock the size of a wagon that fell from the Sun in Thrace in the 5th century BC. In 1492 a meteorite fell in a wheat field in Alsace, France, but was seen as a miracle rather than extraterrestrial. One sacred meteorite has survived in the form of the Ka'aba in a Temple in Mecca: a holy black stone revered by followers of Islam. There is also an iron dagger, believed to be of meteoritic origin, found in the burial chamber of the Egyptian Pharaoh Tutankhamen in the Valley of the Kings.

The study of meteorites became a science after 1803 when a massive fall of 3,000 stones in south-western France was witnessed by hundreds of people. Since then there have been many 'finds', as they are known. Hot and cold deserts, such as the Sahara and Atacama, are prime areas to search since meteorites survive for long periods without weathering. Antarctica is also a fruitful region. They are usually named after the town nearest to where they were found, and age-dating confirms that almost all are 4.5 billion years old and, therefore, date to the formation of the Solar System.

Meteorites are divided into three main types based on their composition: stones (94%), irons (1%), and stony-irons (5%). All provide evidence of the cosmic slot machine. Stony meteorites are subdivided into chondrites and achondrites. Chondrites, the most common, contain rounded objects 1mm or so in size known as chondrules; these droplets were once suddenly heated and then rapidly cooled – perhaps in the solar nebula around the proto-Sun or in impacts on the surface of planetesimals. Around 4% of chondrites are carbonaceous (resembling or containing carbon), thought to have the most unaltered chemical composition of all meteorites and thus providing the signature of Solar System formation.

Mundrabilla Meteorite: this 100-pound meteorite sample was found 36 years ago in Australia.

'Block Island': an iron-nickel meteorite on Mars measuring 60 cm across taken by NASA's Mars Exploration Rover *Opportunity* on Martian day (Sol) 1961 of its mission.

Achondrites are stony meteorites that have melted, suggesting a source large enough to retain heat – a giant asteroid, perhaps even the Moon or Mars. Likewise, iron and stony-iron meteorites are also thought to have the same source whereby sufficient heat created an iron-rich core and rocky outer layer. Core collisions created iron fragments. Some stony-irons, pallasites, are composed of minerals embedded in iron-nickel metal, indicating they originated from the boundary between an iron core and rocky mantle; and other stony-irons, mesosiderites, perhaps formed during a collision between two asteroids of differing composition.

Many millions of meteorites land on Earth every year, most unseen in oceans or remote areas. Among recent 'falls' are the Carancas Meteorite in Peru in September 2007, when windows shattered over a kilometre away and the resultant crater was 13m across and 4.5m deep. Thousands saw a ten-tonne rock flaming through the sky in Saskatchewan, Canada, in November 2008. Although a crater was not found, fragments from the Buzzard Coulee Meteorite, weighing as much as 13kg, are still being discovered.

A meteorite weighing more than a few hundred tonnes and travelling at speeds between 10 and 20mi/s (15–30km/s) will form a crater on impact. The largest found to date, Hoba West, fell in 1920 near Grootfontein in Namibia and weighed in at an estimated 60 tonnes. Kinetic energy is converted to heat causing a tremendous explosion that vaporises both the meteorite and some of the rock beneath and forms a hole at least ten times the size of the impactor. If the impactor lands in the sea, a 'crater' will be produced in the water that will rapidly fill to form a tsunami.

A crater-forming impact occurs on Earth roughly every 5,000 years, causing localised devastation. The largest impacts have global consequences, the best documented of which is the impact that took place 65 million years ago in the Yucatan region

Barringer/Meteor Crater near Flagstaff, Arizona, 1km wide, 170m deep: formed 50,000 years ago when an iron meteorite 50m in diameter (Canyon Diablo) smashed into the ground.

of Mexico. The crater is named Chicxulub after the small town near its centre. It is not known if it was a comet or asteroid but it had a diameter of at least six miles and weighed more than a trillion tons. The crater is 110 miles (177km) wide and the impact energy equalled five billion Hiroshima-size nuclear bombs, generating an earthquake exceeding 12 on the Richter Scale. Millions of tons of dust surged upward to form a cloud that rapidly encircled the planet – evidenced by a layer of iridium-rich clay sediment discovered by geologists in 1978 – and led to a runaway greenhouse effect, as seen on Venus, that lasted for centuries. Sunlight was eradicated for a month and Earth was plunged into darkness for several more. Debris rained down, creating more impacts, more explosions, and sending temperatures soaring. Melted rock sprayed everywhere igniting global firestorms. Great tsunamis engulfed the coasts of Mexico and Florida. Sulphuric acid rained down. It is believed that there were subsequent mass extinctions, wiping out 80% of all species, including the dinosaurs.

Such impacts are ubiquitous in the Solar System. In July 1994 fragments of Comet Shoemaker-Levy 9 (after co-discoverers Carolyn and Dr Eugene Shoemaker and Dr David Levy in 1993) plunged into the giant planet Jupiter. It had been in orbit around the gas giant for over 60 years but, after a close approach in 1992, shattered into a 'string of pearls' comprising more than 20 fragments. All the impacts occurred on the far side but, as the planet rotated, astronomers witnessed 'bruising' in the form of dark spots, some 2,000 miles across, which were easily visible even in amateur telescopes. Had this happened on Earth craters measuring some 37 miles (60km) across would have gouged out its surface. And impacts still continue, the most recent caught by the Hubble Space Telescope in July 2009.

And then there are the giant impacts: the one that possibly created our Moon, as described earlier, and the enormous impacts with Mercury. One or more of these may have torn away a fraction of its mantle and crust, leaving a body dominated by an iron core. Craters, such as the Caloris Basin on Mercury, or the Victoria crater on Mars, bear testimony to the violent epoch of our Solar System's creation.

Victoria Crater: an 800m-wide crater in Mars' Meridiani Planum region, imaged by NASA's *Mars Reconnaissance Orbiter* as it flew 300km overhead.

Comet Shoemaker-Levy: impacts Jupiter in July 1994.

When the comet was observed, its train of 21 icy fragments stretched across 710 thousand miles (1.1 million km) of space.

A view of the barred spiral galaxy NGC1672 captured by the Hubble Space Telescope (HST): barred arms funnel gas and dust towards the 'hungry' supermassive black hole at its core.

Our perspective

Where are we in space? Where is our yellow dwarf, type G2 star with its retinue of planets? Are there other stars 'out there'? Do they play host to planets and life? What does the Universe look like? How did it begin? When, where and how will it end? So many questions... so many answers... leading to further questions demanding further answers, our enquiry as ravenous, far-reaching and restless as the cosmos itself. We cannot know. We may never know... but we must always ask... all of us... always...

'No more or less important than a bug on an aphid's back.'

Cheryl R. Lutring, b.1948

Star stuff

For the most part, the cosmos is an ocean of seemingly empty space, its darkness occasionally relieved by giant gatherings of stars. Our Sun resides in one of these galaxies – the Milky Way – so to our eyes in our minuscule region of the Universe, space is stuffed with stars. As already seen, with the exception of the Moon and planets, every fixed point of light, including our Sun, is a nuclear reactor – a star – and they are innumerable and diverse. There are stars resembling our Sun, red dwarf stars, white dwarfs, giants, supergiants and even hypergiants, Wolf-Rayet stars, and more. To our earthbound eyes they are strewn across the night sky, their twinkling (scintillation) caused by their light interacting with atoms in Earth's upper atmosphere. Some are immensely hot, pumping out the power of 100,000 of our Sun, or more. Some are cool. Some are 'close' and others unimaginably distant. Like our Sun, they are born, they live, and they die, but all stars have lives longer than 10,000 of our human lifetimes. Half consist of two or more stars locked in a gravitational embrace, but there are triple, quadruple, and quintuple star systems, and then there are the beautiful multiple systems where hundreds, to thousands, to millions of stars are gravitationally bound in open or globular clusters.

APPARENT BRIGHTNESS

Some stars are apparently brighter than others due, in part, to their distance. Imagine looking along a road of equally powered streetlights: those closest appear brighter. Those more distant appear dimmer, but they are the same. Gaze along the same street to more distant floodlights, a sports stadium say, and these appear brighter still; that's because they are. This impression of brightness is known as apparent magnitude. Recognising this in the 2nd century BC, the Greek astronomer Hipparchus divided stars into six brightness classes, or magnitudes, the faintest being sixth magnitude and the brightest first. The scale was refined in 1856 by Norman Podgson of Radcliffe Observatory. A star of apparent magnitude one is exactly 100 times brighter than a star of magnitude six (see table opposite). These are all apparent magnitudes as they appear from Earth. At absolute magnitude – at a standard distance of ten parsecs (32.6 light years) – the Sun would be an unremarkable +4.8 and Sirius a modest +1.4.

TOP TEN BRIGHTEST OBJECTS

Magnitude	Object	Detail
-26.8	Sun	As seen from Earth
-12.6	Moon	When at full Moon
-4.7	Venus	When at its brightest
-2.9	Mars	When at its brightest
-2.9	Jupiter	When at its brightest
-1.9	Mercury	When at its brightest
-1.4	Sirius	Brightest star in night sky
-0.7	Canopus in night sky	Second brightest star
-0.3	Saturn	When at its brightest
-0.01	Alpha (α) Centauri	Leading star in constellation Centaurus

+3.0 = faintest stars visible in light-polluted urban sky.
+6.5 = faintest stars visible at dark-sky site.
+9.5 = faintest stars visible with 10 x 50 binoculars.
+30 = faintest stars visible with the Hubble Space Telescope.

Omega (ω) Centauri NGC 5139: the largest globular cluster in the Milky Way galaxy's halo; home to some ten million stars.

STELLAR DISTANCES

If stars are so unimaginably far away, how can astronomers calculate their distance?

The ancient Greeks invented star patterns/pictures for the prominent nearby stars – the 88 constellations. Unaware of their distances, they 'placed' them on an imaginary transparent sphere of uncertain size and accepted they were farther away than the Sun. Early attempts to measure distances proved unsuccessful because stars did not show any detectable shift in position as Earth orbited the Sun – a requirement for trigonometric calculation. But in 1685, Isaac Newton (1642–1727) used an indirect method to gauge the distance to the brightest star, Sirius, basing his calculations on the brightness difference between it and the Sun. He worked out Sirius was nearly a million times farther away than the Sun. He was wrong, but it was a revelatory indication of the enormity of the cosmos.

PARALLAX

The key to measuring the nearest stellar distances came with the technique of trigonometric parallax, first successfully used in 1838 by the German astronomer Friedrich Wilhelm Bessel when he determined the parallax of the star 61 Cygni. Parallax is the shift in the position of a nearby star with respect to more remote background stars and is caused by the change in our vantage point as we orbit the Sun – similar to holding a pencil steady in front of your nose, then closing one eye, opening it and closing the other. As you view with each eye the pencil appears to move against the background even though it is stationary. The diameter of Earth's orbit provides a similar baseline and the size of the parallax shift correlates directly to the distance of the star. As telescope precision improved, microscopic parallax shifts could be measured for stars at even greater distances. It was soon shockingly realised that stars were so far away their light took years to reach us. The concept of the light year (ly) was born.

Parallax: the shift in position of a nearby star with respect to background stars caused by the change in our vantage point as we orbit the Sun.

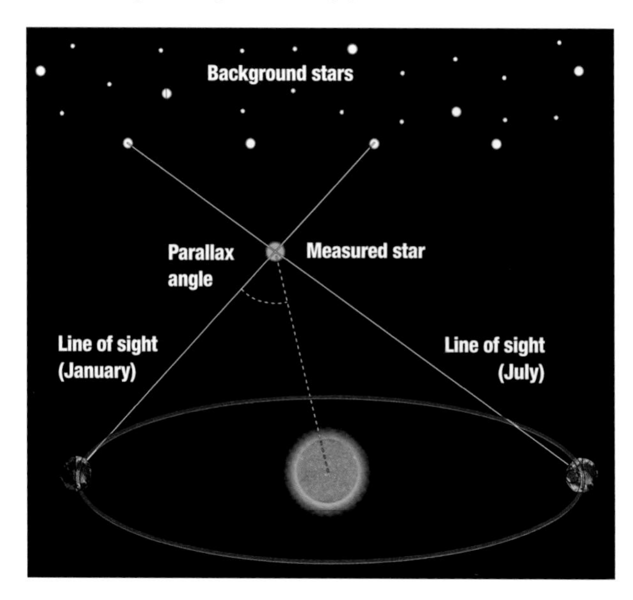

Sirius the Dog Star: by comparing its brightness with the Sun, Isaac Newton calculated that it was nearly a million times more distant.

THE LIGHT YEAR

When contemplating objects beyond our Solar System the word 'distance' assumes an entirely new, and bewildering, meaning and never fails to elicit gasps of awe from the uninitiated. The concept of the Earth 'mile' is inconsequential and irrelevant. To convey distances 'out there' astronomers talk in terms of the light year (ly). This is a vital unit of distance (not time) and simply equates to how far a beam of light can travel in one calendar year. Light travels at the fastest known speed in the Universe – a staggering 186,000 miles a second (300,000km/s) – covering a mind-numbing distance of 670 million miles an hour, or 300 million metres every second. A light year is, therefore, equivalent to 5.9 million million miles (9.5 million million kilometres) – the distance it will travel in a year. It takes light eight minutes to reach us from our Sun, an average distance of some 93,000,000 miles (1AU). It takes 43 minutes to reach us from Jupiter (5AU) and roughly seven hours from the dwarf planet Pluto (40AU). A related unit of distance is the parsec, the range at which a star would have a parallax of one second of arc (1/3,600 of a degree), and this is equal to 3.26ly.

It was soon realised that the light from Sirius takes 8.6 years to reach Earth, even though it covers a distance of 186,000 miles per second. It is enormously far away. The star 61 Cygni's parallax placed it even farther – 10.3ly. Once a star's distance from the Sun was known, astronomers could gauge its apparent magnitude (brightness) to how bright it really was – its absolute magnitude – the brightness it would have at a standard distance of 10 parsecs (32.6ly).

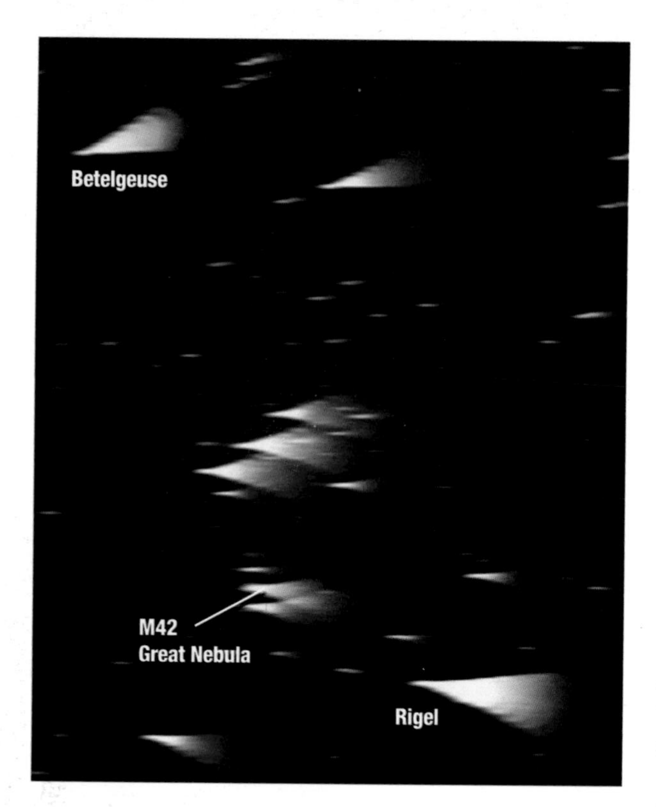

The constellation of Orion, the Hunter: temperature determines a star's colour. Imaged using a time exposure method to create star trails, the 'cooler' supergiant star, Betelgeuse, glows red among its hotter, bluish companions.

STELLAR MOTION

Although stars appear fixed in their constellation patterns, astronomers know from precise measurements, over enormous timescales, that they do move relative to each other – transverse proper motion. Barnard's Star, a red dwarf only 6ly away, has the largest proper motion across the sky, travelling the apparent width of the Moon in 180 years. Stars also move away from us or towards us, their speed being known as radial velocity. These motions are a result of the movement of the stars around the centre of our Galaxy. Consequently, over many millions of years the appearance of the constellations as we know them will change.

STELLAR CLASSIFICATION

The size of a star is obtained indirectly by studying its temperature and luminosity, and temperature is determined by colour. The colours of the visible 'surface' of stars can be equated to placing an iron poker in a fire; at first it will appear dull, but, as it heats up, it will glow orange, then yellow, and ultimately white. Likewise, the hotter a star, the brighter its colour. Yellow stars, like our Sun, have surface temperatures in the region of 6,000°C (10,800°F). A cooler star like the supergiant Betelgeuse in the constellation of Orion has a temperature of 3,650°C (6,600°F) and appears orange/red, emitting most of its energy as infrared radiation. The star Rigel, at the foot of Orion, dazzles blue-white since its surface temperature is over 11,000°C (20,000°F).

A star's basic properties denote its 'spectral type'. Each star is designated one of seven letters: O, B, A, F, G, K, M, (remembered by the mnemonic 'Oh, Be A Fine Girl/Guy Kiss Me'). This is a sequence of ever-decreasing surface temperatures, and also (for main sequence stars), of ever-decreasing luminosity, mass, and radius. Sub-classes 0 to 9 were also introduced to allow finer distinction. Using this scheme, the Sun is classified as a G2 star. The table below details the main sequence stars at the start of each division and includes the Sun.

STAR TYPES AND THE MAIN SEQUENCE

There is a link between the absolute magnitude and types of stars. The Hertzsprung-Russell (HR) diagram is fundamental to our understanding of stars and their life cycles and was created in 1913 by Henry Norris Russell from the USA and, later, Ejnar Hertzsprung from Holland. It plots surface temperatures (or spectral types) against luminosities (absolute magnitudes), making allowances for distance. Ninety per cent of stars lie along the band that runs from top left to bottom right, called the main sequence. Our spectral type G2 Sun lies midway. Main sequence stars are often called dwarfs, although they can be 20 times larger and 20,000 times brighter than our Sun.

Stars populate specific areas of the diagram and this gives an indication of their physical nature and their stage of evolution. At the cool, faint end of the main sequence (lower right) are the most common red dwarfs. These are smaller than the Sun and have longer lives – tens of billions of years – since they require less fuel and they use it more slowly. If these stars were visible to us, then the night sky would be rich with them, but since they appear so dim we can only observe the closest, like Proxima Centauri, 4.22ly distant.

To the lower left of the main sequence are the white dwarfs. These are smaller than red dwarfs, typically the size of Earth but with the mass of the Sun. They are dying stars, rich in heavy materials such as iron and consequently enormously dense; their mass equates to squeezing a Boeing 747 aircraft into a matchbox! Having exhausted their fuel, they will cool and die over millennia.

STELLAR CLASSIFICATION

Type	°C	Colour	Radius (1 = Sun)	Mass (1 = Sun)	Luminosity (1 = Sun)
O	38,000	Blue	18	40	500,000
B	30,000	Blue-white	7.4	18	20,000
A	10,800	White	2.5	3.2	80
F	7,240	Yellow-white	1.4	1.7	6
G	6,000	Yellow	1.05	1.1	1.26
G2	5,920	Yellow	1	1	1
K	5,150	Orange	0.85	0.78	0.40
M	3,920	Red	0.63	0.47	0.063

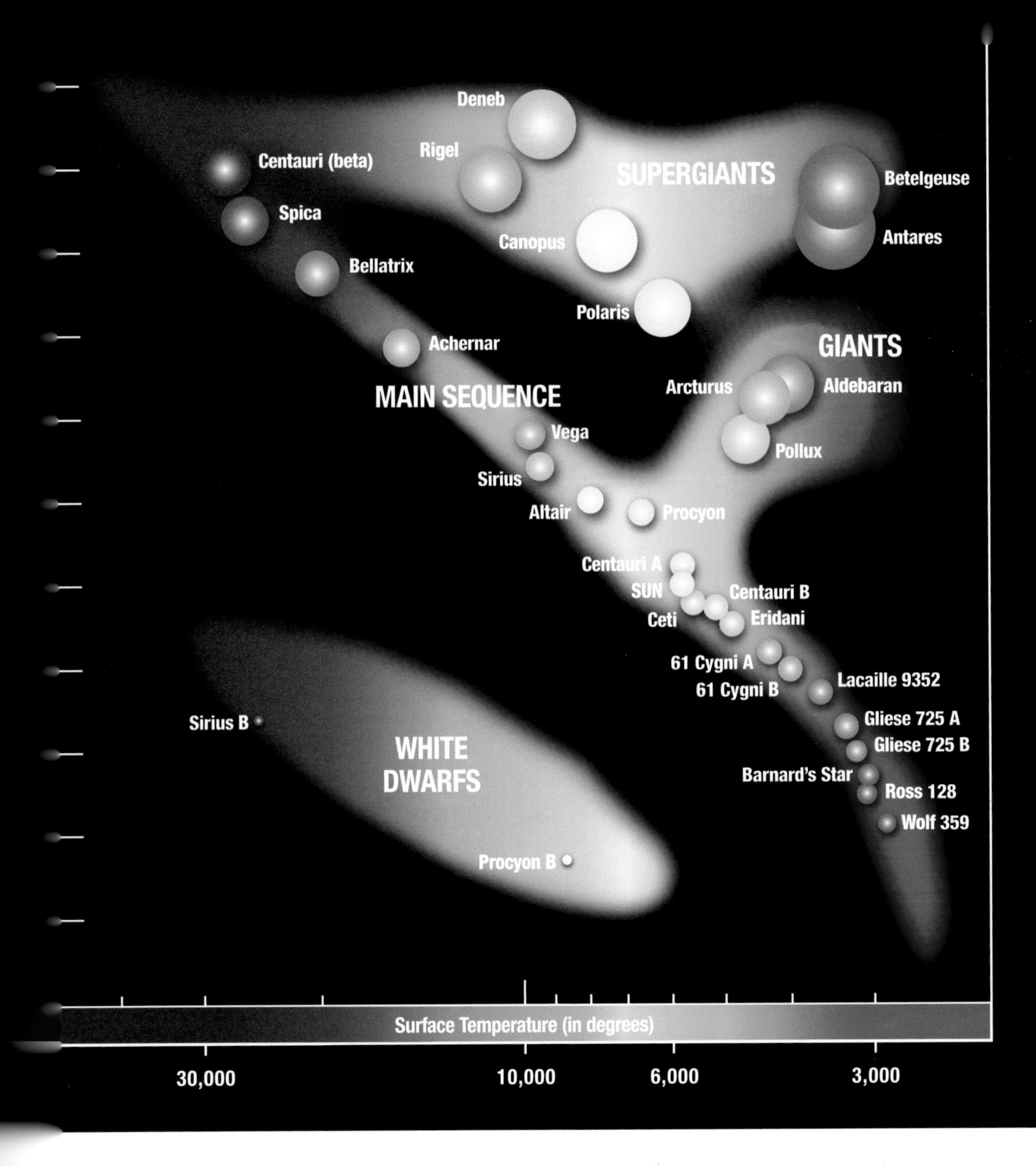

usually orange in colour and are the most after those on the main sequence. They have ce temperature as red dwarfs but are much e luminous. Consequently, they appear above ence (upper right of centre). These stars have s as our Sun but their atmospheres are huge es sufficient to engulf the inner planets of our

pergiants are located at the upper right of the HR as their name suggests, they are monsters with at more mass than our Sun. Examples are the red d blue Rigel mentioned earlier. Betelgeuse has 5.5 times the Earth–Sun distance: if it usurped our Sun, its atmosphere would stretch out to the (5.2AU). Rigel, although one-tenth the size of the 8.5-million-year-old Betelgeuse, is still almost 100 times the size of our Sun! These stars normally have short lives – mere millions of years – since they require an enormous amount of nuclear fuel to sustain them. They use up their energy reserves very quickly: hydrogen is fused in a set of reactions known as the CNO cycle where the nuclei of carbon (C), nitrogen (N), and oxygen (O) act as catalysts. This process induces enormously high temperatures and tremendous amounts of energy.

If the mass of a proto-star (baby star) is less than about 80 times the mass of Jupiter, the gas at its core will never become sufficiently hot or dense to ignite nuclear burning. These 'failed' stars are known as brown dwarfs with low surface temperatures of 1,000–2,000°C (1,000–3,600°F), detectable at infrared wavelengths. If the gaseous ball has a mass of only around ten Jupiters it will be classed as a planet.

■ STARLIGHT

Stars radiate their light as electromagnetic radiation. The colour of this light depends on its wavelength (see also Chapter 4). White light is a mixture of wavelengths and can be separated by passing it through a prism or diffraction grating. The band of colours produced is known as a spectrum and the study of spectra is known as spectroscopy. The spectrum can contain bright (emission) or dark (absorption) lines, indicating chemical elements present in the star or the intervening medium. The patterns of these lines, unique to each star like a stellar fingerprint, is determined by many factors, dominant ones are temperature and density.

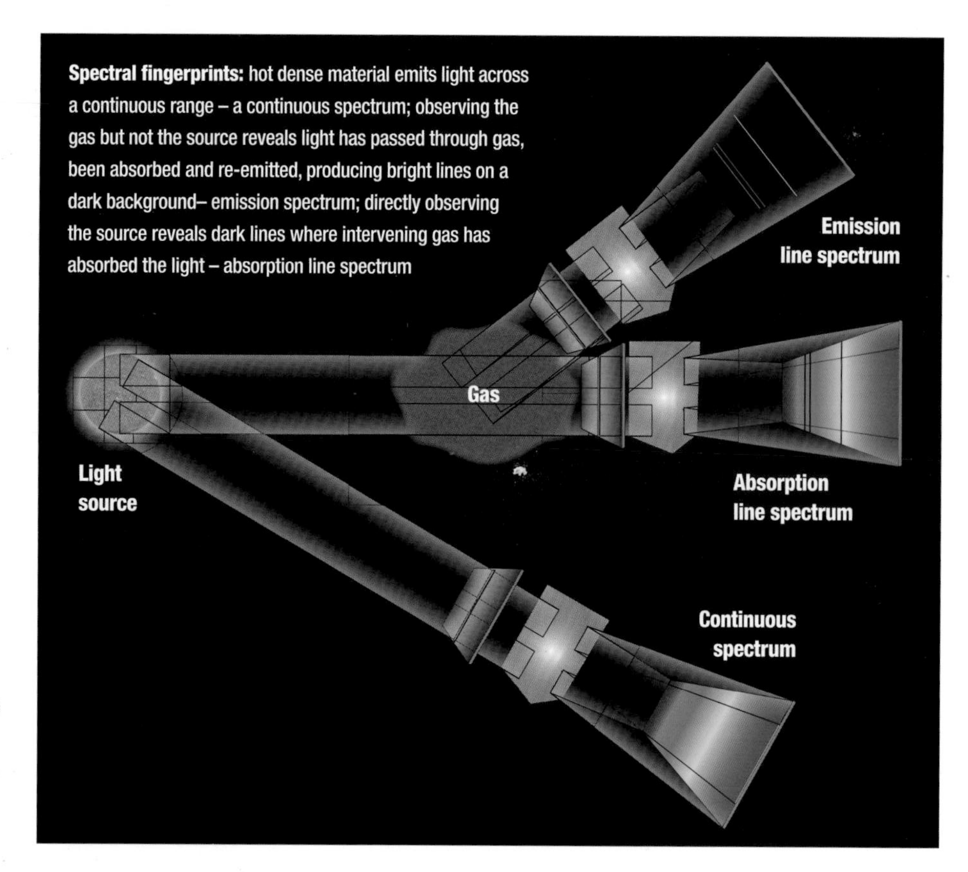

Spectral fingerprints: hot dense material emits light across a continuous range – a continuous spectrum; observing the gas but not the source reveals light has passed through gas, been absorbed and re-emitted, producing bright lines on a dark background– emission spectrum; directly observing the source reveals dark lines where intervening gas has absorbed the light – absorption line spectrum

■ STAR DEATH

We have seen how our Sun was born and, like all the stars out there, it will die. The manner of death for a star is determined by its initial mass; the bigger a star, the more spectacular its demise.

RED GIANTS

When our Sun's core supply of hydrogen runs out, hydrogen burning will spread to its outer layers and helium burning will take over in the core. More energy will be released and our Star will 'bloat' into a red giant, engulfing the inner planets of Mercury, Venus, and possibly Earth. Eventually, unable to sustain this energy release, it will, along with all stars of a similar mass, shed its outer layers into space, forming beautiful, glowing planetary nebula. At their centre will be a dense dying white dwarf, slowly cooling into invisibility.

Butterfly Nebula, NGC 6302: a striking example of a bipolar planetary nebula 2,100 light-years away in the constellation Ophiuchus, formed when the central object is not a single star, but a binary system.

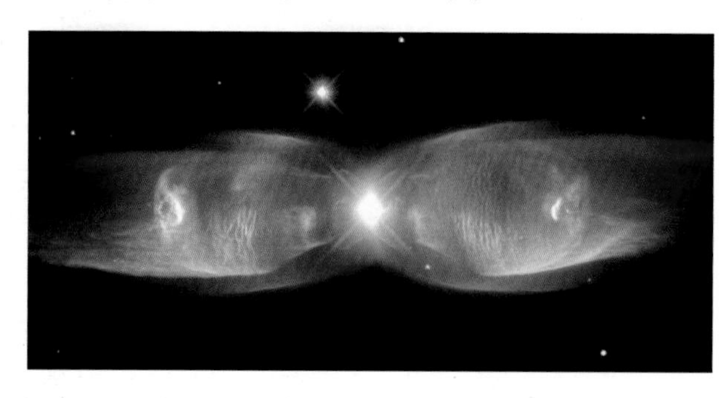

SUPERNOVA

The death of the giant and supergiant stars, those with at least 20 times more mass than our Sun, involves successive cycles of nuclear reactions, producing increasingly complex chemicals, ending up with iron which is unable to fuse to create any further energy. As a result, the core collapses and rebounds in a fraction of a second; a temperature in excess of 50 billion degrees is created and a catastrophic explosion – a supernova – results. It is estimated up to five supernovae occur every second somewhere in the Universe, and in our Galaxy perhaps two or three each century, the last observed by Johannes Kepler in 1604. Light from the explosion can be as bright as billions of stars put together. Material in the form of clouds is blasted into space, over several centuries, at a staggering 20,000mi/s to 6,000,000mph: the reflected light, in the form of 'echoes', is detected by astronomers and used to yield clues as to its origin. As this material hurtles through space, it smashes into other giant molecular gas clouds causing compression: more compression leads to greater gravity and the birth of more stars – sequential star formation. Supernovae also act like snowploughs; the initial shockwave sweeps up more material to feed star birth, utilising the heavy elements created in the explosion to generate new planets and life.

At the core of a supernova atomic protons and electrons will have been crushed together to form neutrons. The result will be a highly magnetised neutron star – a tiny 12-mile wide decaying star. These are so dense that, using an earlier analogy, it equates to squeezing 36 million Boeing 747s into a matchbox, or packing all the world's cars into a thimble.

Atomic particles focus into beams along the star's magnetic poles. When observed from Earth these beams appear to flash as the neutron star spins sometimes up to a thousand times a second. The star is then known as a pulsar. The neutron star at the centre of the Crab Nebula, M1 – a supernova remnant from an explosion seen by Chinese astronomers in the constellation of Taurus in 1054 – is known as the Crab Pulsar and rotates at 30 times a second. Over thousands of years this will eventually slow down. Even though it has the same mass as our Sun, it measures a mere six miles across!

Some stars are simply so gigantic that when they die in a supernova explosion what remains is a black hole. The central mass is so great that the core cannot counteract the inexorable squeeze of its own gravity. Its ultimate collapse produces 'singularity' – a point in space where gravity is so intense that even light cannot escape – the surrounding 'surface' of which is known as an 'event horizon'. Astronomers cannot see a black hole directly; its presence is indicated by the effect it has on its surroundings.

Supernovae are generally described as being two types: Type I, which do not reveal hydrogen in their spectra and take one of three forms; and Type II, which have strong hydrogen emission lines and are the result of the collapse of a star with at least eight times the mass of our Sun (see table at right).

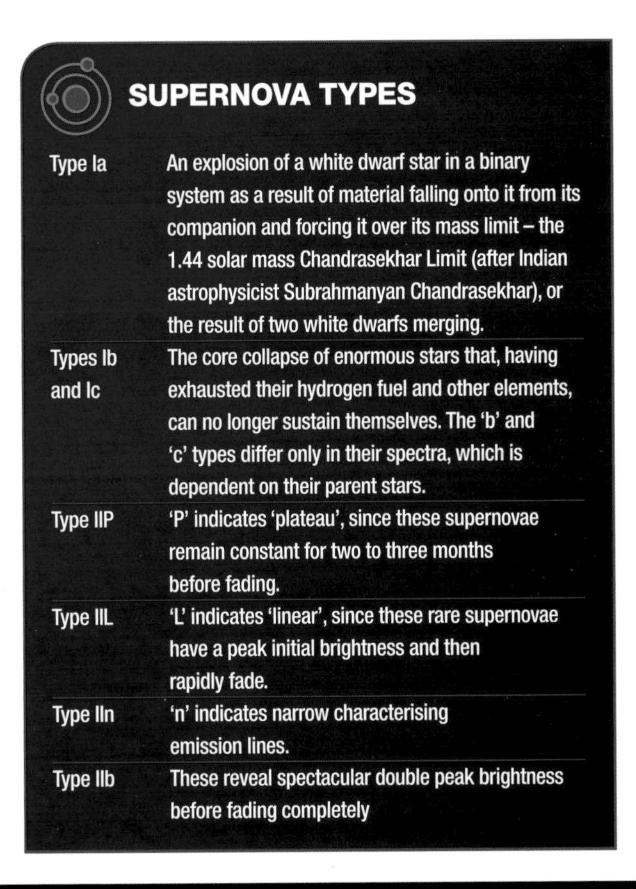

SUPERNOVA TYPES

Type Ia	An explosion of a white dwarf star in a binary system as a result of material falling onto it from its companion and forcing it over its mass limit – the 1.44 solar mass Chandrasekhar Limit (after Indian astrophysicist Subrahmanyan Chandrasekhar), or the result of two white dwarfs merging.
Types Ib and Ic	The core collapse of enormous stars that, having exhausted their hydrogen fuel and other elements, can no longer sustain themselves. The 'b' and 'c' types differ only in their spectra, which is dependent on their parent stars.
Type IIP	'P' indicates 'plateau', since these supernovae remain constant for two to three months before fading.
Type IIL	'L' indicates 'linear', since these rare supernovae have a peak initial brightness and then rapidly fade.
Type IIn	'n' indicates narrow characterising emission lines.
Type IIb	These reveal spectacular double peak brightness before fading completely

EVOLUTION OF STARS

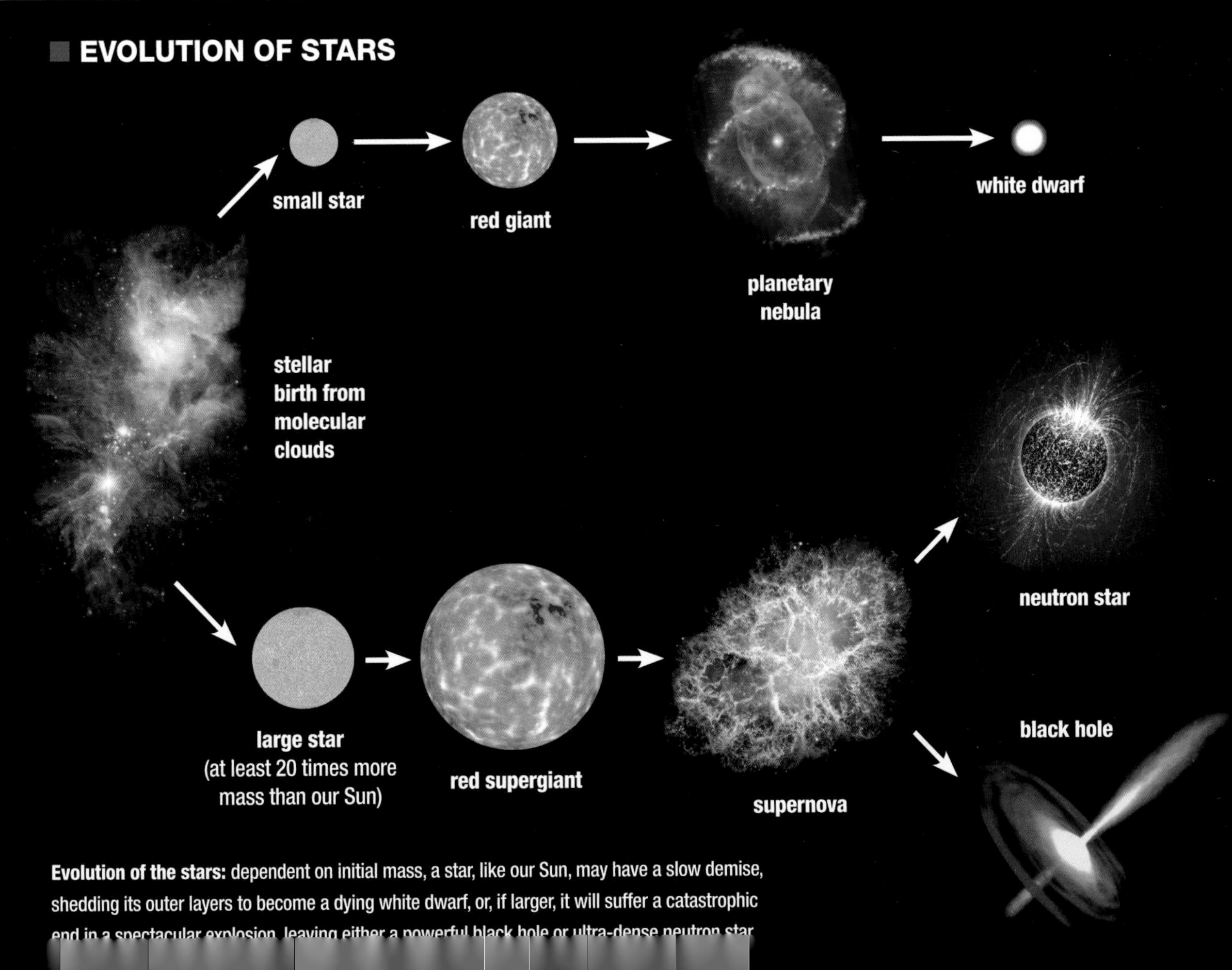

Evolution of the stars: dependent on initial mass, a star, like our Sun, may have a slow demise, shedding its outer layers to become a dying white dwarf, or, if larger, it will suffer a catastrophic end in a spectacular explosion, leaving either a powerful black hole or ultra-dense neutron star

VARIABLE STARS

Like us in our brief lifetimes, stars are not always well behaved or predictable prior to their demise! Most exhibit a form of 'irritability' at some stage; it is a natural part of their evolution. Such instabilities can be temporary or permanent and take many forms:

PULSATING VARIABLE STARS

Pulsation – a puffing out and subsequent collapse of their outermost layers – is a common source of change in stars. Perhaps the most important category of pulsating stars is the Cepheids (spectral type F and G), named after Delta (δ) Cephei, the first to be discovered. Like Mira stars (see *Red variables*), a Cepheid's fluctuations result from a cycle of changes inside the star, causing it to expand and contract. These stars are yellow supergiants that oscillate with clockwork regularity on anything from a daily to monthly basis. As they do so, their visible light changes by as much as two magnitudes, their temperature by over 1,000°C and their diameter by about 10%. The northern pole star, Polaris in Ursa Minor, is a Cepheid star with minor fluctuations.

They are rare but highly luminous and can be seen at enormous distances. In 1784, a deaf-mute English teenager, John Goodricke, discovered Delta (δ) Cephei's changes. A century later, Harvard's Henrietta Leavitt (1868–1921) studied the cycles of around 25 Cepheids in the Small Magellanic Cloud, a satellite galaxy of the Milky Way, and realised that if their average magnitudes were brighter, then the periods of their variation were longer. Harlow Shapley (1885–1972) later realised that if two Cepheids had the same period of variation, the one with the brighter average magnitude would be closer to us. This he called the period-luminosity relationship and it became the Rosetta Stone for measuring distances in space.

Also used as distance indicators are the RR Lyrae stars found in great numbers in old globular clusters. They are similar in brightness to the Cepheids but their periods are shorter than a day and all have about the same absolute magnitude, +0.5. If a star is recognised as an RR Lyrae, its apparent magnitude can be used to calculate distance. Then there are the almost undetectable Delta (δ) Scuti stars that oscillate with periods of just a few hours or less, changing in brightness by only a few per cent.

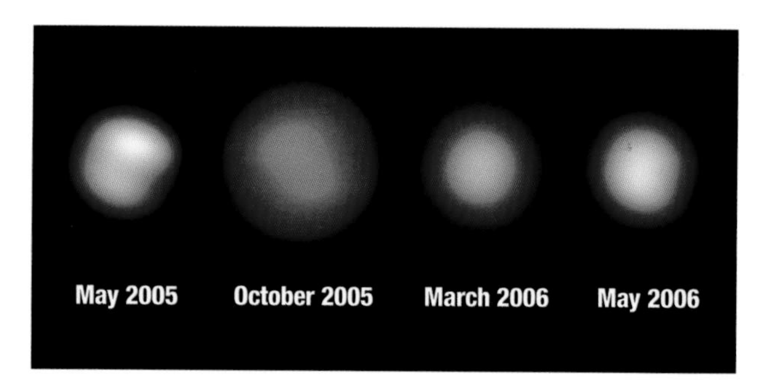

Red giant star Chi Cygni: four different times of the pulsation are shown, and hot spots on the star's surface can be seen. At its smallest, the star's diameter is about 300 million miles.

RED VARIABLES

Many red giants and supergiant stars pulsate to some degree. In the constellation of Cetus there is an old red giant known as Mira (Omicron [o] Ceti) with a mass roughly the same as our Sun. Over a period of 11 months, it expands and contracts, brightening sufficiently to become visible to the naked eye and then fading until it is only visible through a telescope. The process then reverses. There are several thousand of these stars – known as Mira stars after the first discovered. Their pulsations typically occur over periods of hundreds of days, hence they are known as long-period variables. Stars with less predictable behaviour are known as semi-regular variables, such as Betelgeuse in Orion whose radius has shrunk by 15% during the last 15 years. Those whose fluctuations do not show any detectable predictability are termed irregular variables. Several billion years from now our Sun may well endure the same pulsations before its ultimate slow demise.

PULSATING VARIABLES

The brightness, or spectrum, of most stars will vary during their lifetimes as part of their natural evolution. Such changes, or pulsations, can by cyclical and their periods of short or long duration.

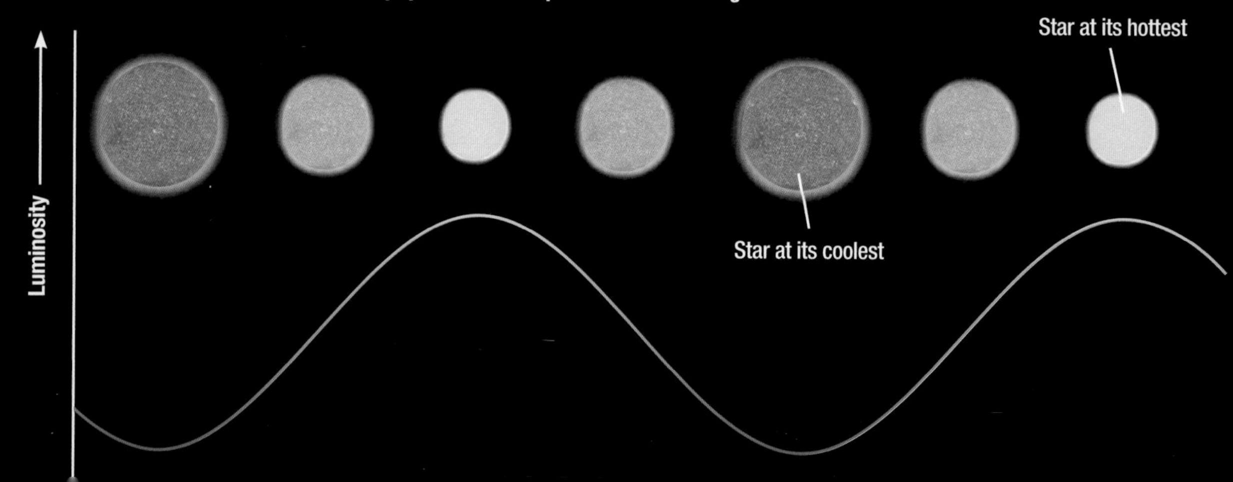

R Coronae Borealis star: such stars show erratic variability that is thought to arise from the presence of large clouds of dust in their envelope.

R CORONAE BOREALIS STARS

These are very bright supergiant stars (spectral type F or G) with odd behaviour. Every few years, instability in their atmospheres causes them to eject great clouds of carbon into space. This turns to soot and shields the star's light – like a reverse supernova. R Coronae Borealis itself is normally a near sixth magnitude star but can nosedive unpredictably by as much as nine magnitudes and take months to return to its original brightness.

LUMINOUS BLUE VARIABLES (LBV) – ETA (η) CARINAE

The most massive stars (spectral type O) are constantly disgorging copious quantities of gas into space. When they slide from their steady main sequence state, violent instabilities release huge amounts of mass every few hundred or thousand years. As their outer shell expands, they brighten by several magnitudes to change their spectral type classification from a blue supergiant (B) to a cooler super-supergiant (A). Perhaps the most remarkable of these LBVs is the southern hemisphere's Eta (η) Carinae. It has an estimated mass of around 120 Suns, with five million times the luminosity, and pumps out as much energy in six seconds as our Star does in an entire year. In the mid-19th century there was an enormous eruption lasting 20 years during which it brightened to a peak magnitude of -1 and then declined by seven magnitudes. A smaller eruption followed in the 1890s. The star itself is embedded between two enormous shells of material expanding at nearly 500 miles per second.

ECLIPSING BINARIES

Some variables are double stars. Seen from Earth, one star passes in front and then behind the other. The result is a periodic variation in the emitted light. The most famous of these is Algol in Perseus, which drops in magnitude from 2.1 to 3.4 every 2 days 20 hours and 48 minutes in an eclipse lasting ten hours.

CATACLYSMIC BINARIES AND NOVAE

If the component stars in a binary system are sufficiently close (less than 600,000 miles/1,000,000km, or twice the distance from Earth to the Moon) and orbit each other in less than half a day, they can interact and ignite spectacularly. These are the cataclysmic variables. Here, a larger cool red dwarf star (spectral type K or M) is partnered by a gravitationally superior white dwarf. The white dwarf draws gas from the red; this gas spirals on to the white dwarf's accretion disc and creates hot spots. These spots create instabilities in the disc. Consequently, every few days or weeks, there are variations in brightness, sometimes by as much as six magnitudes. But when the gas hits the surface of the white dwarf, the result is a thermonuclear explosion or 'nova'. Nova is the Latin name for 'new' – a term used in ancient times to describe stars, or explosions, that were previously unseen. The explosion blasts off the star's outer layers and the star's brightness flares by a massive 11 magnitudes. Dozens of these occur in our Galaxy each year, many recurrent, although rarely do they become visible to the naked eye. In extreme cases, the nuclear explosion may be so severe it triggers the Type Ia supernova mentioned earlier.

So, these stars offer wondrous diversity. It is a 'stellar zoo' astronomers seek to understand ... see Chapter 4 to find out how.

Eta Carinae: a blue hypergiant star on the brink of destruction. The dumb-bell shaped clouds – the Homunculus Nebula – is the remnant of an earlier eruption, the 0.5ly-wide vista captured by the Hubble Space Telescope (HST).

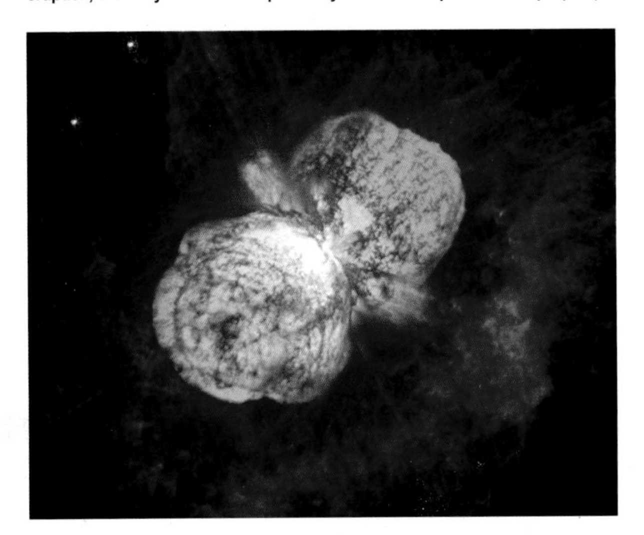

Cataclysmic binary: the red giant star dumps enough hydrogen gas onto its companion white dwarf star to set off a brilliant thermonuclear explosion on the white dwarf's surface.

Exoplanets

We know our Sun, our Star, has an entourage of eight planets, dwarf planets, asteroids, comets and a plethora of other orbiting objects, but what about the other stars 'out there'? Do they support planets – known as extrasolar, or exoplanets – and are they rich with life?

DISCOVERY

In 1990 astronomers were confident exoplanet systems existed: stars were not necessarily Sun-like but they could have planets mirroring our Solar System – close-orbiting small rocky worlds and giant outer gaseous bodies. Furthermore, they would have accreted, like ours, from an infant solar nebula disc of gas and dust. This assumption began unravelling in 1991 when American radio astronomers Aleksander Wolszczan and Dale Frail discovered the first exoplanets circling an unexpected rapidly spinning pulsar – a remnant body of a massive supernova. Four years later, Swiss astronomers Michel Mayor and Didier Queloz discovered a Jupiter-size planet circling a star, 51 Pegasi. Shockingly, this massive satellite was located only one-hundredth the distance from its star as Jupiter is from our Sun and, furthermore, orbited every 4.2 days. By comparison Mercury takes 88 days and Neptune 165 years. American astronomers Geoff Marcy and Paul Butler found two more circling Sun-like stars: one around 47 Ursae Majoris and the other around 70 Virginis. Both stars are yellow dwarfs similar to our Sun. In 2011 it was confirmed that three Jupiter-like extrasolar planets orbit 47 Uma. The standard planetary system theory was annihilated and the search for eclectic exoplanets began in earnest.

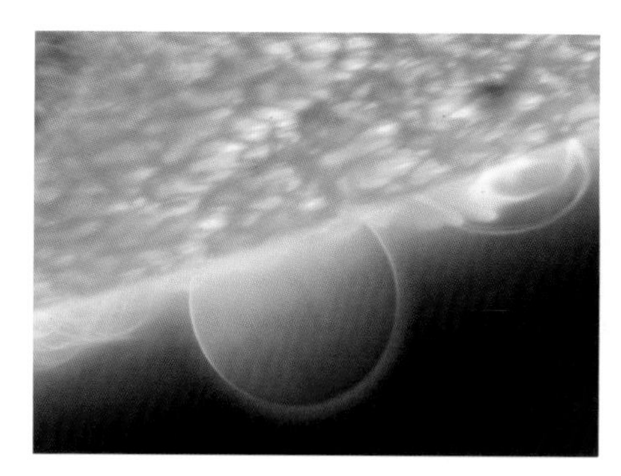

Jupiter-size extrasolar planet, HD 189733b: an artist's impression of the planet being eclipsed by its parent star.

SEARCHING FOR OTHER WORLDS

The discovery of a super-Earth-sized planet orbiting a Sun-like star brings us closer than ever to finding a twin of our own watery world. But NASA's Kepler space telescope has captured evidence of other potentially habitable planets amid the stars of the Milky Way Galaxy. It helps to zero in on the "habitable zone". This is the band of congenial temperatures for planetary orbits - not too close and not too far. Too close and the planet is fried, like Venus. Too far and it's in deep freeze. But in the habitable zone, a planet can have liquid water on its surface - just right. Scientists call this water-friendly region the "Goldilocks zone." Recently discovered, Kepler-452b, is the closest match to our Earth-sun system. This artist's conception of a planetary line-up shows habitable-zone planets with similarities to Earth: from left, Kepler-22b, Kepler-69c, Kepler-452b, Kepler-62f and Kepler-186f. Last in line is Earth itself.

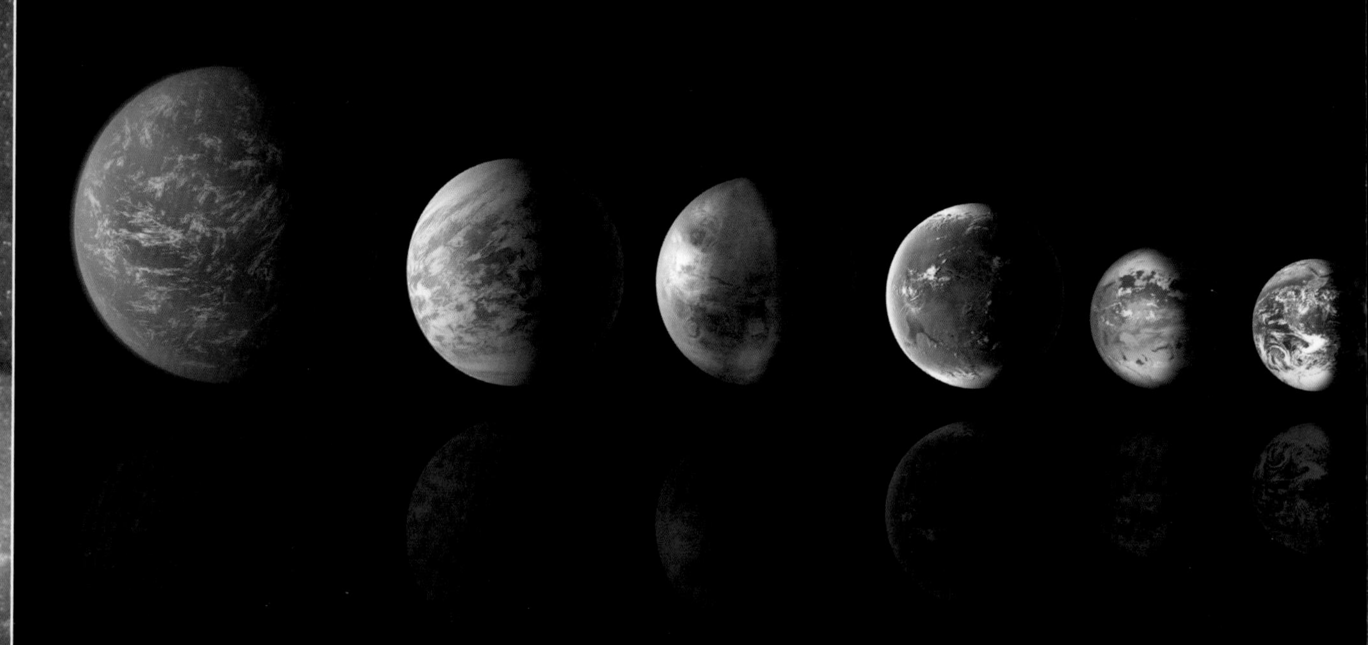

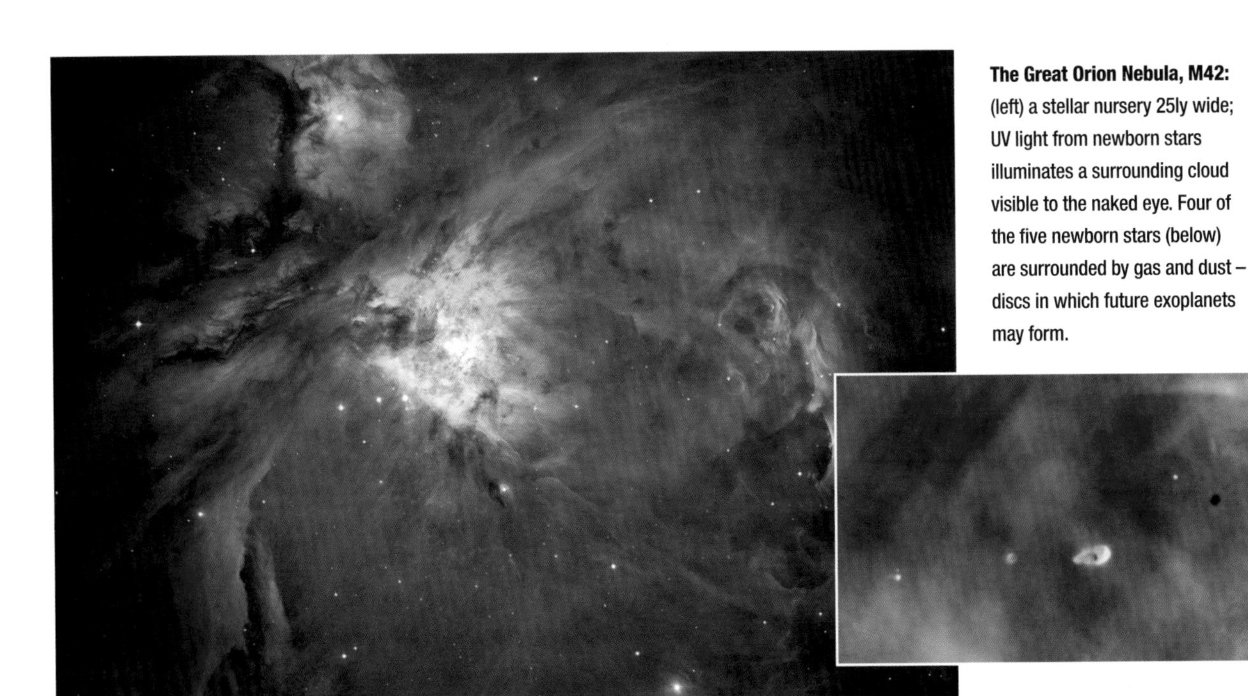

The Great Orion Nebula, M42: (left) a stellar nursery 25ly wide; UV light from newborn stars illuminates a surrounding cloud visible to the naked eye. Four of the five newborn stars (below) are surrounded by gas and dust – discs in which future exoplanets may form.

EXOPLANET TYPES

At time of going to press, there are over 2000 known exoplanets. As observing techniques improve and searches are expanded over longer periods, both high and low mass planets, and those in larger elliptical orbits, will undoubtedly be discovered.

As we have seen, space is replete with giant molecular clouds of gas and dust – the raw material of our Solar System and other solar systems. One example, among many, is the Great Orion Nebula (M42). Within this giant stellar nursery are countless protoplanetary discs, all rich in gas and dust, all able to coalesce into diverse planetary systems.

NOTABLE EXOPLANET TYPES

HOT JUPITERS

Gas giants orbiting less than a tenth of the Earth–Sun distance (1AU). Their year lasts around ten Earth days. They possibly formed far from their parent star but migrated inwards: a disc of gas could have sapped the planet's angular momentum or the disc's material swirled onto the star and took the planet with it.

SUPER-EARTHS

Low-mass planets orbiting parent stars in ten days or less.

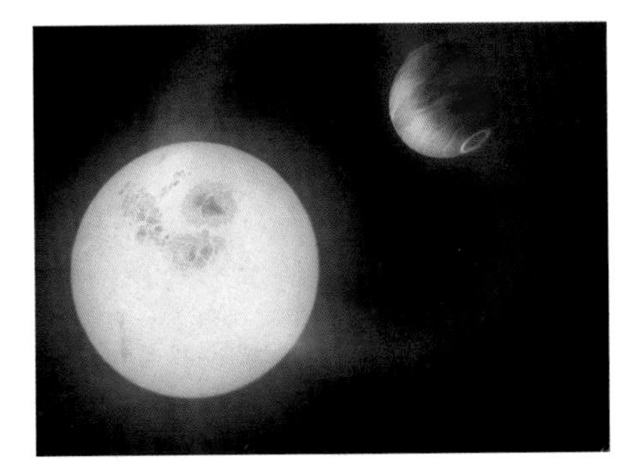

SWEEPS-10: the planet, roughly 1.6 times heftier than the gas giant Jupiter, circles the star to which it is gravitationally anchored in just over ten hours.

Earth-like planets: OGLE-2005-BLG-390Lb, is around 25,000 light years away from Earth. It has an orbital period of ten years and is about three times as far from its parent star as the Earth is from our Sun.

FREE-FLOATING PLANETS

Also known as interstellar, nomad or orphan planets. These rogue planets orbit within a galaxy and have either been ejected from the planetary system in which they formed or have never been gravitationally bound to any star or brown dwarf. They may not even be planets but possibly brown dwarf stars: lightweight bodies, with masses possibly 5 to 13 times that of Jupiter 'floating' around without parent stars. In 2000, a dozen such orphans were located in the Trapezium star cluster in the Great Orion Nebula, M42.

PULSAR PLANETS

Multiple planets orbiting pulsars (spinning neutron stars): rich in heavy elements and possessing strong magnetic fields.

ORDINARY GIANTS

Long-period planets, like Jupiter, with sedate orbits: their orbits are so large it can take at least a decade to find them.

■ EXOPLANET DETECTION

But how are exoplanets found when their host stars are so incredibly distant and consequently infinitesimal? There are a number of methods:

RADIAL VELOCITY METHOD

This is currently the leading method and accounts for around 90% of exoplanets found. A planet's gravity tugs on its parent star, inducing a wobble in the star's motion. As a star moves closer to Earth its spectral lines (its unique fingerprint) shift towards the blue; as the star moves away, the lines shift to red – an effect called the Doppler Shift. The heavier the planet, the greater its star's wobble. Studying these wobbles with sophisticated spectroscopy equipment, and allowing for 'jitters' such as coronal mass ejections and star quakes, astronomers can calculate a planet's minimum mass and its orbital period – quite a thought, when stars can be millions of miles wide, some 600 million, million miles distant, and astronomers are measuring movements of perhaps only one to three metres a second … walking speed!

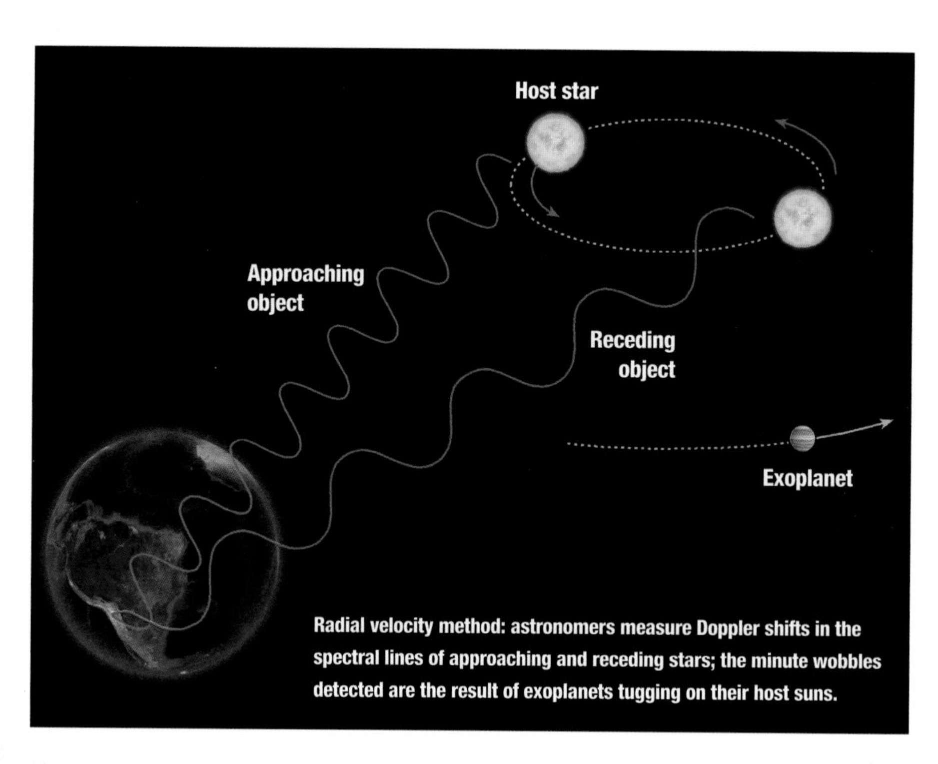

Radial velocity method: astronomers measure Doppler shifts in the spectral lines of approaching and receding stars; the minute wobbles detected are the result of exoplanets tugging on their host suns.

TRANSIT METHOD

It is rare, but if a planet's orbital plane happens to coincide with our line of sight, then it passes in front of its star each time it orbits. Such a transit was first observed in November 1639, and later witnessed by Captain James Cook (1728–79) and astronomer Charles Green (1735–71) on *Endeavour* in 1768, when Venus passed in front of the Sun. When a planet 'transits' its tiny black disc causes a minuscule dip in the star's brightness and this is recorded by a diagrammatical light curve. If the dip is regular and of equal dimming each time, and if the star is not a known periodic variable, then the suspect could be a planet. The frequency of the dimming reveals how long the planet takes to orbit its star and, therefore, its speed. The time taken for a planet to edge in front and fully cover the star reveals the planet's diameter, and the total time it takes for the planet to cross from one side of the star to the other discloses the star's diameter. Additional radial velocity measurements will confirm the planet's exact mass.

Transiting extrasolar planet: the planet's transit results in a decrease in the perceived luminosity of the host star.

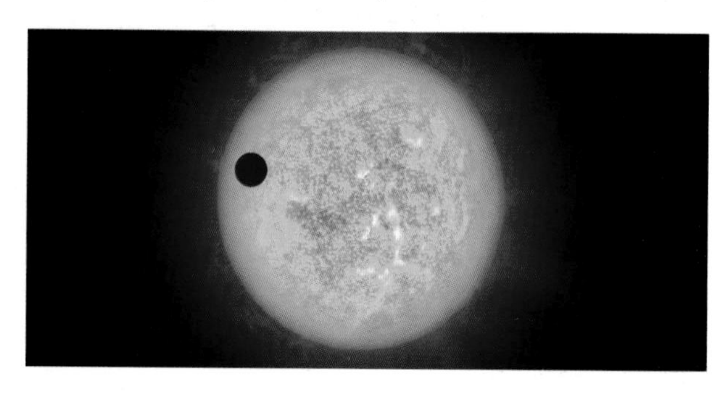

MICROLENSING METHOD

Despite the fact that it is a chance alignment and, therefore, not repeated, if a star or planet passes in front of a more distant star, then gravity distortion from the intervening object acts as a magnifying lens causing the distant star to brighten. Robotic telescopes monitor an enormous number of these background stars for fluctuations since a smooth rise and fall in brightness indicates a dim star and a blip on a smooth light curve indicates an orbiting planet. Such technology is enabling astronomers to find Earth-mass planets.

DIRECT-IMAGING METHOD

Since around 2008, a new planet-detection method has been gaining significance: direct imaging. In this process, astronomers use starlight-suppression techniques, such as ground-based coronagraphs or space-based starshades – masks placed at one of a telescope's focal planes which, combined with adaptive optics, blocks a star's bright light in order to directly image small dots – planets – orbiting around them. Although direct imaging cannot be used to estimate a planet's size (as per the transit method) or its mass (as per the radial-velocity method), both size and mass can be inferred from a planet's brightness and spectrum – or from just its colour. Excitingly, by observing different wavelengths during transits or eclipses of a star, direct imaging can capture the spectrum of an exoplanet's atmosphere. At time of going to press, nine planets have been directly imaged. The dots are currently unresolvable because their discs are orders of magnitude smaller than the resolution of even the

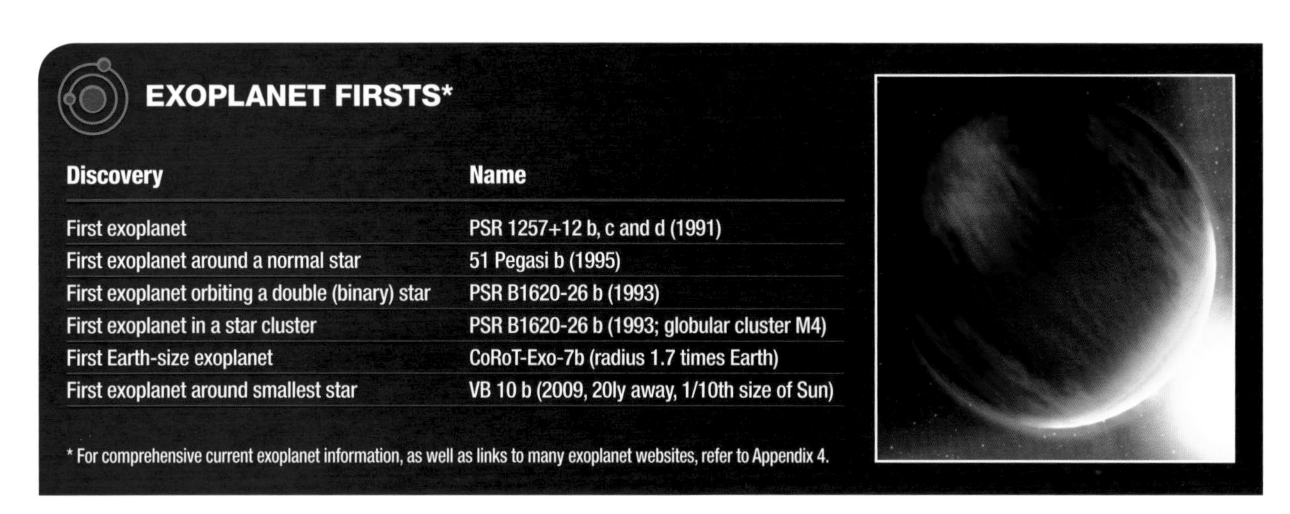

EXOPLANET FIRSTS*

Discovery	Name
First exoplanet	PSR 1257+12 b, c and d (1991)
First exoplanet around a normal star	51 Pegasi b (1995)
First exoplanet orbiting a double (binary) star	PSR B1620-26 b (1993)
First exoplanet in a star cluster	PSR B1620-26 b (1993; globular cluster M4)
First Earth-size exoplanet	CoRoT-Exo-7b (radius 1.7 times Earth)
First exoplanet around smallest star	VB 10 b (2009, 20ly away, 1/10th size of Sun)

* For comprehensive current exoplanet information, as well as links to many exoplanet websites, refer to Appendix 4.

largest telescopes. Yet, whilst astronomers cannot currently resolve continents, oceans or polar caps on any Earth-like habitable worlds, it will not be long before advancing techology triumphs. The next generation of instruments, collectively known as 'extreme adaptive optics', are already pushing the envelope: Project 1640, Subaru Coronagraphic Extreme Adaptive Optics (SCExAO), Gemini Planet Imager (GPI) and Spectro-Polarimetric High-contrast Exoplanet Research (SPHERE).

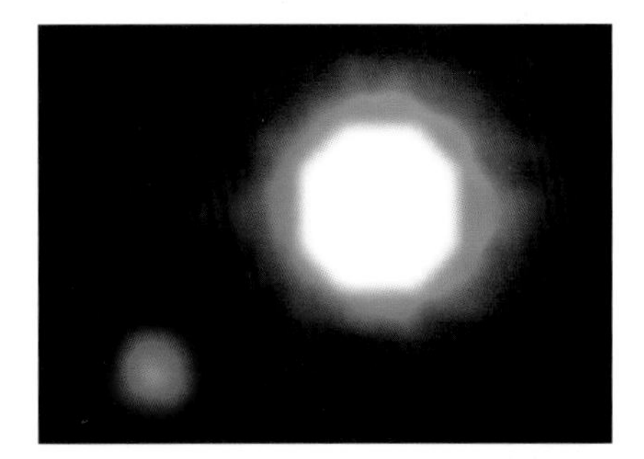

2M1207A: the first directly imaged planet (at left).

Fomalhaut b: observing dust disturbances reveals hidden exoplanets.

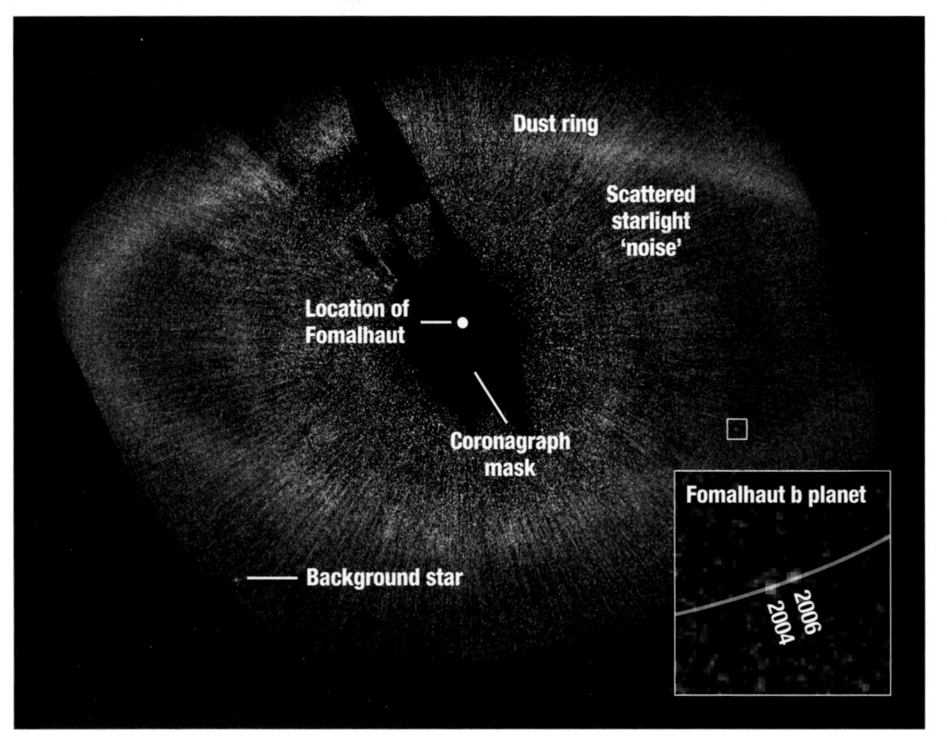

PULSAR TIMING

Planets have been discovered around pulsars – rapidly spinning neutron stars that emit polar radio beams. These beams sweep across Earth hundreds of times each second and can be measured by radio telescopes. Irregularities in the beep rate can signal the presence of a planet whose gravity tugs the neutron star back and forth. This incredibly sensitive technique can even detect small asteroids.

ASTROMETRY METHOD

This is the study of the motions and positions of stars: the gravity of a large planet tugs its star from side to side sufficiently far enough to enable the movement to be measured against background stars.

Another method used by astronomers is the observation of dust disturbances in the proto-planetary discs surrounding young stars.

The hunt for exoplanets continues – see Chapter 4. Nor does the search stop here: astronomers now seek single, or multiple, exomoons – moons orbiting exoplanets – using similar transit methods. The gravity of a large moon circling an exoplanet will cause the planet to wobble around the centre of mass of the planet/moon system. This, in turn, causes tiny but detectable variations in the timing and duration of the exoplanet's transit around its host star. The discovery of a habitable rocky moon may not be too far away and would be a landmark in our search for life elsewhere in the Universe.

The Milky Way Galaxy

We have briefly mentioned 'Galaxy' and 'galaxies', but what are they? Where is our Sun and its entourage of planets? Where are the stars and exoplanets? Where and what is the great 'out there'? We approach the grandest enquiries of all. Look up on a dark, moonless night, allow your eyes 30 minutes to acclimatise to the dark – become light-adjusted – and you will see a faint swathe of 'cloud' sweeping overhead from horizon to horizon. This is an arm of the Milky Way Galaxy, the Galaxy in which our Sun with its family of planets, and all the other objects mentioned thus far, are embedded. It is our home Galaxy, and it is also home to perhaps over 400 billion other suns – other stars – a stunning realisation that transforms perspective forever.

STRUCTURE AND MOTION

What does a galaxy look like? It is a great gathering of billions of stars immersed in enormous collections of gas, dust, and mysterious dark matter, bound by gravity. The word 'galaxy' comes from the Greek word for 'milky' or 'cloudy' (galaxias), since this is how the swathe of billions of stars appeared to the ancients. Our home Galaxy, the Milky Way, is accorded a capital 'G' to distinguish it from other galaxies and is a construction on a truly massive scale. If we could board Captain Kirk's *Enterprise*, and view the Galaxy from above at 'one-quarter impulse power', we would see four major Catherine wheel-like spiral arms highlighted by the vivid young hot stars, open clusters and glowing nebulae nestled within them. Along with giant clouds of gas and dust, these arms coil around a central bar that appears end-on down its long axis from our viewpoint on Earth.
The size of the bar is uncertain but could extend to

Whirlpool Galaxy M51: note its distinct spiral structure and dazzling pink areas of massive star formation.

THE MILKY WAY: AN EDGE-ON VIEW

Inset: area surrounding Antares (yellow) with globular clusters, dark nebulosity and colourful Rho Ophiuchi complex. 500 ly distant, 10 ly size.

The California Nebula (NGC 1499): amidst stars of Perseus constellation.

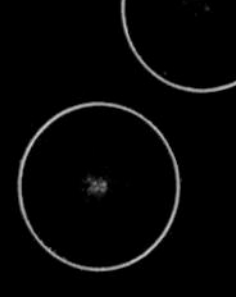

The Pleiades (M45): 440ly distant, 7.25 ly size.

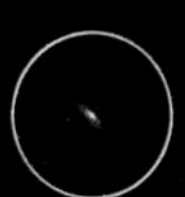

The Andromeda Galaxy (M31): 2.9 million ly distant, 140 thousand ly size.

around 10,000ly from the centre. The differing orbit of spiralling gas and stars over varying distances creates 'density waves'. Like traffic slowing on a highway, this leads to 'bunching': where clouds bunch together new stars are born.

However, if we could travel at the fictional 'warp factor one', the speed of light, and whiz across the Galactic diameter, then even though travelling at a mind-blowing speed of 186,000mi/s, it would still take us over 120,000 years to travel from one side to the other! The disc itself is some 2,000ly thick. It is staggeringly immense.

Since everything 'out there' is moving, so, too, the entire Milky Way galaxy rotates. By studying radio observations of hydrogen gas, astronomers have discovered it takes a mere 240 million years to make just one revolution – a cosmic blink of the eye. This speed of rotation remains, surprisingly, the same to at least about 150,000ly from the centre. According to Kepler's laws, the speed should lessen with distance. The fact that it does not implies another force is in play – dark matter, whose composition remains a mystery. Dark matter is 'dark' in the sense that it emits no light and could comprise faint, low-mass stars too dim to be seen, or dying white dwarfs, or even black holes. It may even be exotic particles as yet unknown to science. But by observing the motions of the stars and galaxies, astronomers have discovered they are being tugged by the gravity of this mysterious invisible 'material'. It is also thought that the Galaxy's overall mass (weight) is dominated by a halo of it, possibly extending some 750,000ly beyond the central disc, and it could weigh at least ten times more than all the visible stars in the Galaxy combined.

SOLAR SYSTEM LOCATION

Our insignificant Solar System lies roughly 14ly above the central plane of the Galaxy and is some 27,000ly from the core in one of the spiral arms known as the local or Orion Spur. The arm closer to the centre is the Sagittarius Arm, while the one farther out is the Perseus Arm, both named after the constellations in which they appear. We, and the planets in our neighbourhood, are orbiting our 'tiny' Sun that itself follows a roughly circular orbit around the centre of the Galaxy at a speed of around 140mi/s (220km/s). Recent evidence suggests that our Star even 'bobs' along, dipping 230ly above and then 230ly below the Galactic plane following a 64-million-year cycle.

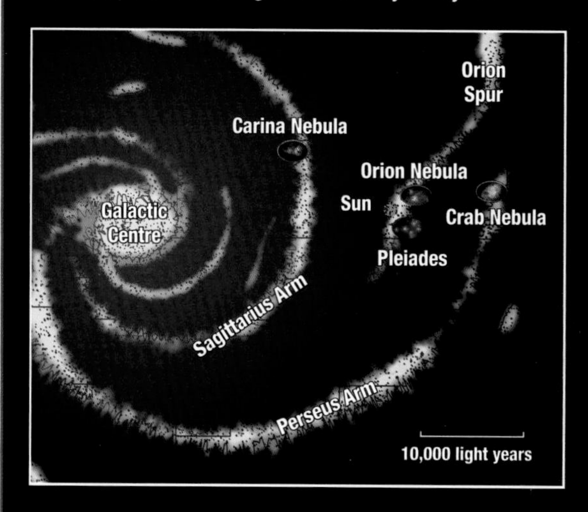
Orion Spur
Carina Nebula
Orion Nebula
Sun
Crab Nebula
Galactic Centre
Pleiades
Sagittarius Arm
Perseus Arm
10,000 light years

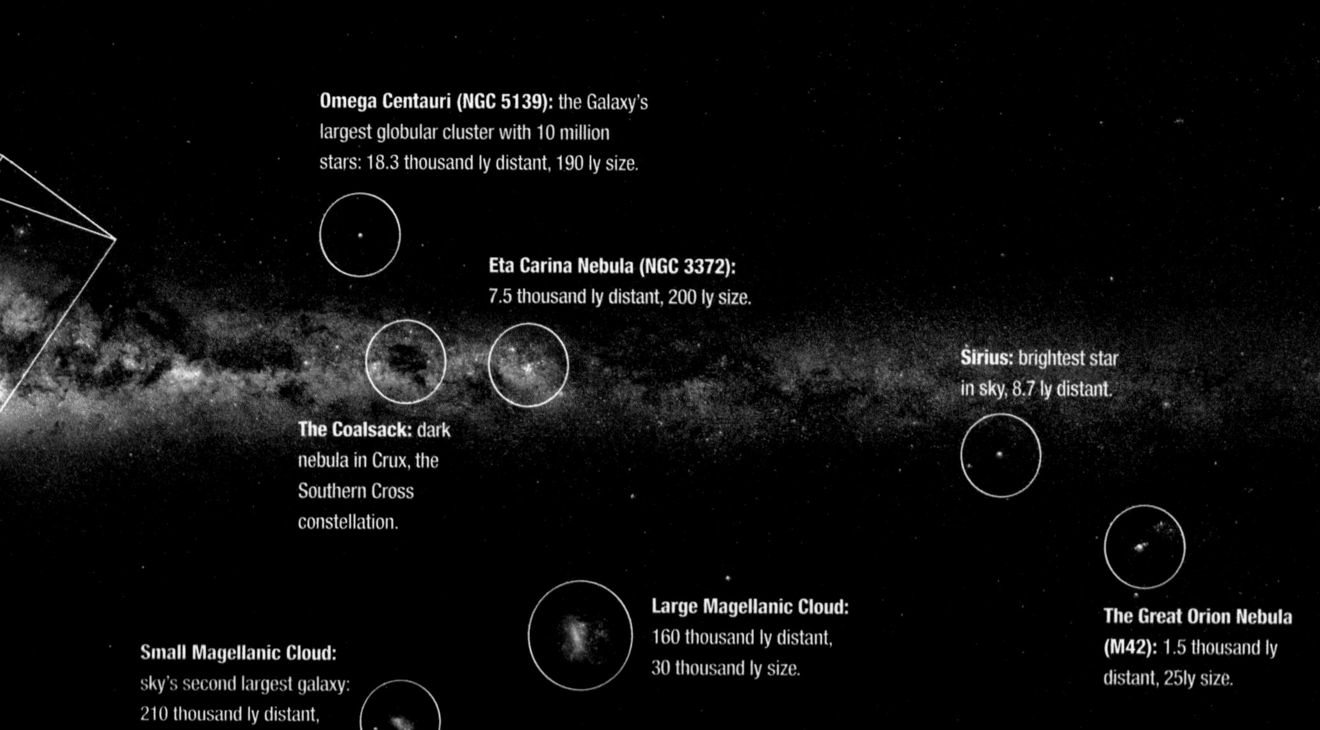
Omega Centauri (NGC 5139): the Galaxy's largest globular cluster with 10 million stars: 18.3 thousand ly distant, 190 ly size.
Eta Carina Nebula (NGC 3372): 7.5 thousand ly distant, 200 ly size.
Sirius: brightest star in sky, 8.7 ly distant.
The Coalsack: dark nebula in Crux, the Southern Cross constellation.
Large Magellanic Cloud: 160 thousand ly distant, 30 thousand ly size.
The Great Orion Nebula (M42): 1.5 thousand ly distant, 25ly size.
Small Magellanic Cloud: sky's second largest galaxy: 210 thousand ly distant, 10 thousand ly wide.

ANATOMY OF THE MILKY WAY

Marking dead centre of the Milky Way, around which the entire Galaxy turns, is an unusual radio source, Sagittarius A* (pronounced 'A-star'). Its diameter is about the same as that of Earth's orbit around the Sun but it has a mass some 4.5 million times greater than our Star. With such enormous mass crammed into a 'minuscule' volume, Sagittarius A* is probably the site of a supermassive black hole, where gravity is so powerful not even light can escape and around which stars race at phenomenal speeds.

Surrounding this Galactic centre is a bulge of some ten billion stars. Looking edge-on 'into' the Galaxy, we see it resembles a flattened rugby ball shape with the central bulge shaped like a peanut. Invisible in optical light because of the enveloping obscuring dust, and with the exception of an area known as Baade's Window, the centre is nevertheless 'visible' at penetrating infrared and 21cm (8in) radio wavelengths (see Chapter 4).

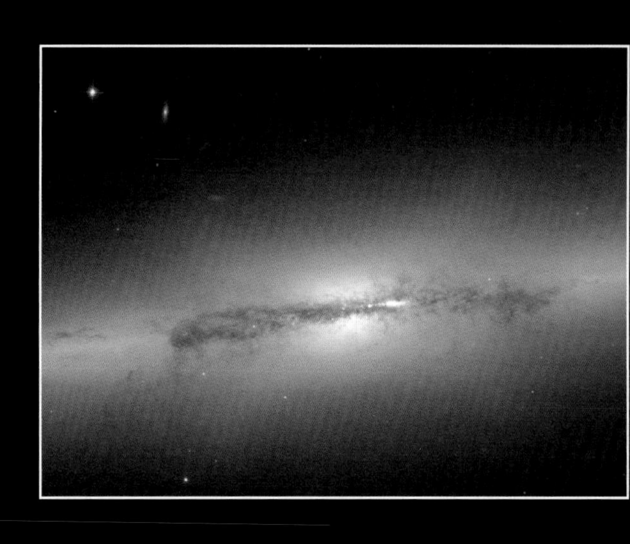

ESO 510-G13: an edge-on view of a spiral galaxy with a warped dusty disc.

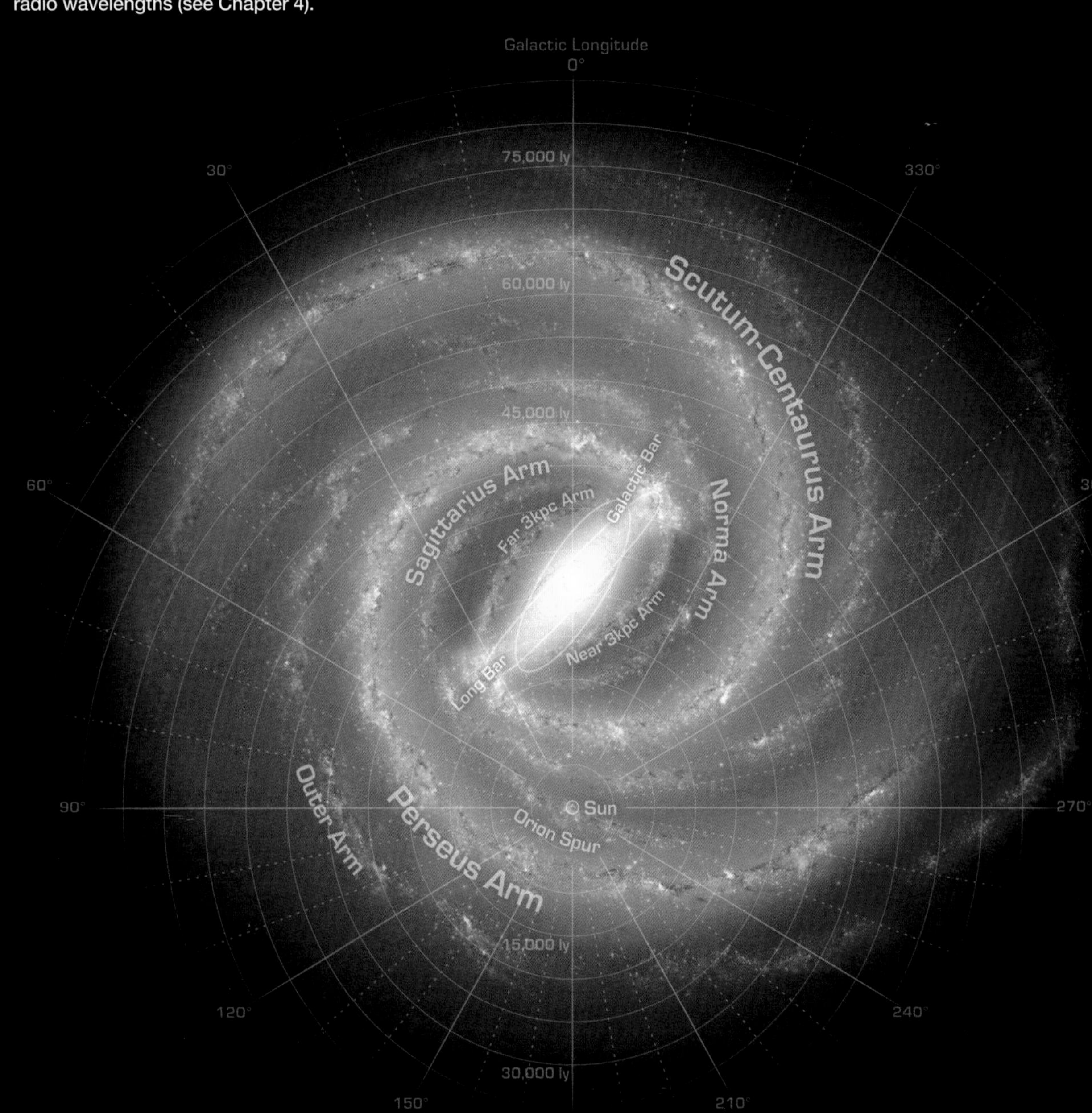

Infrared edge-on view of Milky Way galaxy: revealing copious cold dust clouds (around 20 degrees Kelvin) hidden from visible cameras.

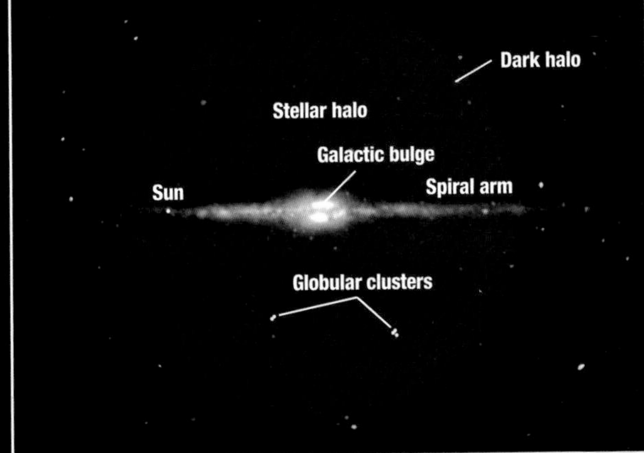

The Milky Way galaxy: edge-on view. A central galactic bulge with spiral arms stretches out to form a flattened disk. From our viewpoint, near our Sun, the Galaxy appears as a band of starlight because we see more stars looking 'through' the plane (left to right) than we would looking at right angles (up, down).

Our Galaxy contains two distinct populations of stars. Those in the spiral arms, including our Sun, are classified as Population I: stars of varying ages, including the very young, short-lived hot blue stars, and all rich in elements heavier than hydrogen and helium. Population II stars are yellow-coloured, long-lived and old, almost as old as the Universe itself, and are located in the Galactic halo and central bulge. These formed within the first billion years or so of the Galaxy's history.

Closer to the centre of the Galaxy is the stellar halo that differs from the dark matter halo. Within this, there are around 150 globular clusters and many individual stars containing few heavy elements. These individual stars are thought to be among the oldest in the Galaxy, born before they could be enriched with the material from a succession of other explosive dying stars. Unlike those in the disc, these stars, and the globular clusters, swing out on highly elliptical orbits, plunging wildly through the centre of the Galaxy as they do so.

The sporadic globular clusters are the largest individual components of our Galaxy, all around ten billion years old. They are enormous spheres measuring around 100ly across and containing hundreds of thousands of stars but with little or no gas or dust. Of those known, the most distant are over 300,000ly from the Galactic centre.

Some of the brightest stars in the night sky reside in a broad band inclined at about 16° to the plane of the Galaxy, called Gould's Belt after the US astronomer who studied it in the late 19th century. These young stars frequently belong to 'OB associations' or open clusters – loose collections of stars all formed from the same giant molecular cloud. Open clusters are smaller and more dense than the associations and can contain up to a few thousand stars in a volume less than 50ly wide. They clearly delineate the spiral arms, examples being the Pleiades and Hyades in the constellation of Taurus, both visible to the naked eye.

Ancient Globular Clusters: swing around the Galaxy on enormous elliptical trajectories.

INTERSTELLAR MEDIUM AND NEBULAE

Between the stars is the interstellar medium, or ISM. Empty? Certainly not; this area is rich in gas, dust, magnetic fields, and cosmic rays. Some of the gas, found in giant molecular clouds where stars are born, is very cold, dark, and dense, but there are clouds with temperatures of thousands of degrees. The latter are the areas of hot, ionised gas, known as HII regions, and appear as spectacular bright clouds, or emission 'nebulae' – Latin for 'mist'. A fine example is the Great Orion Nebula where the hot young stars are irradiating the surrounding cloud. And then there are the enormously hot, low-density clouds, their temperatures measuring millions of degrees. These contain the gas heated by the blast waves from supernovae explosions and comprise two-thirds of the volume of the ISM. If there is gas within the ISM then there is mostly interstellar dust – likened to soot or very fine sand and comprising minuscule grains of carbon, silicon, and small ice crystals. These dust grains absorb and scatter starlight with the result that they can dim, or redden, the light from distant stars. When stars illuminate dense, dusty clouds, the scattered light can be seen as a bluish reflection nebula, as with the Pleiades seen earlier. See right for the different types of nebulae. See also Chapter 3 for observing tips.

ABSORPTION (DARK)

Definition: Dim and dense: they block light from background stars.

Examples:
- Barnard 72 – Snake Nebula: 650ly away, size 2ly.
- Barnard 33 – Horsehead Nebula: 1,500ly away, size 4ly.
- NGC 2264 – Cone Nebula: 2,500ly away, size 2.5ly.
- Thackeray's Globules: 6,000ly away, size 1.5ly.

Horsehead Nebula (Barnard 33): an absorption (dark) nebula: a tower of dust resembling a horse's head silhouetted by background irradiated clouds.

The Eagle Nebula, M16 'Pillars of Creation': an emission nebula.

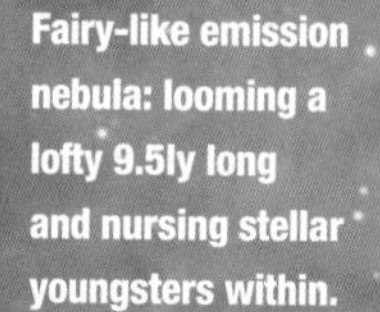

Embryonic stars race to emerge from dark dusty clouds before ultraviolet light erodes their dark nursery (top left of image).

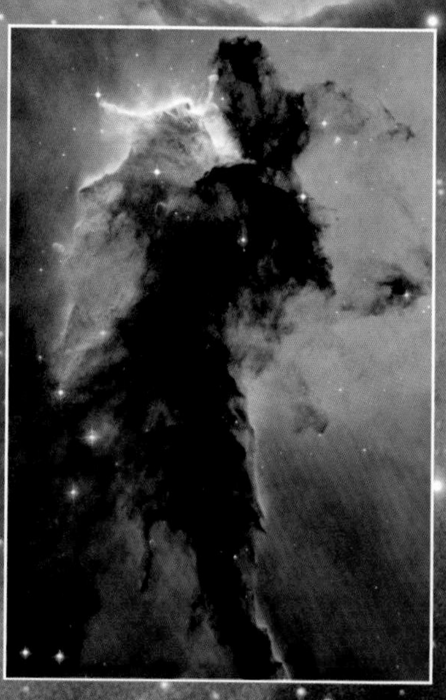

Fairy-like emission nebula: looming a lofty 9.5ly long and nursing stellar youngsters within.

EMISSION

Definition: Clouds illuminated by the UV light from stars within them.

Examples: NGC 2070 – Tarantula: 180,000ly away, size 1,000ly.

M42 – Great Orion: 1,500ly away, size 25ly.

W5 – Mountains of Creation: 7,000ly away, size 50ly.

M16 – Eagle: 7,000ly away, size 40ly.

Mountains of Creation (W5): the towering pillars are ten times the size of the 'Pillars of Creation' in the Eagle Nebula.

PLANETARY

Definition: Clouds jettisoned into space by the death of a red giant star; a central dying white dwarf remains.

Examples: NGC 2440: 4,000ly away, size 1.3ly.

NGC 6543 – Cat's Eye: 3,000ly away, size 1.2ly.

NGC 6537 – Red Spider: 3,000ly away, size 1.3ly.

NGC 7293 – Helix: 650ly away, size 5.7ly.

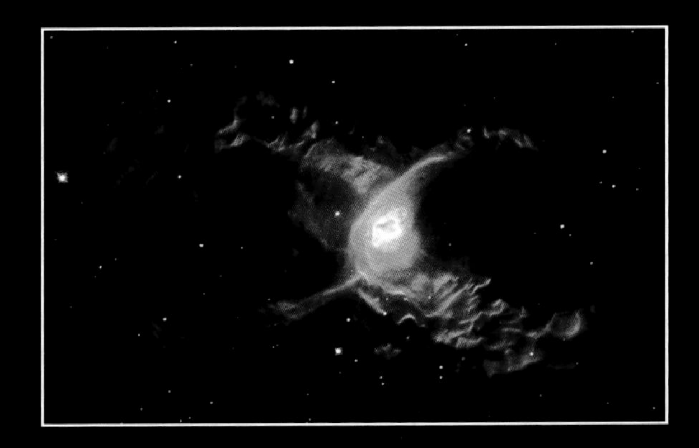

Red Spider Nebula, NGC 6537 - a planetary nebula: at its core dwells one of the hottest white dwarf stars with a temperature of 500,000 degrees Kelvin.

REFLECTION

Definition: Hidden dark dust clouds where stars are too weak to illuminate them. Instead, radiation is scattered – reflected – off the constituent dust and gas.

Examples: M20 – Triffid: 5,500ly away, size 35ly.

IC 2118 – Witch Head: 660ly away, size 34ly.

Rho Ophiuchi Complex: 500ly away, size 10ly.

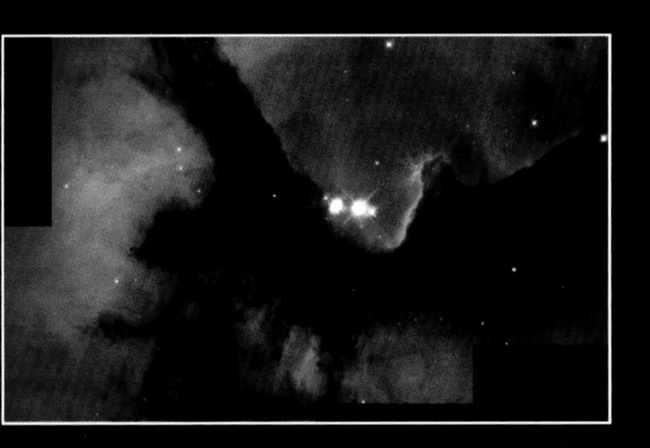

The Triffid Nebula, M20, NGC 6514: a close-up of the core where radiation from the giant stars illuminates the entire nebula.

SUPERNOVA REMNANT

Definition: Remnant clouds of a spectacular explosion signifying the death of a giant or supergiant star.

Examples: M1 – Crab: 6,500ly away, size 7ly.

NGC 6960 – Veil: 1,500ly, size 50ly.

IC 443 – Jellyfish: 5,000ly away, size 72ly.

3C 461 – Cassiopeia A: 10,000ly away, size 10ly.

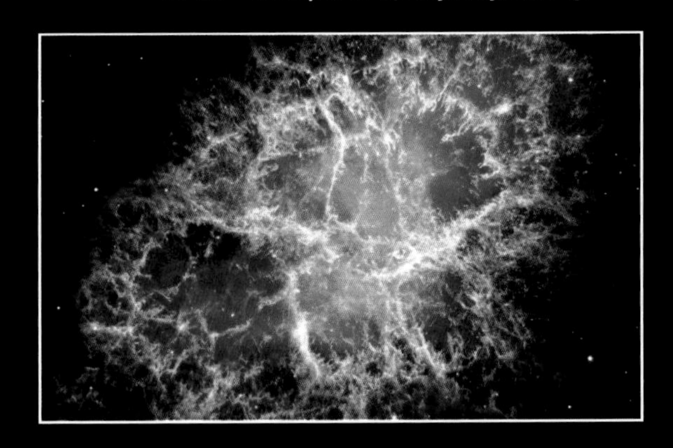

Crab Nebula, M1 – a supernova remnant: at its core beats a rapidly spinning neutron star, a pulsar.

THE LOCAL GROUP

Are there other galaxies out there? Of course! One, amazingly only discovered in 1995, is on our cosmic doorstep, the Sagittarius Dwarf. This is a gathering of stars some 80,000ly from the Sun that is slowly unravelling as a result of our Galaxy's gravitational influence. And there are other recently discovered 'gatherings' that are closer still.

Next closest are the Large and Small Magellanic Clouds (LMC and SMC), around 170,000 and 200,000ly from the Sun respectively. Both these galaxies are visible to the naked eye in the southern hemisphere. The LMC has about a hundredth the mass of our Galaxy and is about 64,000ly wide, with the SMC having about a tenth the mass of the LMC. These are both irregular galaxies (see *Galaxy classification* page 74), each enveloped in a common cloud of hydrogen gas invisible to the naked eye. However, radio astronomers have been able to 'see' the Magellanic Stream – an enormous swathe of gas extending some 100° across the southern sky and probably formed as a result of 'close encounters' with our Galaxy some 300 million and five billion years ago. Perhaps the SMC was even ripped from the LMC during one such close approach. It is thought that both the LMC and SMC will spiral into our Galactic centre some ten billion years from now. It will be accretion like that which formed our Solar System, but on a much grander scale.

But where are these galaxies, including our own? Well, they form part of what is known as the Local Group, a name given to a cluster of around three dozen galaxies all bound by gravity and covering a cosmic arena some five million light years wide. Two spiral galaxies dominate: our own, and the larger Andromeda galaxy (M31), which is around 220,000ly wide. The Andromeda galaxy is the nearest spiral galaxy to us but is still a staggering 2.9 million light years away. With the exception of M33 in Triangulum, which requires excellent seeing and good eyesight, the Andromeda Galaxy is the most distant object that can be seen with the naked eye and is perhaps home to a supermassive black hole at its core with the mass of some 30 million Suns. The light we see now left this galaxy at the dawn of humanity. And, for a sobering thought, the massive Andromeda is actually moving towards us. If astronomers subtract the Sun's orbital motion around our Galaxy (140mi/s), the Andromeda is still approaching us at 75mi/s. It is thought that, in around three billion years time, the two will merge: accretion on an even grander scale!

SUPERCLUSTERS AND BEYOND

The Local Group is travelling through space and is itself constrained by the gravity of an even larger cluster some 60 million light years away – the Virgo Cluster, its 1,300 galaxies feathering an area some 20 million light years wide. This Virgo Cluster is itself gravitationally pulled by other 'nearby' clusters of galaxies, each dotting a region some seven million light years wide. There is even a Great Wall of clusters measuring some 500 million light years long, 200 million light years wide, and 15 million light years deep. In 2003 another wall stretching over 1.37 billion light years was found … and there may be more.

But, defined in September 2014, the Laniakea Supercluster (also called the Local Supercluster or Local SCl) is the galaxy supercluster that is home to our Milky Way Galaxy. This gargantuan supercluster encompasses 100,000 galaxies stretched out over 520 million light years, and consists of the Virgo Supercluster, where the Milky Way resides, the

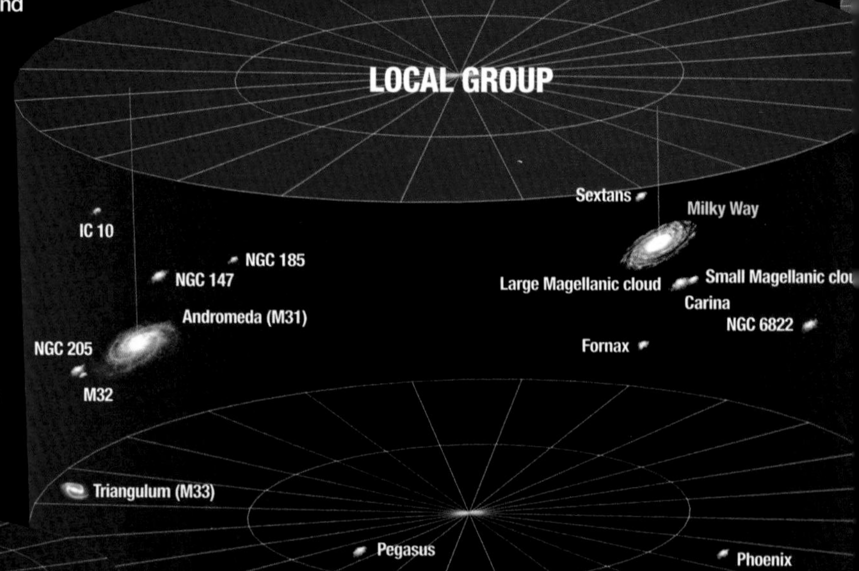

Our Galactic Realm, the Local Group and the Virgo Supercluster, all of which are dwarfed by the realms of the recently discovered Laniakea (Hawaiian for 'Immeasurable Heaven') Supercluster.

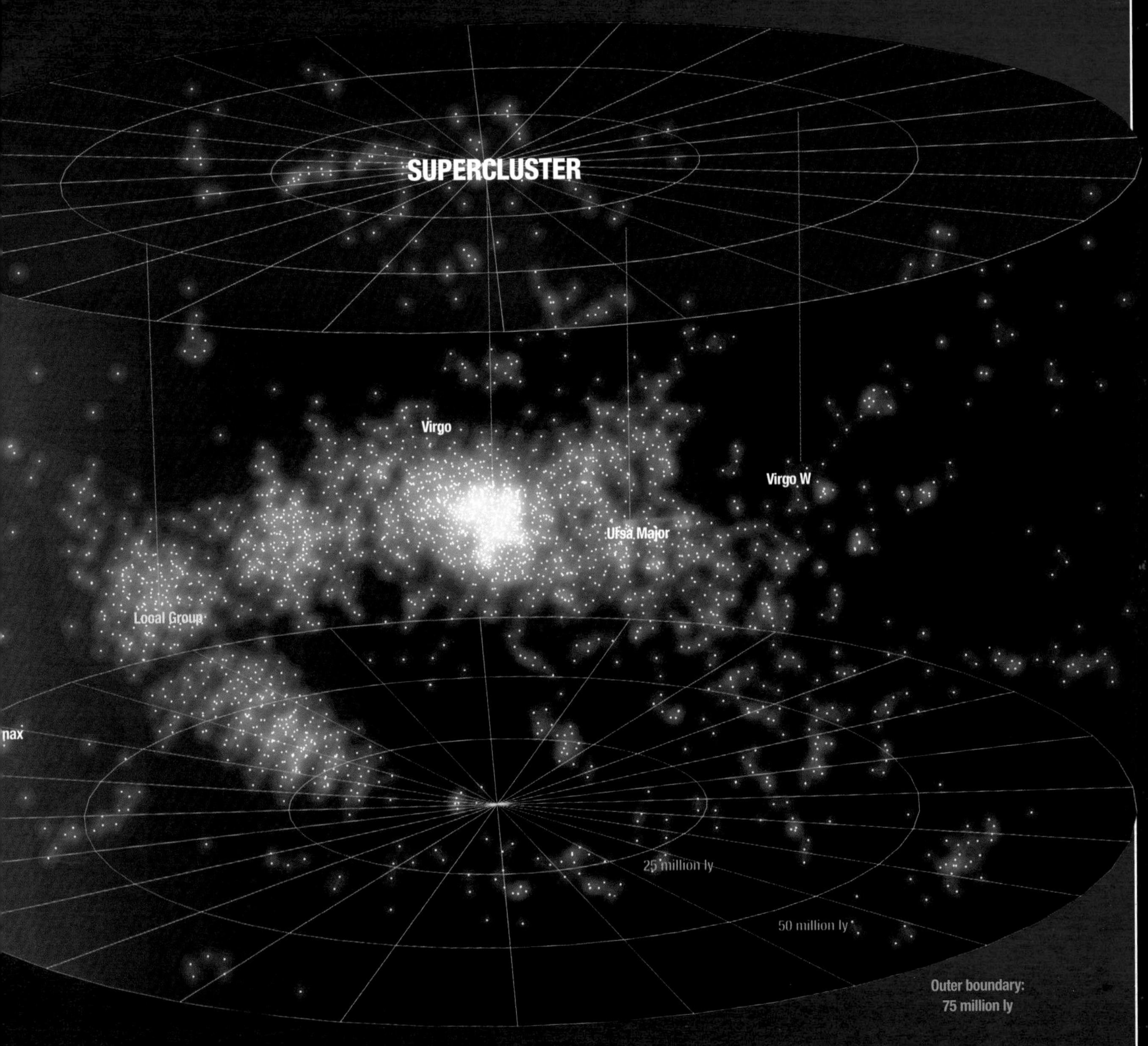

Hydra-Centaurus Supercluster, the Pavo-Indus Supercluster and the Southern Supercluster. Within the vicinity of the Hydra-Centaurus Supercluster, at the centre of the Laniakea Supercluster, is the Great Attractor, a gravity anomaly in intergalactic space with a localised concentration of mass tens of thousands of times more massive than the Milky Way. The immense gravity generated by the Great Attractor is luring galaxies and their associated clusters over regions of millions of light years of space.

Beyond these, computers simulate a 'super structure' almost impossible to convey in two dimensions. Cosmologists compile maps of 'slices' of the sky. These show superclusters arrayed along filaments hundreds to millions of light years long. In between, like giant bubbles, are enormous voids around which the filaments thread: superclusters gather where these surfaces join. Each supercluster, when compared to the whole, appears like a particle of dust on a vast celestial cobweb.

When simulating the observable Universe as seen from its 'edge' some 13 billion light years away, astronomers see a 'worldwide web' on the grandest scale of all: a cosmos of 'matter' mapped on a 'froth of whisked egg whites'. Here, the clusters are but minuscule dots akin to 'bugs on the back of an aphid'. And beyond that …

Astronomers know there are over 250 billion galaxies out there, each home to billions of stars. To visualise just one galaxy cluster, or supercluster, in its motion through space is like imagining raisins in a fruitcake bound for baking. The cake mixture represents space, the raisins the clusters. Place the mixture in an oven: it expands, taking the raisins along with it. Similarly, as the fabric of space itself (the mixture) expands, so the clusters (raisins) are carried with it.

Some way into the 20th century, astronomers were divided as to the nature of these 'island universes', or 'spiral nebulae', as galaxies were known. Despite a Great Debate in 1920 in Washington DC between proponents of opposing theories as to whether the spiral nebulae were close or distant, or whether they were moving away from us, the conclusion remained elusive. It took Edwin Hubble (1889–1953) to demonstrate that these spiral nebulae did indeed exist way beyond our Galaxy and that their recession could be explained by the expansion of the Universe itself.

GALAXY CLASSIFICATION

Over the last half-century astronomers have reached a greater understanding about the classification of galaxies, the main three types being spirals, barred spirals, and ellipticals, but there are others:

Sombrero Galaxy, M104: a spiral galaxy, tilted nearly edge-on, with an estimated 800 billion stars, 28 million ly away, size 50,000ly.

SPIRALS

Whether viewed face-on or edge-on, these galaxies exist in many varieties, from swirling 'grand designs' with two spiral arms unwinding from a central bulge, to the 'flocculant' (fleecy) systems. Viewed face-on, they display a characteristic whirlpool appearance indicating rotation. A spiral's rotation draws stars into a flat disc similar to the protoplanetary disc (ecliptic) that spawned our Solar System. They have masses from a billion to a million million solar masses, diameters from 10,000 to 200,000ly and a central bulge, or nucleus, like a flattened sphere. Typically, the stars in the arms are short-lived with very luminous young supergiants, clusters, and star-forming molecular clouds, all delineating the spiral structure.

BARRED SPIRALS

Almost half of the known spiral galaxies, including our own, have a central cigar-shaped bar consisting of stars, gas and dust, from the ends of which radiate arms. These bars can be up to five times as long as they are broad and have varied orientations. Occasionally, a single galaxy can even host several bar-like structures of varying sizes. They work almost like a food processor, mixing together stars at differing distances. Dust accumulates along the leading edges whereas star formation and gas clouds are found near the ends.

NGC 6217: an example of a barred spiral galaxy, 6 million ly away in the constellation of Ursa Major.

ELLIPTICAL GALAXIES

These are featureless, slowly rotating conglomerations of mostly old stars in spherical or elliptical shapes. Notably, they do not contain clouds of cold gas from which future stars can form and are distinguishable only by their overall appearance, which can be round, squeezed, or elongated, dependent on viewing angle. Their brightness also varies: dwarf ellipticals contain only a few million stars and resemble giant globular clusters – like Omega (ω) Centauri – while supergiant ellipticals can be 20 times more luminous than our Galaxy.

NGC 1132: an example of an elliptical galaxy, 318 million ly away in the constellation Eridanus, the River: the result of a possible merger.

LENTICULAR

These lens-shaped (hence lenticular) galaxies appear to be midway between spirals and ellipticals. Like spirals, their stars reside in discs around a large central bulge and there can be a central bar. Like ellipticals, however, they contain mostly old stars, have little gas or dust, lack spiral arms, and are featureless. Indeed, if viewed face-on, lenticulars would probably resemble round ellipticals. Perhaps they were once spiral galaxies that now have a dearth of gas or perhaps they formed just as they are today.

NGC 5866: an example of a lenticular galaxy, 44 million ly away in the constellation Draco. It is tilted edge-on to our line of sight.

STARBURST GALAXIES

As has already been seen, galaxies can be involved in collisions or near-misses with other galaxies, despite the enormous distances between. The result is cloud compression and subsequent massive star formation. Astronomers study these galaxies in infrared wavelengths – heat energy generated by the dust from the hot, newly born stars. A fine example of a starburst galaxy is M82 in the constellation of Ursa Major; this may well have been disturbed by a close encounter with one of the other galaxies in the cluster or, indeed, suffered a collision with the nearby M81 spiral some 200 million years ago.

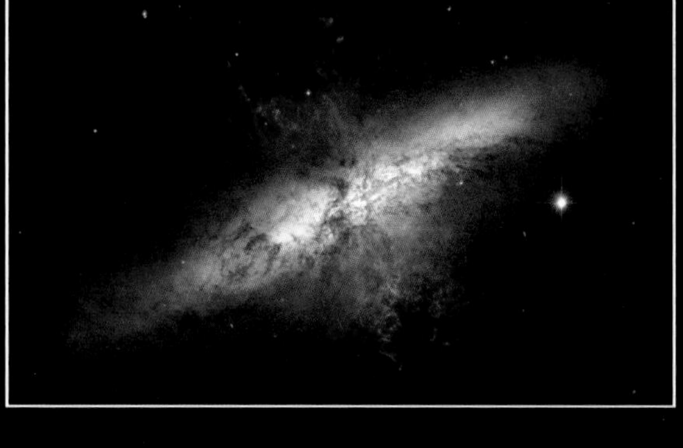

The Cigar galaxy, M82 in Ursa Major: two enormous plumes of hydrogen gas are sucked from the galaxy by the gravitational near-miss of another.

DWARF SPHEROIDAL GALAXIES

These difficult to detect galaxies are the most common in the Universe; half the members in the Local Group are of this type. Typically, they are only a few thousand light years wide and have a mass of just a few million Suns. Appearing as a loose gathering of stars, they have little or no gas, hence there is no current star formation.

NGC 1569: a dwarf galaxy, 7 million ly away; the bubble structure is sculpted by super-winds created by colossal energy from supernova explosions.

IRREGULAR GALAXIES

Most galaxies fit into the categories already mentioned. Those that do not are deemed irregular! They tend to have a ragged appearance and are smaller and fainter than other galaxies. Since they are not eye-catching nor aesthetically pleasing they are not 'popular', and yet, surprisingly, they account for more than a third of all known galaxies, including the Large and Small Magellanic Clouds.

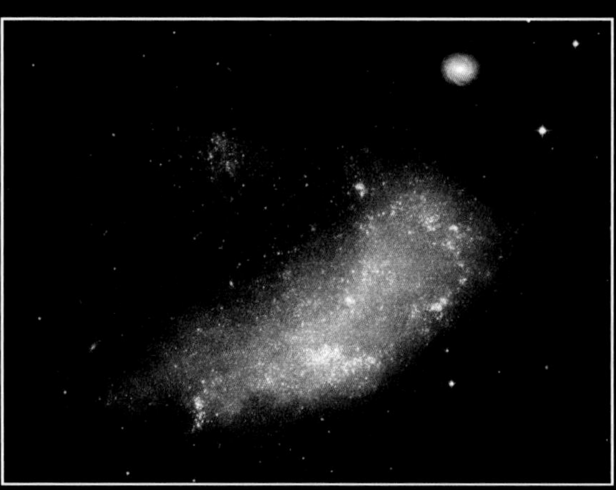

NGC 1427A: pulled through space at 600km/s by the combined gravity of the Fornax cluster of galaxies, this irregular will gradually tear apart.

HUBBLE CLASSIFICATION

The galaxies were studied by Edwin Hubble in 1926: the result was a diagram resembling a musician's tuning fork, which forms the basis of a system still used today. He started with the ellipticals, classifying them from the roundest to the flattest – E0 to E7 – and followed these with the 'midway' lenticulars – S0. The diagram then divided into ordinary spirals (S) and barred spirals (SB). Within these two branches each galaxy type was categorised according to the openness of their arms and the relative size of the bulge. Types Sa and SBa have tightly wound arms and a relatively large bulge. Types Sc and SBc have loosely wound arms and a relatively small bulge. Those intermediate between Sa and Sb are termed Sab. The irregulars were added at the ends of the spiral galaxy sequences.

CIRCULAR

E0

NGC 4486 (M87)

E3

ELLIPTICAL GALAXIES

NGC 4406 (M86)

E5

NGC 205 (M110)

S0

NGC 7217

BARRED SPIRALS

SPIRALS

NGC 1291

SBa

Sa

NGC 2775

SPIRAL GALAXIES

NGC 1300

SBb

Sb

NGC 2841

NGC 7741

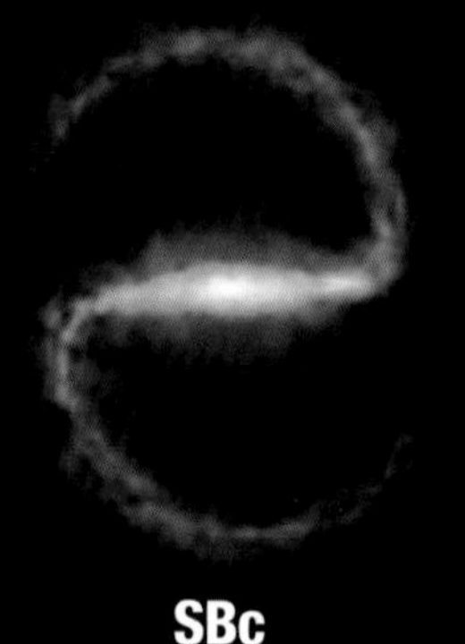

SBc

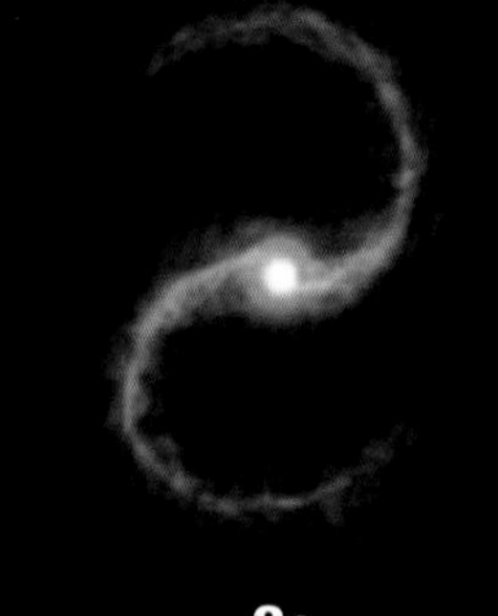

Sc

NGC 5194 (M51)

QUASARS AND ACTIVE GALACTIC NUCLEI

Today, it is thought that a quasar – once deemed a strange, hugely distant star-like object, also known as a quasi-stellar source – is just one form of active galactic nucleus (AGN). This is a term used to describe various violent phenomena apparent at the centre of some galaxies. These AGNs are highly luminous, emitting enormous amounts of energy across the electromagnetic spectrum from radio to X-rays.

Many galaxies are thought to host central black holes with masses of millions to billions of Suns. A supermassive black hole at the centre of any of these galaxies could easily be transformed into an AGN simply by gas collecting into an accretion disc around it. This gas could come from the intermittent tidal disruption of surrounding stars which, drifting too close, were captured by the black hole's gravity. All galaxies may have black holes, but since the supply of gas would be in short bursts it would explain why only a few galaxies seem to have AGNs. Moreover, when galaxies interact with each other more stars and gas could fall into the black hole, fuelling more activity.

These AGNs are among the most distant objects yet detected in the Universe. Astronomers 'see' them as they appeared when the Universe was only about 10% its present age. It is now agreed that all the various forms of AGNs stem from a central source; their description is dependent upon their orientation to an observer. This source can vary in size: anything from one light day in diameter to a light month. When observed at radio wavelengths, AGNs comprise three key features: a compact source, two large symmetrical radio-emitting lobes, and a single jet.

In the model opposite, a small source is surrounded by a torus (a ring) of gas and dust. Narrow jets of gas are emitted at right angles to the ring at light-speed velocities; these may be caused by differential rotation in the disc generating magnetic fields. Where the jets point towards the line of sight, the observer sees a blazar (A). Where the observer looks towards the central black hole over the edge of the torus, a quasar is seen (B). Where the torus obscures the central black hole, a radio galaxy is seen (C).

Centaurus A, NGC 5128: colour composite image revealing lobes and jets emanating from the active galaxy's central black hole, the result of

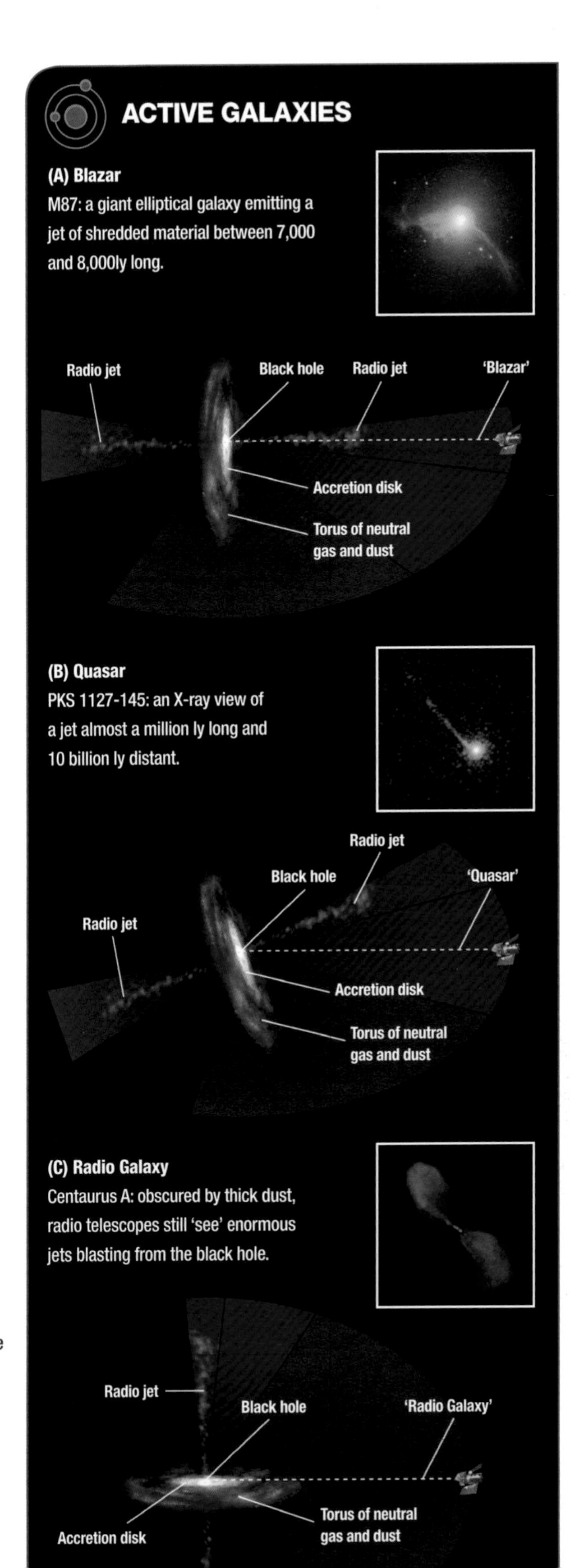

The big picture

How did these 250 billion known galaxies originate? Where, when, and why? Where are they heading? What and where is the Universe, and how did it originate? When, why, and where will it end, if indeed it does? Many, many questions …

THE BIG BANG

The past is where the Universe began … from a single dense hot state producing a 'big bang explosion' … except that this primordial 'explosion' did not happen in a fixed location, it happened everywhere, at every point in space. From this event came all the matter that surrounds us as well as space and time itself, and since time started in this 'explosion' it is impossible to scientifically ask what went before. Quite a concept! The Universe underwent a rapid expansion in the first flash of its existence, known as inflation, and it has been expanding, albeit at a slower rate, ever since. During the first few hundred thousand years it was a primordial cauldron of highly energetic colliding particles and radiation. Electrons constantly escaped from atoms. Photons of light were consequently caught and trapped. Light could not escape. The result was an impenetrable 'fog' of plasma that was opaque to electromagnetic radiation – like cloud obscuring sunlight on an overcast day.

After around 380,000 years, as the Universe continued to expand, the wavelengths of energy lengthened, the temperature subsequently cooled to about 3,000°K, and the energetic collisions decreased; matter and radiation could separate, electrons could be caught by atomic nuclei to create stable atoms – a period known as recombination. But the Universe was then engulfed by vast clouds of neutral hydrogen, and it would take hundreds of millions of years before ultraviolet radiation from hot newborn stars in infant galaxies could split the fog – reionisation. The epoch between recombination and reionisation is called the 'dark ages', whose end is signalled by the birth of the first generations of stars and galaxies – 'Renaissance'. Since the time of recombination the Universe has expanded by a factor of 1,000, so the most distant galaxies detectable by present-day telescopes are being seen at a time when the Universe was around 1,000 million years old!

The 'Big Bang' left a signature residual background radiation, an afterglow that initially escaped into space around 379,000

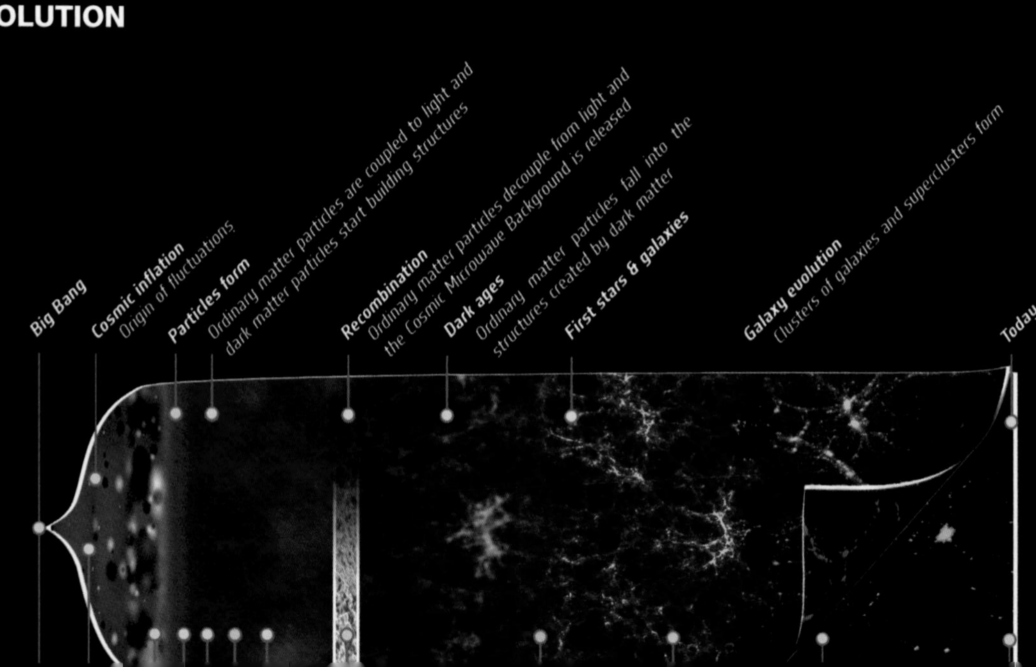

Big Bang
Cosmic inflation
Origin of fluctuations
Particles form
Ordinary matter particles are coupled to light and dark matter particles start building structures
Recombination
Ordinary matter particles decouple from light and the Cosmic Microwave Background is released
Dark ages
Ordinary matter particles fall into the structures created by dark matter
First stars & galaxies
Galaxy evolution
Clusters of galaxies and superclusters form
Today

COSMIC EVOLUTION

This illustration summarises the almost 14-billion-year long history of our Universe. It shows the main events that occurred between the initial phase of the cosmos, where its properties were almost uniform and punctuated only by tiny fluctuations, to the rich variety of cosmic structure that we observe today, from stars

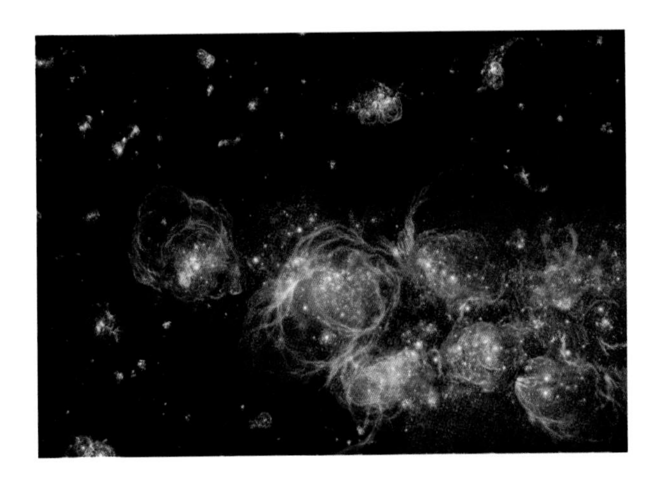

Celestial fireworks: an artist's impression of the Universe during the first billion years. Devoid of dust, it glitters with myriad starburst galaxies, massive star formation, hot gaseous bubbles and spectacular stellar demise.

years after the Big Bang and has only now reached Earth – the Cosmic Microwave Background (CMB). Its cosmic 'hiss' was discovered in 1965 by independent researchers Arno Penzias and Robert Wilson of Bell Laboratories, New Jersey. In 1989, NASA's COBE (Cosmic Background Explorer) satellite accurately measured the radiation at 2.725°K. COBE also found tiny temperature variations (anisotropy): warmer regions indicated lower density and cooler regions indicated high. It was thought these could have arisen from quantum fluctuations during or prior to inflation but, however formed, they 'seeded' the galaxy clusters.

Since COBE, various ground and balloon-based experiments have measured the CMB on smaller scales, but the Wilkinson Microwave Anisotropy Probe (WMAP), launched at the start of the 21st century, resolved structures that were only a fraction of a degree wide. Furthermore, it revealed that everyday atoms accounted for just 4% of the cosmos as we know it, with a further 23% being dark matter and the final 73% a mysterious, repelling dark energy. ESA/NASA's Herschel Space Observatory and ESA's Planck spacecraft (both no longer operational) have unveiled the coldest, dust-shrouded and most distant objects in the Universe within the regions of infrared and microwave wavelengths. Thanks to Planck, cosmologists are now able to study the most detailed map of the CMB ever made.

■ PRE BIG BANG?

Where did the elements that would produce the material for stars originate? Since astronomers cannot see any further back than around 300,000 years after the Big Bang, they are reliant on theoretical models as to what may have happened before. How did the Big Bang arise? Was it a quantum mechanical fluctuation in the realms of the infinitesimal where interaction of fundamental particles takes place? How did general relativity – the effects of gravity on a truly large scale – interact? What happened during its earliest existence for which current physics fails? A unifying theory for quantum mechanics and general relativity has yet to be developed. Theorists have yet to account for how the Universe originated from a singularity, with zero size and infinite density and temperature. Theories abound.

When radiation was produced in the earliest moments after the Big Bang, it should have produced a symmetry – equal amounts of particles (protons, neutrons, and electrons) and antiparticles (antiprotons, antineutrons, and antielectrons, also known as positrons). These should have annihilated each other and we should not be here! But, clearly, matter is here. It is a conundrum on a truly colossal, and yet minuscule, scale. Astronomers search for this antimatter but are hampered: they study spectral lines from atoms but the photons produced from antimatter are the same as those produced from matter. Only the antineutrino is a distinct particle, but these too are elusive. If matter annihilates antimatter, then a giant flux of high-energy gamma rays should result, but gamma-ray telescopes have failed to find any such evidence.

However, it is now generally accepted that most of the mass in the Universe is an invisible form of mysterious dark matter which, though yet to be detected in particle accelerators, perhaps resides in the halo of galaxies; even though stars exist in the halo, they do not account for its entire mass. Additionally, galaxy clusters appear to move much faster than should be expected: perhaps there is invisible dark matter in the space between the galaxies providing the source of the gravitational pull. The visible matter in the Universe, in the form of stars and gas, is relatively insignificant when compared to the whole.

Universal expansion: according to the Big Bang model, the Universe expanded from an extremely dense and hot state and continues to expand today. A common analogy explains that space itself is expanding, carrying galaxies with it, like raisins in a rising loaf of bread.

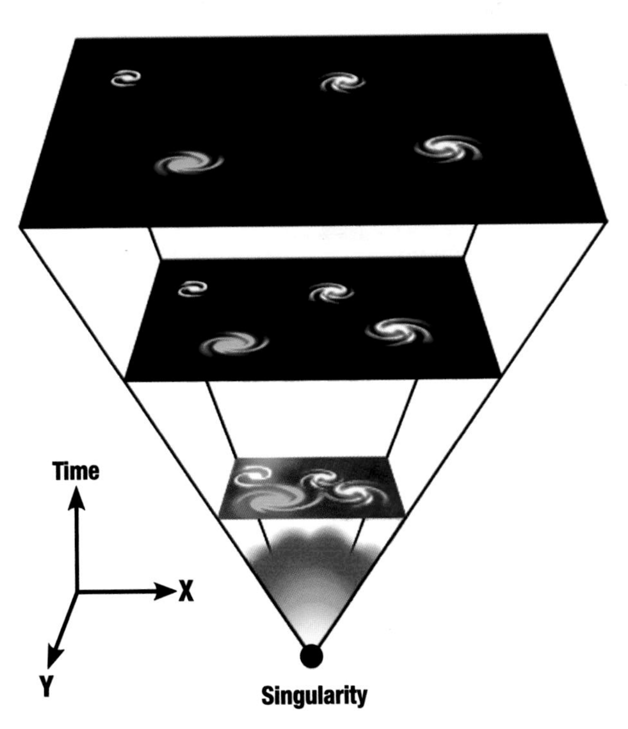

EARLY GALAXY FORMATION

In the 1960s, astronomers held that galaxies formed from a giant cloud of primordial gas which collapsed under its own gravity; a process mirroring the birth of stars but on a much grander scale. This was known as Monolithic Formation. As the cloud contracted it created small lumps of gas that then spawned a halo of globular clusters. Whether the galaxy became spiral or elliptical was dependent on its speed of rotation; if sufficiently fast, the cloud could spread out into a disc and later form spiral arms.

This theory has now been usurped by the Hierarchic model: large galaxies present today formed from mergers of smaller galaxies – proto-galaxies – which themselves formed from the merger of other smaller objects now mind-numbingly distant and shrouded in dust. Perhaps all galaxies started as ellipticals but after amassing gas they transformed into spirals. Vast 'empty' spaces exist between stars so they are able to merge freely, their attraction induced by the mutual gravitational forces of their parent galaxies. The results of these mergers are magnificent tidal tails that stream hundreds to thousands of light years out into space.

When gases merge in galaxies, massive star formation ensues and luminosity increases. If the star formation explodes throughout the entire galaxy we see the dazzling starburst galaxies or, if sufficient stars and gas cascade on to the central black hole, the quasars – both mentioned earlier. After hundreds of millions of years of mergers the result is an elliptical galaxy. But astronomers are now witnessing 'mature' galaxies: these must have formed in isolation and existed in the first billion years of the Universe's formation. The picture is incomplete. With rapid developments in telescope and computer technology enabling studies of long-term behaviour and small-scale intricacies, astronomers are rapidly gaining a better understanding of galaxy evolution (see Chapter 4). It is an astro-archaeological survey on the largest scale of all.

The Antennae Galaxies in a slow hundred-million-year collision: dark dust pillars are massive molecular clouds being compressed in the encounter and causing the explosive birth of millions of stars and star clusters.

RELIC 'AFTERGLOW' MICROWAVE RADIATION

'Captured by ESA's Planck spacecraft, this image is a full-scan of the entire Universe sky – a cosmic blueprint, the smoking gun of Creation. Two CMB anomalous features hinted at by Planck's predecessor, NASA's WMAP, are confirmed in new high-precision data. One is an asymmetry in the average temperatures on opposite hemispheres of the sky (indicated by the curved line), with slightly higher average temperatures in the southern ecliptic hemisphere and slightly lower average temperatures in the northern ecliptic hemisphere. There is also a cold spot that extends over a patch of sky that is much larger than expected (circled). In this image the anomalous regions have been enhanced with red and blue shading to make them more clearly visible.

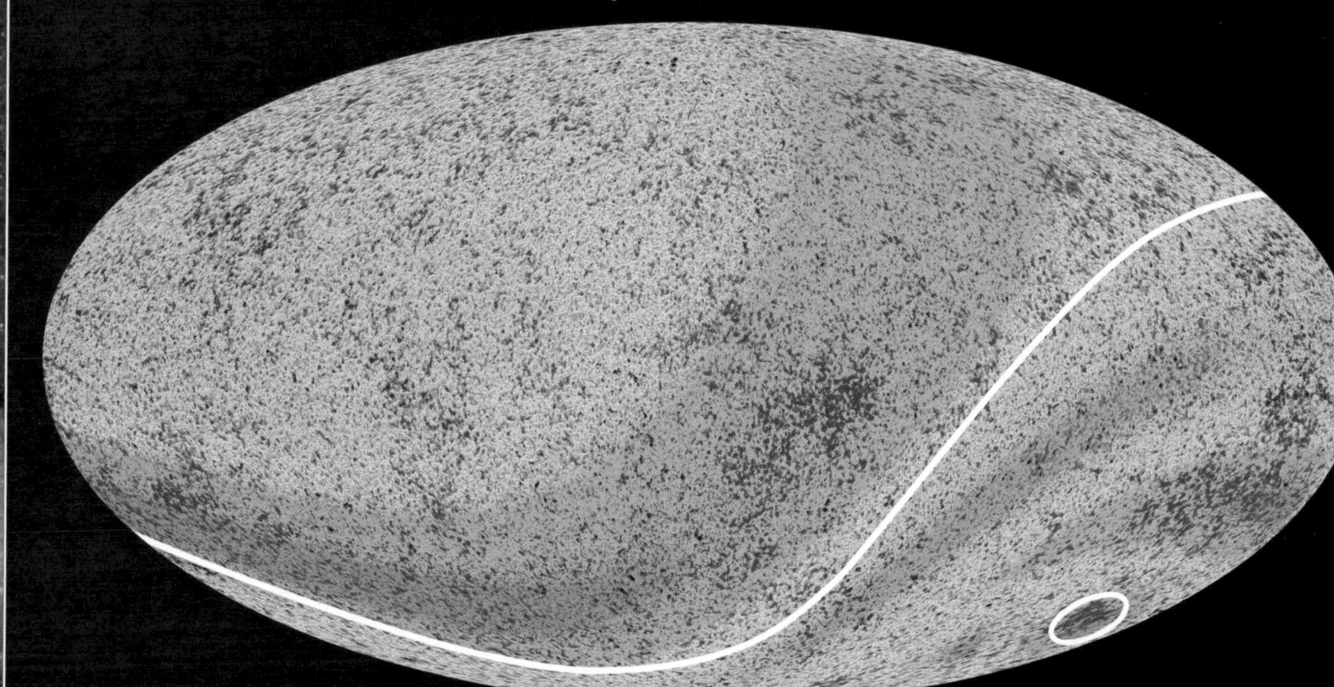

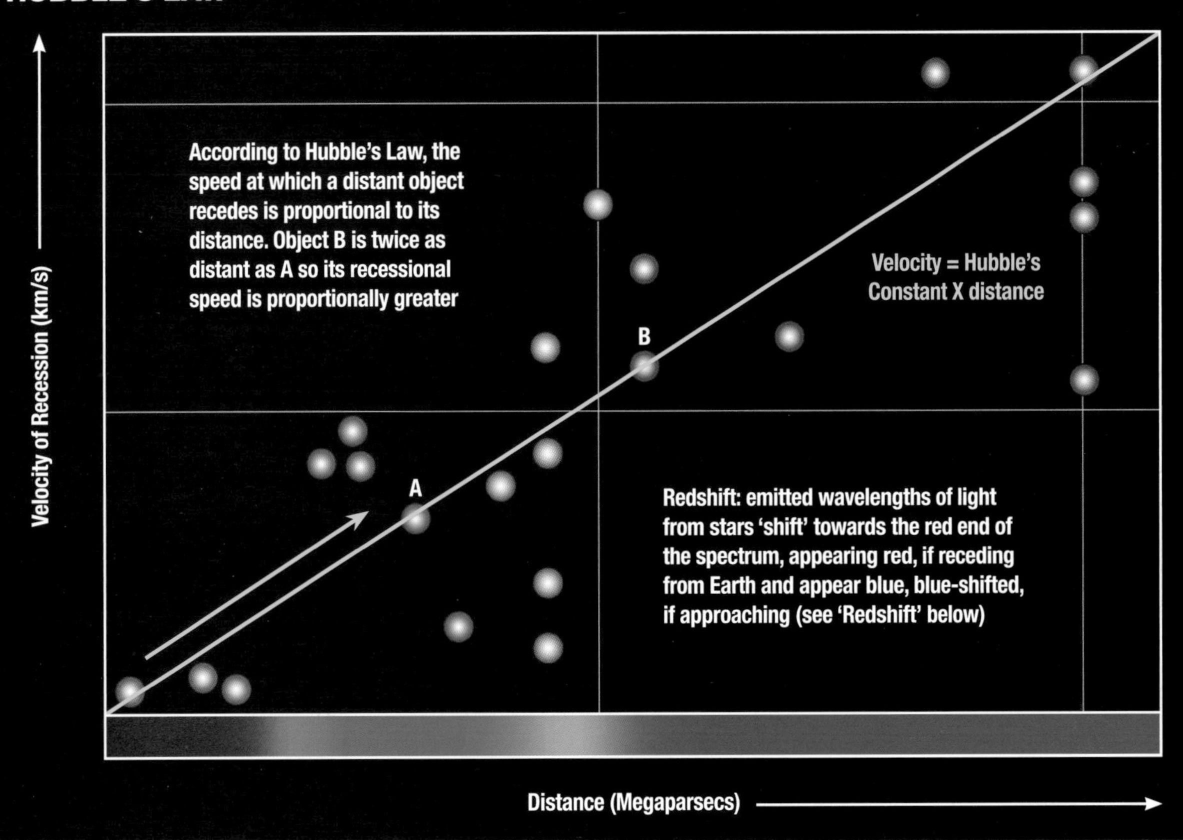

HUBBLE'S LAW
According to Hubble's Law, the speed at which a distant object recedes is proportional to its distance. Object B is twice as distant as A so its recessional speed is proportionally greater
Velocity = Hubble's Constant X distance
A
B
Redshift: emitted wavelengths of light from stars 'shift' towards the red end of the spectrum, appearing red, if receding from Earth and appear blue, blue-shifted, if approaching (see 'Redshift' below)
Velocity of Recession (km/s)
Distance (Megaparsecs)

SIZE OF THE UNIVERSE

In 1781, Charles Messier drew up a catalogue of 100 nebulae. These were diffuse, often enigmatic, objects visible in small telescopes and with the naked eye. For years it was argued that they were gas clouds in our Milky Way galaxy, whilst others believed they were other galaxies, like ours, but hugely distant. Following on from US astronomer Harlow Shapley's (1885–1972) work on distant globular clusters, the first galaxy proven to lie beyond the Milky Way was the great Andromeda (M31), and this was proved in 1924 by the American Edwin Hubble (1889–1953), using the 100-inch Mount Wilson telescope in California. He knew the measurements of Cepheid variable stars (see page 60) within our Galaxy and used them as 'standard candles' to calculate the distance of the same types of stars in Andromeda; the galaxy was found to be ten times more distant than the diameter of our Milky Way. Furthermore, allowing for M31's angular size and brightness, the galaxy had to be much larger too. It was a revelation.

REDSHIFT AND HUBBLE'S LAW

With the 19th century discovery of emission and absorption lines in the spectra of gases (see Preface), astronomers were able to determine chemical composition. Subsequently, these spectra were also used for an even greater revolution – redshift. Redshift is a change in the wavelength of light as a result of cosmological processes. All astronomical objects emit radiation (energy) in the form of waves, the distance between each wave crest being the 'wavelength'. These lengths vary: the number of times a crest passes a fixed point is known as the 'frequency' – the shorter the wavelength the higher the frequency. It is similar to the Doppler effect of sound waves on Earth: the pitch of a train whistle changes as the train passes a stationary listener because the sound waves ahead of the moving train are compressed and, therefore, pitched at a higher frequency than those which are more spread out behind and, therefore, pitched lower. Light waves in space are the same. When light is shifted (compressed) towards the blue end of the spectrum the object is approaching us, when shifted towards the red, it is receding. In cosmology, this redshift of light from distant galaxies is not caused by the galaxies rushing away but by the very expansion of the fabric of space itself (see *Superclusters and beyond*, page 72). As the Universe expands it stretches the wavelengths of light received from the objects within it, just as the 'cake mixture' with the 'raisin galaxies' stretches in our fruitcake. Using this knowledge, Edwin Hubble clearly saw that nearly all the galaxies were rushing away in all directions at a rate that was proportional to their distance. This concept became known as Hubble's Law.

Astronomers were soon able to measure the speed of this line-of-sight expansion – radial velocity – and could see that the greater the separation of the objects, the faster their recession. Indeed, recent results from ESA's Planck spacecraft posit that the expansion rate of the Universe (called the Hubble parameter or Hubble Constant) which sets not only the time scale for cosmic expansion but also the scale for the Universe's size and age is 67.8 +/- 0.9 km/s per mega-parsec (a parsec is 3.26 light years). So for every 3.26 million light years separating two galaxies, the galaxies are being driven apart by 67.8 km per second.

DESTINY

Where these galaxies are rushing to is the ultimate unanswered question, one limited by the finite speed of light itself. The farthest astronomers can see – the edge of the observable Universe – equates to the distance light has had time to travel during the billions of years of the Universe's existence. LIght from stars beyond this observable threshold, which, because the Universe has been constantly expanding, is 46.5 billion light years away, has not yet reached us. This highlights a phenomenon known as 'look-back' time whereby, owing to the finite velocity of light, the more distant an object being observed, the older the information received from it. A galaxy one billion light-years away, for instance, is seen as it looked one billion years ago. The look-back time cannot exceed the age of the Universe (13.8 billion years) but since the expansion of the Universe has been accelerating, the observable radius – the distance to the most remote objects – may well exceed 13.8 billion light years.

AGE OF THE UNIVERSE

How old is the Universe? Knowing Hubble's Constant, and if the expansion of the Universe has remained steady, then astronomers could, in theory, easily calculate the elapsed time since the Big Bang. It works out to around 15 billion years in Hubble time ... but the expansion may not have been constant. As galaxies rush apart from each other, they resist a mutual gravitational pull that slows their expansion. Expansion could have been faster in the past. In support, astronomers study the oldest globular clusters too. By observing and calculating how long it has taken for their young hot blue stars to redden and die, astronomers know they are around 12 billion years old, and, since they formed around a billion years after the Big Bang, it seems the best current estimate of the age of the Universe is 13.8 billion years. In consolidation, astronomers also use 'forensic' evidence: radioactive elements created in old stars from past supernova explosions – nucleosynthesis. By comparing these elements with those that arise from their known decay, an estimate of age can be achieved in a process similar to the dating of rocks on Earth. This technique seems to indicate that the oldest stars in our Galaxy are between 10 and 20 billion years old.

GEOMETRY OF THE UNIVERSE

Who knows for sure? Since the moment of the Big Bang, the Universe has been expanding and its legacy radiation surrounds us (see page 80). Perhaps we can better visualise this expansion by picturing painted dots on an expanding balloon: all the dots move away from each other as the balloon expands at a speed proportional to their distance from each other, but no one dot is at the centre. The Universe, on a grand scale, looks the same in all directions – it is described as isotropic. It does not appear to have an edge. Just as none of the dots on the balloon are at its edge, so none of the known galaxies are at the 'edge' of the Universe. Consequently, the Universe, arrayed with its filamentary structure of galaxy clusters and superclusters, looks the same from all points in space and is described as homogenous.

The real breakthrough in understanding the Universe came with Einstein's publication of his General Theory of Relativity in 1916. It was a new interpretation of gravity. In previous Newtonian physics, space was empty and bodies moved through it, but in general relativity space has a substance of its own, with curvature and expansion – an idea temporarily rejected but revived in 1917. It is a study not only of space, but of time and the bodies within it.

But what is the geometry of the Universe? How can it be envisaged? What are the theoretical possibilities? Based on general relativity, modern cosmological models suggest there is an overall curvature of space-time which is more easily understood in two dimensions. It could have positive curvature and be shaped like a sphere: with a closed surface that is finite in extent and without edge. It could be shaped like a saddle: with an infinite negative curvature and, therefore, open since it never closes back on itself. And between the two extremes would be a flat Universe: a plane that is open, lacking either positive or negative curvature and, therefore, infinite.

Inflation has made the observable Universe 'flat'. Remember the balloon: when a small area is stretched, or inflated, trillions of times over, it still looks flat to an insect scrabbling around on its surface. By analysing the small fluctuations in the intensity of the cosmic background radiation, and by comparing the observed size of these with the expected size (as was done in 1998 by balloon-borne detectors flying over Antarctica – the BOOMERanG experiment) it would seem the Universe is flat.

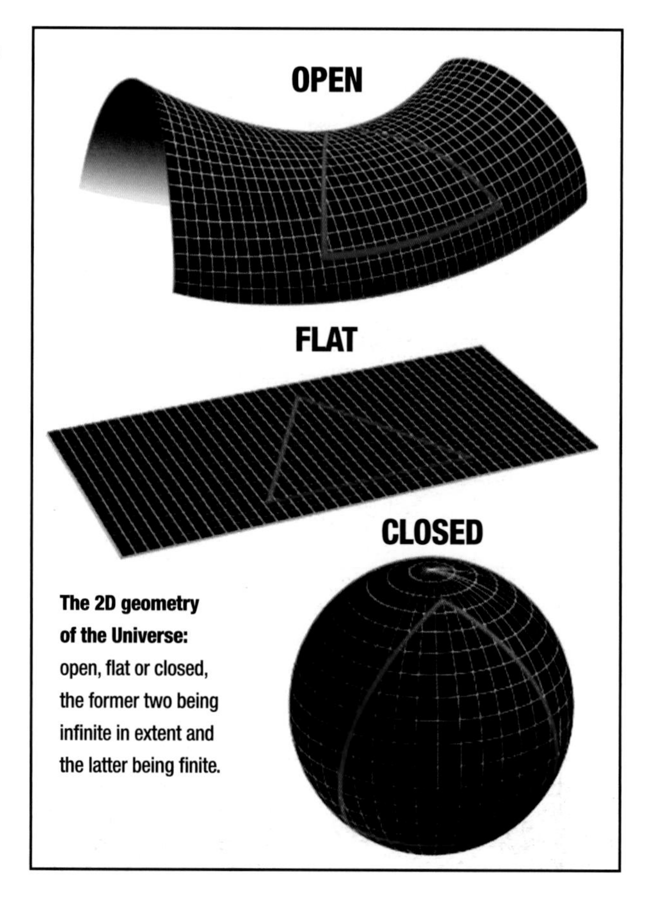

The 2D geometry of the Universe: open, flat or closed, the former two being infinite in extent and the latter being finite.

FATE OF THE UNIVERSE

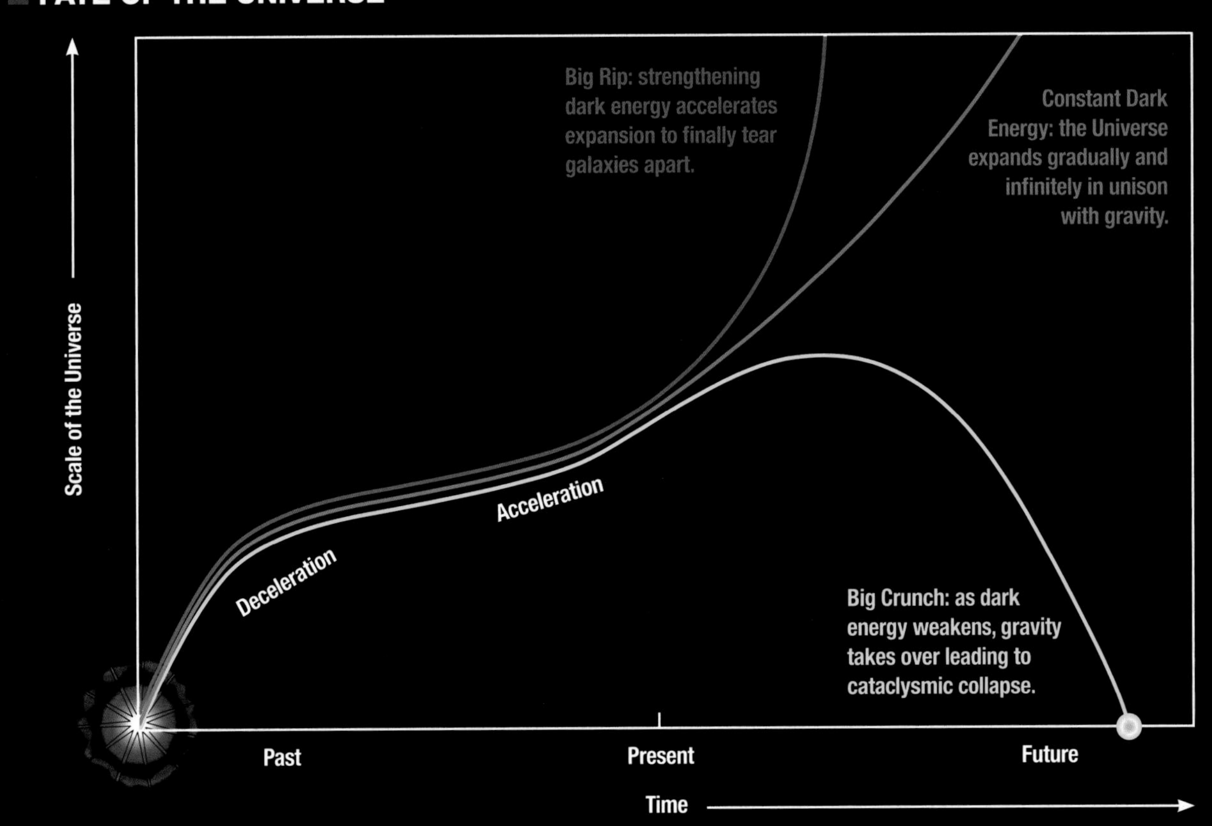
Big Rip: strengthening dark energy accelerates expansion to finally tear galaxies apart.
Constant Dark Energy: the Universe expands gradually and infinitely in unison with gravity.
Scale of the Universe
Acceleration
Deceleration
Big Crunch: as dark energy weakens, gravity takes over leading to cataclysmic collapse.
Past
Present
Future
Time

POSSIBLE DEMISE

Where does the Universe 'end'? Again, astronomers grapple with the unknowable. The question has haunted humanity for millennia. We seek to know and, with increasingly sophisticated technology, and courage, data is being found to address these questions head-on. There appear to be three possible outcomes for the demise of the Universe and all are in some way inextricably linked with the mysterious dark energy that constitutes some 69.2% of the total mass and energy in the Universe (the rest of the 'Cosmic Pie', according to the latest Planck results, breaking down into 25.9% dark matter and 4.9% baryonic matter – protons and neutrons and combinations of these, such as non-emitting ordinary atoms).

Abell 1689 Galaxy Cluster, 2.2 billion ly distant, size 2 million ly: where gravity bends and magnifies light to reveal the oldest galaxies some 13 billion ly beyond.

ETERNAL EXPANSION

Since gravity, generated by matter, should slow down the expansion of space, it follows that the amount of matter – the critical density – should, therefore, determine how slow that expansion could be. But what are the components contributing to the average density of the Universe, and how is this density measured?

There is ordinary matter: protons, electrons, and neutrons found mostly in the stars. This density is measured by comparing the observed cosmic abundance of deuterium (a hydrogen isotope) to the theoretical predictions of what the billion-degree early Universe left as a relic. Secondly, astronomers have weighed intergalactic gas via spectroscopic observations of hydrogen. And thirdly, they have analysed temperature fluctuations in the cosmic microwave radiation, a pattern that encodes the relative densities of ordinary matter, dark matter, and dark energy.

It appears that dark energy, so called not because it is 'dark' but because it is unknown, has a density that remains constant in any given volume of space. It could be a constant energy density produced from the vacuum of space, perhaps like Einstein's Cosmological Constant, an inherent property of space itself. As space expands, the influence of this mysterious, repelling dark energy will increase, ensuring infinite expansion.

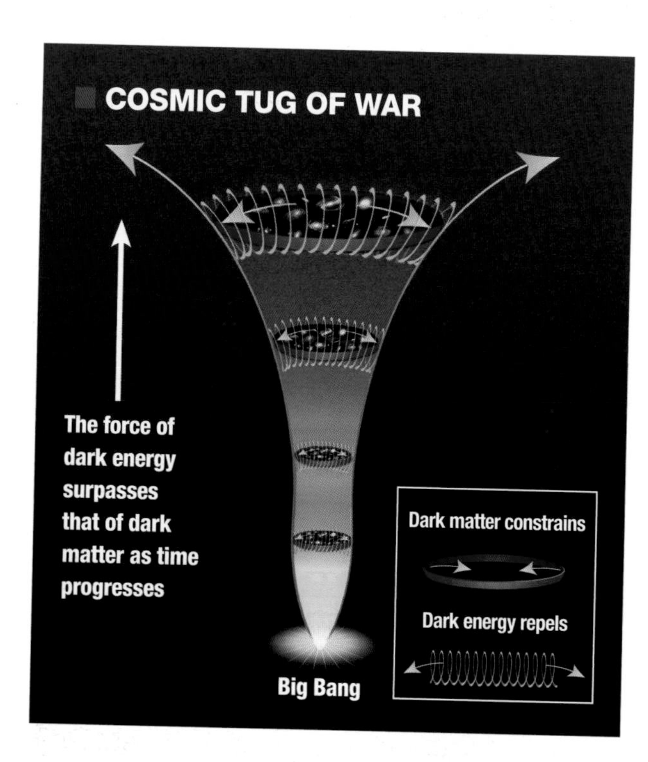

for future star birth. Eventually, however, all the gas will be consumed and interstellar space will consist of dim galaxies, host to decaying white dwarfs, neutron stars, and black holes. These dying stellar systems could, over a long period of time, evolve. Their stellar orbits could change. Stars that approach each other could fling one or the other out of the galaxy. Eventually, all the stars could evaporate. Those that remain will lose energy and gravitate towards the core. A black hole will form and increase in mass until it contains all the mass of the galaxy. Black holes may also lose mass by emitting energy as a result of losing radiation – Hawking radiation, after the renowned physicist Stephen Hawking. As the black hole loses mass, the evaporation rate increases until the hole ultimately disappears in a single explosion of highly energetic gamma rays.

And in the theories of Grand Unification, protons, neutrons, electrons, and dark matter particles themselves may also decay; the fundamental particles will die. What could be left is a Universe that is dark and devoid of anything except photons and neutrinos – fundamental particles without charge and possibly no mass – that are too low in density and energy to ignite matter ever again.

Since Hubble's Constant is a known expansion rate, astronomers could simply measure the expansion rate earlier in the Universe and compare it to the current expansion rate and see whether the expansion is increasing or slowing, regardless of density measurements. Additionally, the finite speed of light enables astronomers to study very distant galaxies. And there are the 'standard candles' mentioned earlier. But there is a problem: these 'candles', such as the Cepheid variable stars and the even more luminous but distant Type 1a supernovae, are either too distant, or too infrequent, to detect. A recent large-scale collaboration between astronomers has, however, meant that a number of supernovae in very distant galaxies have been monitored and it has been found that Edwin Hubble's constant expansion rate was actually smaller in the past, which seems to infer that the Universe is not only expanding but that its expansion rate is actually speeding up! The result will be that galaxies stretch farther apart so that in around 100 billion years their mutual distances will be … incomprehensible!

The Universe itself may not end but the same cannot be said of the objects within it. As we have seen, stars are born, they live, and they die, their ultimate demise dependent on their initial mass. But it is a constant recycling process; when stars die, their elements are returned to interstellar gas and become the material

THE BIG CRUNCH

The expansion of the Universe is accelerating due to dark energy. Cosmologists are agreed that it is unlikely but, if this energy gradually declines, perhaps reverses and becomes somehow negative, then the Universe could begin to shrink. Within around 20 billion years, galaxies could collide and the entire Universe collapse in on itself in a cataclysmic 'big crunch'.

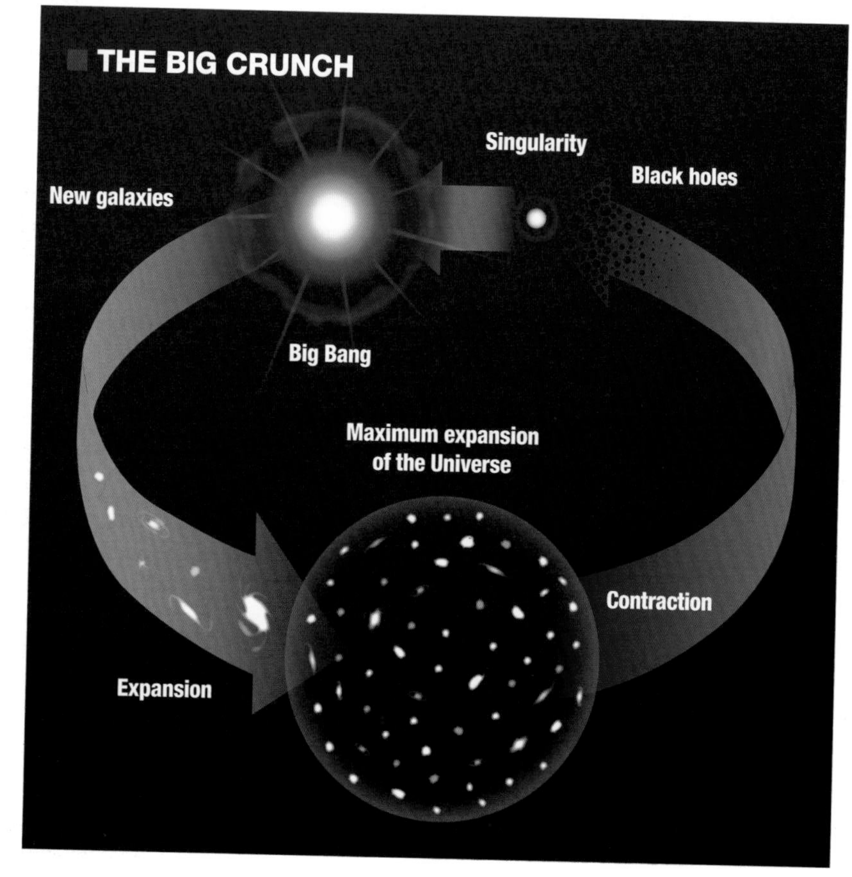

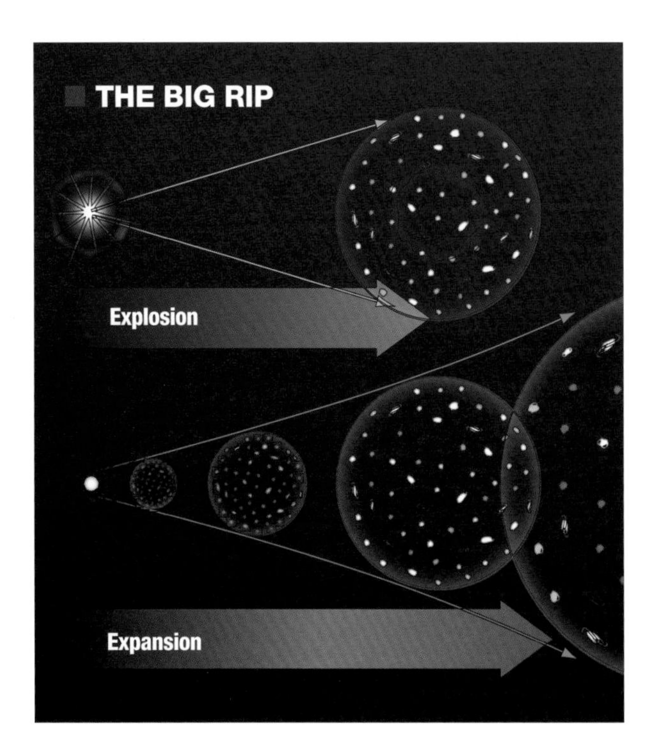

THE BIG RIP

In time, say in around 25 billion years (so not an overnight process!), and if the dark energy becomes even more powerful, it could become progressively violent and even rip galaxies apart. Again, physicists think this is unlikely, but the result would be that planets would explode and even atomic nuclei could disintegrate.

And it does not stop there. Recent calculations have hypothesised an even more powerful dark energy, known as phantom energy, so ferocious it could simply melt omnipotent black holes like sugar cubes in tea, utterly eradicating anything left behind. Even billion solar mass supermassive black holes will not be exempt, their enshrouding galaxies will be shredded like canvas sails in a stormy sea.

Theories and models are proved and disproved. From geological studies, current conjecture is that the Universe is flat. Additionally, the total amount of density within it is around five times smaller than the critical density – the amount needed to maintain a balance – and, therefore, the Universe will continue to expand forever although not necessarily at the same rate.

So where does this leave humanity? Well, as we have seen, our Sun is not exempt from the life-cycle process. In a few billion years, the Sun, having consumed its fuel, will swell to become a red giant that will engulf the planets Mercury, Venus, and possibly Earth. Life on our planet as we know it will consequently cease in around one billion years. Perhaps, in the interim, we will have developed interplanetary, or interstellar, space travel and will colonise elsewhere; perhaps to an exoplanet in a suitably habitable 'Goldilocks' zone. Perhaps our species will have evolved to survive in a dead or dying galaxy… so many seemingly impossible possibilities…

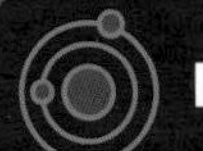

INFINITY

And we have not even tackled the mind-boggling scenario of eternal inflation – the possibility that there could be a never-ending space-time continuum producing multiple 'bubble universes'. Our Universe may not be alone. Indeed, there could be innumerable universes – a multiverse – perhaps interconnected by 'tunnels' or 'wormholes'. Each could have its own physical laws and properties. With the demise of one universe there could be a cyclical resurgence of another – a Big Bounce. Here we transcend to the realms of what seems like science fiction but could perhaps be science fact. There may be multiple dimensions instead of our traditional four.

The popular 'superstring' theory, or 'string theory' – a single theory to unite the forces of nature, including gravity, into a single 'superforce' – moots the idea of the unimaginably minute: if an atom was scaled up to the size of the known observable Universe, then a 'vibrating string' (like a rubber band) would be the size of a tree within it! This theory first originated with Gabriele Veneziano (b.1942) in 1968. A new version emerged in the mid-1980s. But both were superseded in the 1990s by Ed Witten and Paul Townsend's 'M-Theory' – the strings became membranes and the idea of 11 dimensions was suggested. Some physicists consider the 'M' stands for Magic, Mystery, or Mother since it would indeed be the mother of all theories! These theories may resolve the current incompatibility between the world of quantum mechanics and general relativity. There could be parallel universes. Something called the Heisenberg Uncertainty Principle even allows for tiny virtual particles, sited at tiny distances from each other, to jump in and out of existence over tiny intervals, and even disappear – a direct conflict with general relativity! At present, all these theories, and others, are … theories, and will remain so until technology catches up … which it undoubtedly will.

The cosmic web: a detailed computer simulation of dark matter with a bright patch indicating a supercluster of galaxies 300 million ly across. The brighter the colour the deeper the density. Voids are bubbles. It is superstructure on the grandest scale. Infinity appears possible.

The Milky Way: looking towards the dust-shrouded core of our Galaxy with its luminous clouds of billions of stars and pink glowing nebulae.

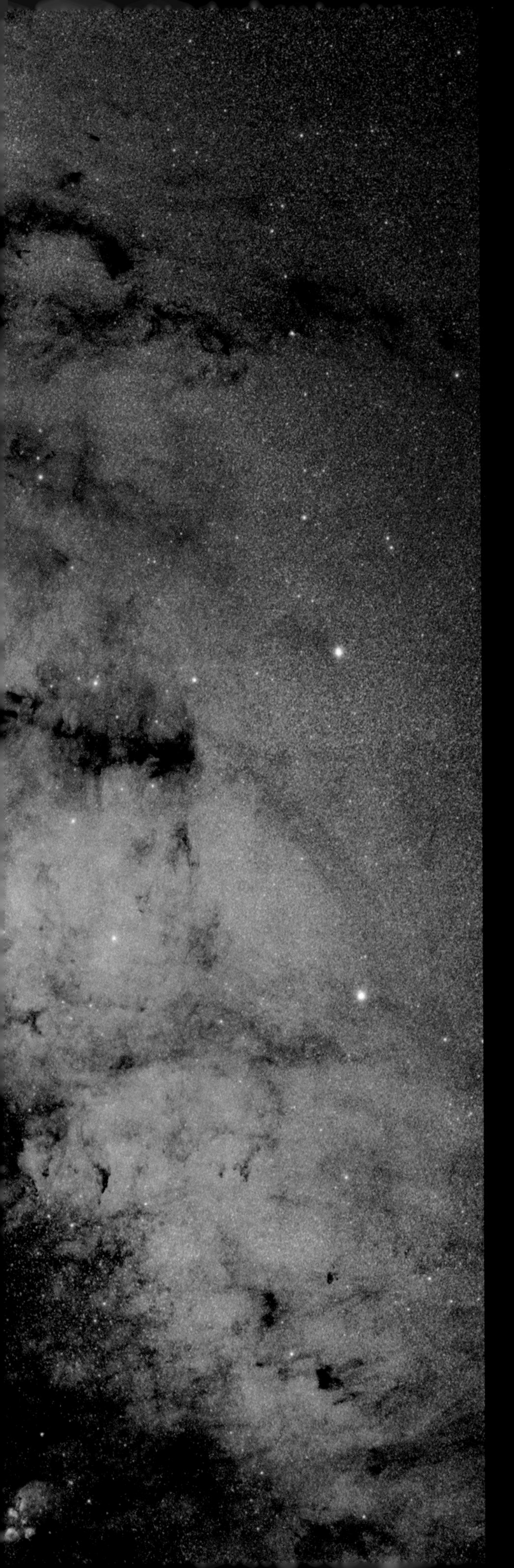

Amateur viewing

What kind of astronomer will you be? An observer, a technophobe, or a fireside buff?

It matters little as one thing will lead to another. In a dusty discarded astronomy book may be a simple diagram of the Sun buried amidst billions of others. An epiphany may follow. You may be drawn to looking up with unaided vision and wondering… perhaps binoculars and star maps will follow… and maybe a small telescope… and larger. You may wish to photographically capture some of the stunning vistas, or be drawn to computer wizardry and the realms of the astronomical internet.

The choice is yours – view it your way.

'Astronomy is looking up'

Anon

Naked eye vision

What better way to enjoy the night sky than simply do as the ancients did: step outside, look up, and absorb the truly 'wide-field' vista. However, if we wish to better understand that 'wide-field' view, whether with the unaided eye, binoculars, or telescope, we need context. Astronomy demands context. Not only do we need to know what we can see, but we also need to know when we can see it, and why. Sounds obvious? It is... and yet not...

■ CELESTIAL SPHERE AND CONSTELLATIONS

When we look up on any given night we can see that stars appear to 'create' shapes in the sky. Since ancient Mesopotamia, mankind has projected these shapes on to an imaginary sphere encompassing Earth. This is known as the celestial sphere. It is a map of the sky, dividing it into 88 distinct areas, or constellations, now recognisable around the world (see Appendix 1 for full list). Each area depicts a pattern or picture representing figures, animals, deities, or other more grounded images. Some bear a resemblance to their asterism, others do not, but each has its attendant customs and stories spanning centuries and about which innumerable maps have been drawn and books written. A particular example is the constellation of Ursa Major, also known as the Great Bear. To the ancient Greeks, this was the legendary Callisto, transformed by Zeus into a bear. For the Hindus, the seven stars represented the homes of seven great sages; the Egyptians saw them as the thigh of a bull and the Europeans saw them as a wagon allied with the tales of King Arthur. Its distinctive shape has inspired Shakespeare, Tennyson, and Homer. With its seven bright stars it is now best known as the Plough or Big Dipper, or simply a 'saucepan with a bent handle'!

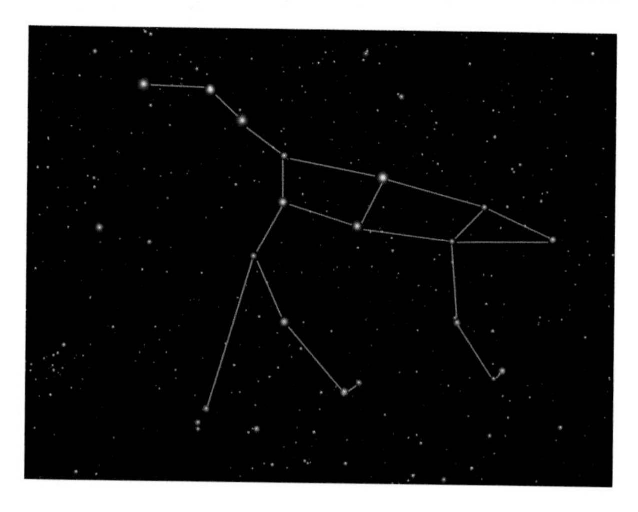

Ursa Major: also known as the Great Bear, the Plough, the Big Dipper or simply a saucepan with a bent handle!

Around the celestial equator on this imaginary celestial sphere is a belt spanning around 18°, in which can be found the 13 astronomical 'zodiac' constellations. Twelve of these form the 30° long equal 'astrological' areas more commonly known as the 'signs of the zodiac'. It is against this backdrop of constellations that the apparent paths of the Sun, Moon, and bright planets appear to pass. The word 'zodiac' has Greek origins and means 'circle of animals' since many of the signs, or star shapes, are creatures. Although there is no scientific evidence to support the idea that the position of the Sun and planets in these zodiacal areas influences everyday affairs, a belief in astrology has played a significant role in the history of mankind.

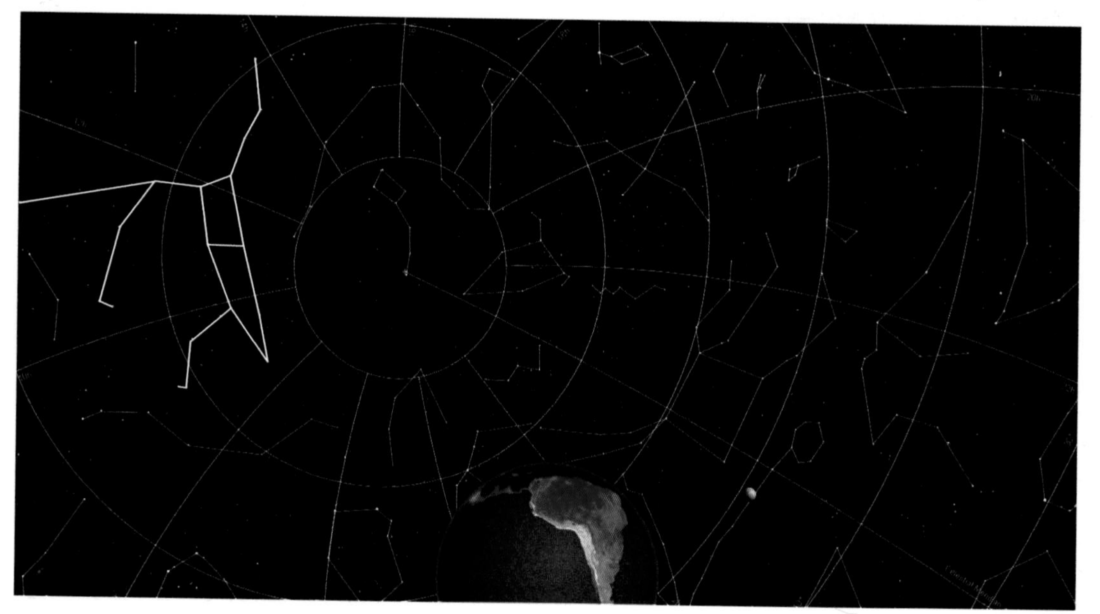

The celestial sphere 'surrounding' Earth: stars are projected on reference lines of longitude (right ascension) and latitude (decination). The constellation of Ursa Major is highlighted.

Also 'within' the 88 constellations, or areas, are the more distant stars, galaxies, and other celestial objects. Of course, these objects are not 'on' this sphere at all: they are all at varying distances. Solar System objects, such as the Moon and planets, can clearly be seen to move because they are, cosmologically speaking, closer. But the more distant stars, nebulae, clusters, and galaxies – the deep-sky objects, as they are known – appear 'fixed', and this is because they are simply so far away their movements are undetectable. But they are moving with the fabric of space just as everything is moving, from the millions of stars within our Milky Way galaxy to the other galaxies out there, millions of light years away. When we gaze skyward, whether with the naked eye, binoculars, or a small telescope, we are seeing only a tiny selection of what is 'out there'.

With so many objects to see, how can we locate them or point them out to others? To do so, astronomers developed a grid system projected on to this celestial sphere, just as we have a map or grid system to locate places on Earth. If Earth could be inflated to fill the celestial sphere, then the reference lines of longitude and latitude on our planet's surface would respectively become the lines of right ascension (RA) and declination (Dec) on the sphere. Earth's equator becomes the celestial equator. Our rotational north and south poles would become the celestial poles and, tipped at an angle of 23.5° to the celestial equator – in line with Earth's 23.5° on its axis – would be the ecliptic. The ecliptic indicates the *apparent* annual path of the Sun around the celestial sphere relative to the stars. This can be a confusing concept for newcomers. The Sun is seemingly immobile at the centre of our Solar System, *but* it does appear to move throughout the year across the celestial sphere (the background of stars) because Earth is orbiting the Sun throughout the year and thus we are seeing it from a different perspective.

Think of it on a smaller scale in terms of our night and day. Earth rotates on its axis, from west to east, every 24 hours, like a spinning top, so the Sun appears to rise in the east in the morning, arc across the sky following a great circle, and set in the west at dusk. It is not moving, of course, we are, both in our annual orbit around the Sun but also on our axis – so the Sun appears to move against the background sky from our viewpoint. Similarly, all objects in the night sky, although fixed in relation to each other (with the exception of the Moon, planets, and 'visitors'), appear to move in the opposite direction to our planet's rotation; they rise in the east after sunset and set in the west towards dawn. As you watch the sky during any evening, you will see that the entire sky moves 15° an hour. It is not the night sky moving (although, as we have seen, everything is moving within the arm of our Milky Way galaxy but over enormous distances and timescales), but our planet's rotation.

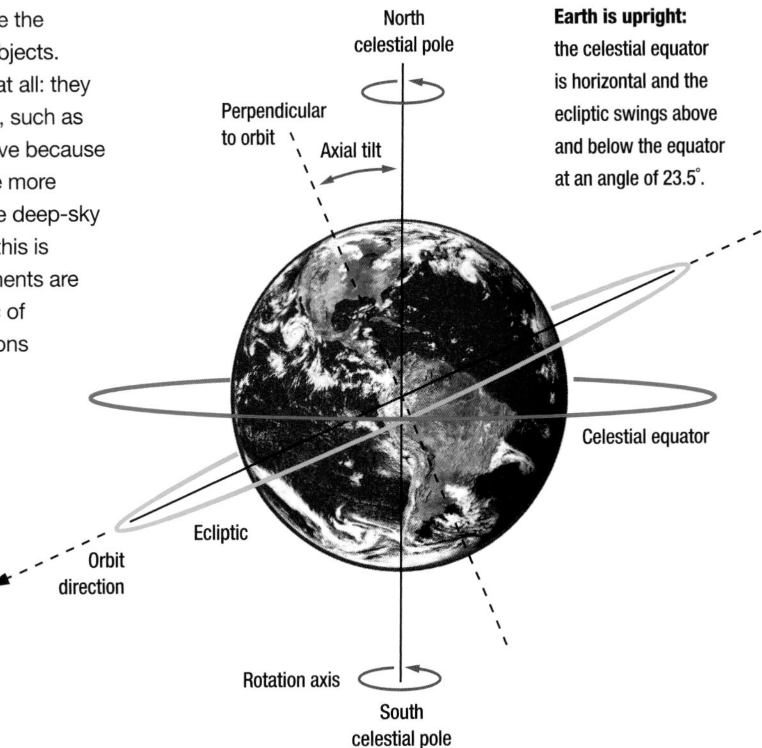

Earth is upright: the celestial equator is horizontal and the ecliptic swings above and below the equator at an angle of 23.5°.

Furthermore, as we know, Earth orbits the Sun, completing one slightly elliptical orbit every year, or 365.24 days – our calendar year. This also influences what we can see at night, and when. Our planet's passage through the heavens is divided into 360°, so Earth travels about 1° a day. As we saw in the Preface, the Babylonians divided circles into 360°. They, too, had a calendar year consisting of 12 months, with each month being 30 days long and amounting to an annual

Earth as it is in space – tilted over by 23.5°: note the position of the celestial equator (red) and the ecliptic (green). The Sun is south of the equator September to March and north of the equator March to September. Where the ecliptic crosses the equator are the equinoxes.

AT THE NORTH POLE, 90°N

Zenith
Apparent motion of stars
Diurnal circle
Horizon
Celestial equator
Stars always invisible

AT THE EQUATOR, 0°

Zenith
Apparent motion of stars
Diurnal circle
North celestial pole
South celestial pole
Celestial equator
Horizon

AT 40°S LATITUDE

Circumpolar region
Zenith
Apparent motion of stars
Horizon
Circumpolar region
Celestial equator
Diurnal circle
Stars always invisible

360 days. As a result, if you gaze at the night sky at the same time over several nights, the entire sky would appear to move an average of 1° to the west each evening (an average of 30° a month); confirmation of our planet's yearly progress around our Star. It goes without saying, therefore, that the position of Earth in its annual orbit around the Sun determines which constellations can be seen, and when. As our perspective changes during a night, so it will change over a year. Certain constellations can only be seen during winter months, such as Orion, the Hunter, since this apparent grouping of stars lies in an area of the Milky Way galaxy only visible when Earth reaches that point in its annual orbit. Likewise, the constellation of Sagittarius, the Archer, can only be seen during the summer months.

PRECESSION

The Earth's axis will not always point in the same direction. It 'wobbles' just like a spinning top, only over a period of 26,000 years – a result of the gravitational pull of the Sun and Moon. Consequently, the entire sky, with its right ascension and declination coordinates, appears to shift. This precession, as it is known, will take the planet's pole to within 0.5° of Polaris around the year 2100 before it starts to move away again.

Circumpolar star trails above Mauna Kea, Hawaii: stars endlessly circle the celestial pole.

Additionally, in keeping with context, what can be seen in the night sky is not only dependent on Earth's 24-hour rotation about its axis and its orbit around our Star, but also on an observer's position of latitude on the planet. For an observer standing at Earth's North Pole (90°N) – marked by Polaris, the Pole Star – the celestial north pole is directly overhead (the zenith). From this position, the celestial equator is parallel with the horizon. Since the stars appear to move parallel to this horizon, the observer will only be able to see stars in the northern half of the celestial sphere. For an observer standing at Earth's equator (0°), the entire sky can be seen. Here, the celestial equator rises vertically from the horizon and runs through the zenith so the stars rise straight up in the east and set straight down in the west. The north and south celestial poles lie on the horizon. For an observer between these latitudes, there will always be parts of the night sky that remain invisible. An observer in Montreal in the northern hemisphere (above Earth's equator) will see pretty much the same as an observer at the same latitude in London. Likewise, those viewing in Santiago, Cape Town, or Perth in the southern hemisphere (below Earth's equator) will share the same sky. All will be unable to see stars in the opposite hemisphere. Anyone observing near either of the Earth's poles will be able to see all the stars in that hemisphere all the time; here the stars do not set but endlessly circle the celestial pole – they are known as circumpolar.

■ LOCATION IN THE SKY

Having established context, let's return to our map of the night sky – the celestial sphere. Locating an object on this sphere relies on using alt-azimuth (altitude and azimuth) coordinates:

Altitude (elevation): The angle of the object above an observer's horizon. The zenith, at 90°, is overhead.

Azimuth: The angle measured clockwise from north along the horizon to the point on the horizon beneath the star. North = 0° or 360°; East = 90°; South = 180°; and West = 270°.

Meridian: An imaginary great circle passing through an observer's zenith from north to south, dividing the sky into an eastern and western half. When an object crosses this line, it is at its highest elevation. The Sun appears to cross this line around noon every day. When it does so, it is said to culminate.

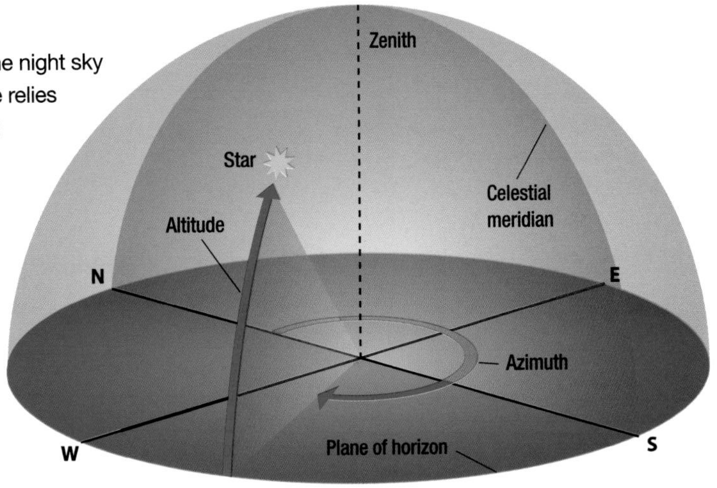

Location in the sky: using altitude and azimuth coordinates stellar objects can be found.

The celestial sphere: a grid for locating stellar objects in the night sky.

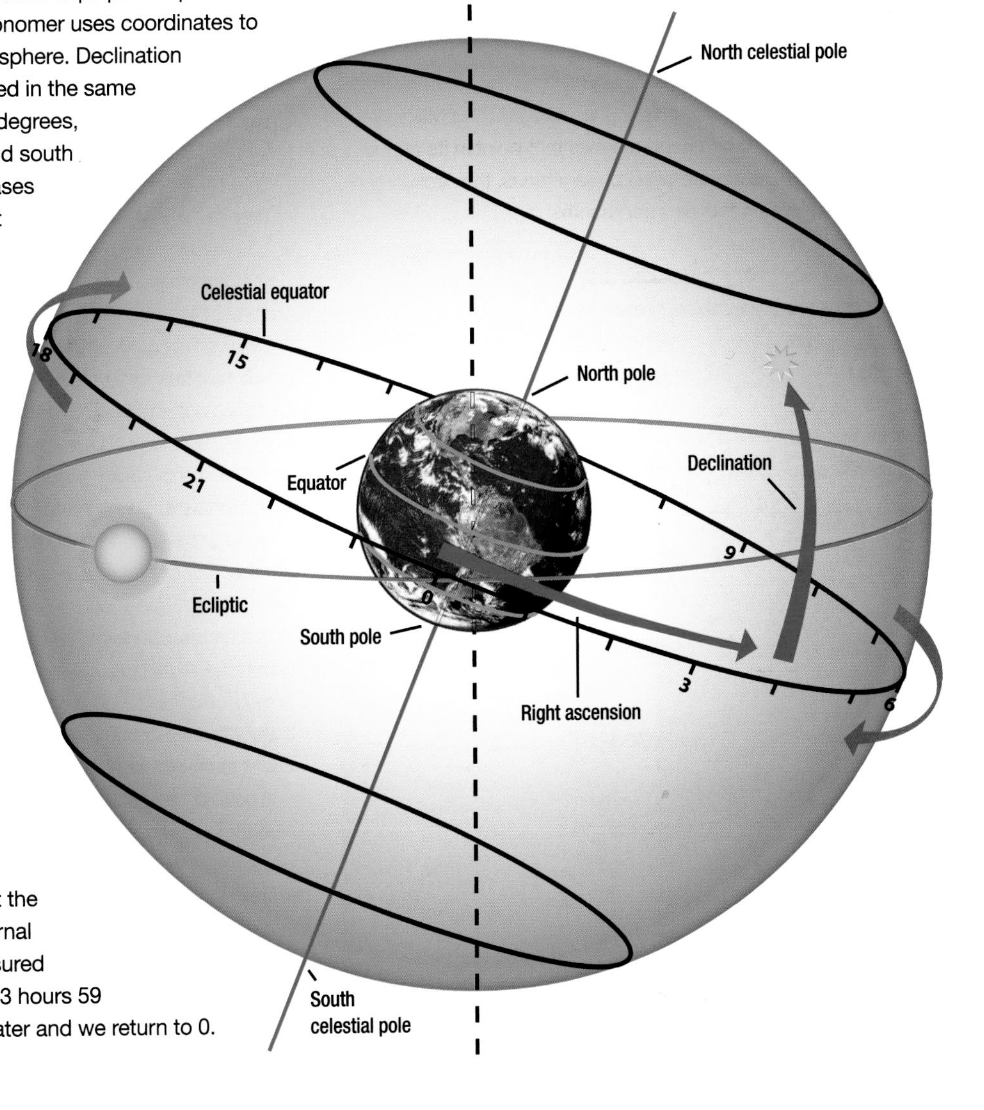

RIGHT ASCENSION AND DECLINATION

Just as a navigator uses coordinates to pinpoint a position on our planet's surface, so an astronomer uses coordinates to locate objects 'on' the celestial sphere. Declination (Dec) in the night sky is measured in the same manner as latitude on Earth: in degrees, minutes, and seconds, north and south of the celestial equator. It increases from 0° on the equator to 90° at the poles. North of the celestial equator is listed as positive (+) and south is negative (–). Right ascension (RA), the equivalent of our longitude, is measured in units of time: hours, minutes, and seconds. Twenty-four hours equals 360° and, as we know, one hour of RA equals 15° of arc. The RA for Polaris, the Pole Star, is thus referred to as 1 hour 49 minutes. The zero (0) line of RA (the celestial equivalent of Earth's Greenwich Meridian of longitude) passes through the point where the Sun, seeming to move northwards, crosses the celestial equator at the first moment of spring – the vernal equinox. Hours in RA are measured eastward from this point until 23 hours 59 minutes is reached. A minute later and we return to 0.

■ OBSERVING TIPS

Having established the significance of our observational viewpoint, let's step outside and start looking. Here are some seemingly obvious, but useful, tips for unaided eye viewing:

Light pollution in Europe: ever-increasing artificial surface lighting obliterates the majority of the night sky.

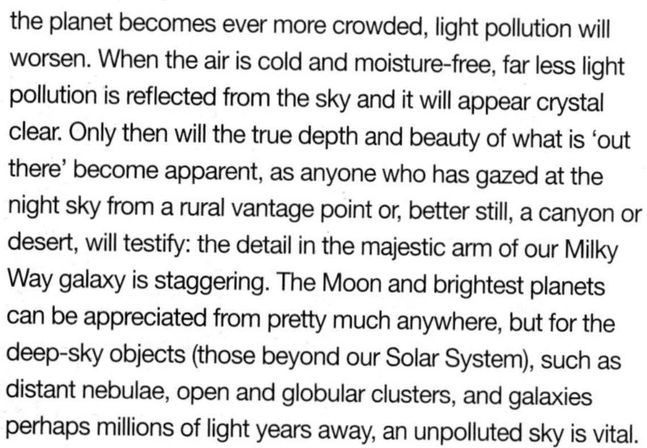

- Whether viewing with the naked eye, binoculars, or a telescope, a 'dark site' as far away as possible from any artificial light – light pollution – is essential. It is staggering how the night sky is washed out by the orange glare of sodium streetlights, vehicle headlights, security lights, house lights, etc. As the planet becomes ever more crowded, light pollution will worsen. When the air is cold and moisture-free, far less light pollution is reflected from the sky and it will appear crystal clear. Only then will the true depth and beauty of what is 'out there' become apparent, as anyone who has gazed at the night sky from a rural vantage point or, better still, a canyon or desert, will testify: the detail in the majestic arm of our Milky Way galaxy is staggering. The Moon and brightest planets can be appreciated from pretty much anywhere, but for the deep-sky objects (those beyond our Solar System), such as distant nebulae, open and globular clusters, and galaxies perhaps millions of light years away, an unpolluted sky is vital.
- Find a clear horizon devoid of obstructions such as buildings and trees, etc. This will optimise viewing objects rising in the east, transiting the zenith, and setting in the west. The higher your elevation the better.

Light pollution: The constellation Orion, imaged at left from dark skies, and at right from within an urban area.

- Ensure the ground in your immediate vicinity is clear of clutter. Sounds obvious, but it is only too easy to stumble over obstacles when carried away viewing the sky!
- Wait at least 30 minutes for your eyes to adjust to the dark – become 'dark-adapted'. During this time your pupils will almost immediately dilate, there will be an increase in sensitivity of the retinal rods, and a regeneration of rhodopsin by which the eye adapts to conditions of reduced illumination.
- Look aside from a target: averted vision sites a target on the more sensitive rod cells surrounding the retina – an unseen object then becomes visible.
- Unless the Moon is your point of interest, avoid viewing unless it has set or is rising later: its glare is notorious for wiping out all other objects in the sky.
- Again, sounds obvious, but dress appropriately. During winter nights, when the skies are darkest and the 'seeing' (a term used by astronomers to measure the steadiness of the atmosphere) crisp and steady, targets are stunning. The downside is that it can be bitterly cold, especially when standing around for long periods of time. Dress accordingly: use layers, headgear, warm socks, boots, and gloves or mittens. Additional comfort, in the form of a vacuum flask of hot drink or soup, and perhaps a sandwich, provides additional warmth. In summer, be prepared for the onslaught of insects: dress suitably and use a good-quality insect repellent, although be sparing in use: these grease up telescope knobs and buttons, and those with high DEET content can decay optical coatings and vinyl.
- Be comfortable: utilise a comfortable chair or, better still, a groundsheet, blanket, sleeping bag, or air mattress for horizontal viewing, especially when observing meteor showers (see *Meteor showers*, page 49). There is nothing like discomfort to draw a night to a premature close!

Items for starting out: a planisphere and basic star charts.

OBSERVING TOOLS AND TECHNIQUES

The following are very useful for naked eye and binocular observing whether a novice or more advanced:

- A star chart.
- A planisphere (star wheel) – an invaluable, inexpensive tool customised for observing at different latitudes at a particular time and date. A good-quality one will help locate stars, constellations, deep-sky objects, and bright planets, as well as indicate sunrise and sunset times.
- A red LED (light emitting diode) flashlight for reading a star chart, or planisphere, without jeopardising 'dark adaptation': bright white light closes the eyes' pupils within seconds. A piece of red plastic or cellophane affixed to a flashlight will do the same job – make sure the light is bright enough to read by, but no brighter.
- An Astronomy App for iPhone or iPad - see page 133.

As we have seen, astronomers use degrees, minutes, and seconds to measure size and distance in the sky. The distance from the horizon to the zenith (overhead) is 90°. The apparent width of the Moon and Sun is around 0.5° (30 arcminutes, since there are 60 arcminutes in a degree). The smallest detail resolvable to the unaided human eye is around one arcminute (60 arcseconds).

An outstretched hand held at arm's length indicates an angular distance of around 20° from the tip of the thumb to the tip of the little finger – about the distance between the first and last stars of the Plough (or Big Dipper) in the constellation of Ursa Major. A clenched fist at arm's length indicates around 10°, and the width of the thumb, again at arm's length, indicates around 2° – sufficient to blot out the apparent width of the full Moon four times over.

These yardsticks are essential for enjoying naked eye astronomy, especially when sharing constellations with others. Invariably, when searching for Orion, the Hunter, newcomers scour the entire sky looking for anything remotely resembling a giant – of which there are numerous candidates. Context is needed. Ask them to hold their outstretched hand at arm's length and search for three stars in a line filling the area behind their thumb and little finger, and they will soon find him. Indeed, the red giant star, Betelgeuse, marking the giant's left shoulder, and the opposite dazzling blue-white Rigel forming his right foot, cover around 20° of the sky. From here, or any constellation, it is easy to 'star hop' to another.

Step outside, find your dark site, wait those obligatory 30 minutes for your eyes to 'dark-adapt', and then look up. Using your red LED flashlight, check your star chart is correct for your latitude and local time – usually indicated as Universal Time (UT) or Daylight Saving (summer) Time (DST). Orientate yourself with north, east, south, and west. Ensure you hold the chart so that it is orientated with the sky: invariably, if you face north, hold the chart so that the label 'northern horizon' is at the bottom, right-side up. The curved edge indicates the horizon and the centre the zenith. A named bright star will be indicated in the north-east, for example – look up, and it will be there. Planets too, when visible, are indicated.

Dependent on the time of evening, your latitude, and the time of year, you can enjoy many of the 88 constellations shown on the map on pages 94 and 95 or on the simple star maps in Appendix 5. For a current customised star chart of constellations visible from your location on any given evening, along with other topical objects of interest, refer to monthly astronomy magazines for circular star charts, or use the Internet (or planetarium software) for printable star-dome maps: see Appendix 4. Finding your way around is fun, rewarding, and – like getting to grips with your ABC when learning to read – fundamental to understanding and appreciating the night sky.

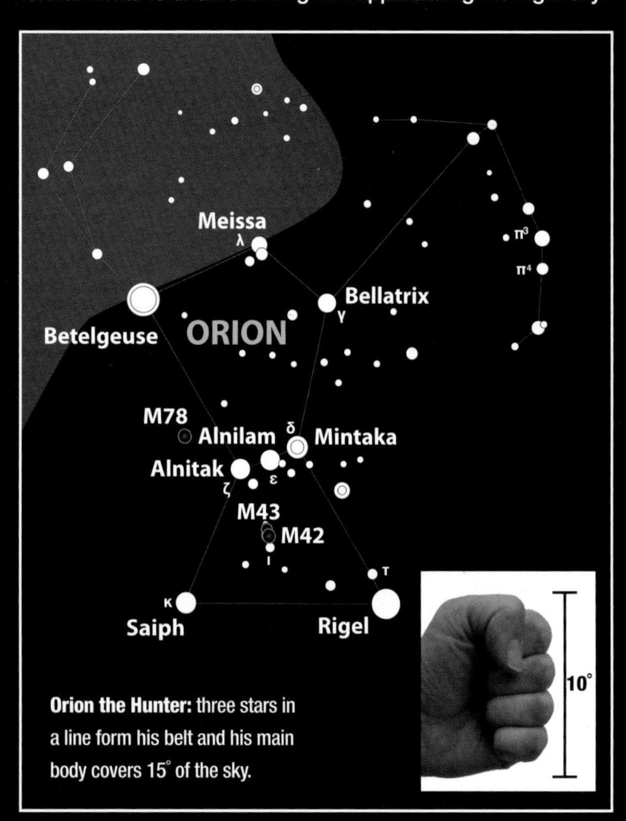

Orion the Hunter: three stars in a line form his belt and his main body covers 15° of the sky.

THE CONSTELLATIONS & NAKED EYE HIGHLIGHTS

Binoculars, or telescopes, are not always necessary to enjoy the wide-field vista of the night sky, especially when starting out. Below are the constellations as visible from the northern and southern hemispheres. Listed right are some objects of interest to the naked eye, numbered as per their location below, and visible at various times of the year, from various locations. Selected detailed descriptions follow on page 97.

1. Andromeda galaxy
2. Praesepe – Beehive Cluster (M44)
3. Sirius
4. Procyon, or Alpha (α) Canis Minoris
5. Gamma (γ) Cassiopeiae
6. Alpha (α) Centauri
7. Omega (ω) Centauri
8. Proxima Centauri
9. Gacrux (Gamma (γ) Crucis)

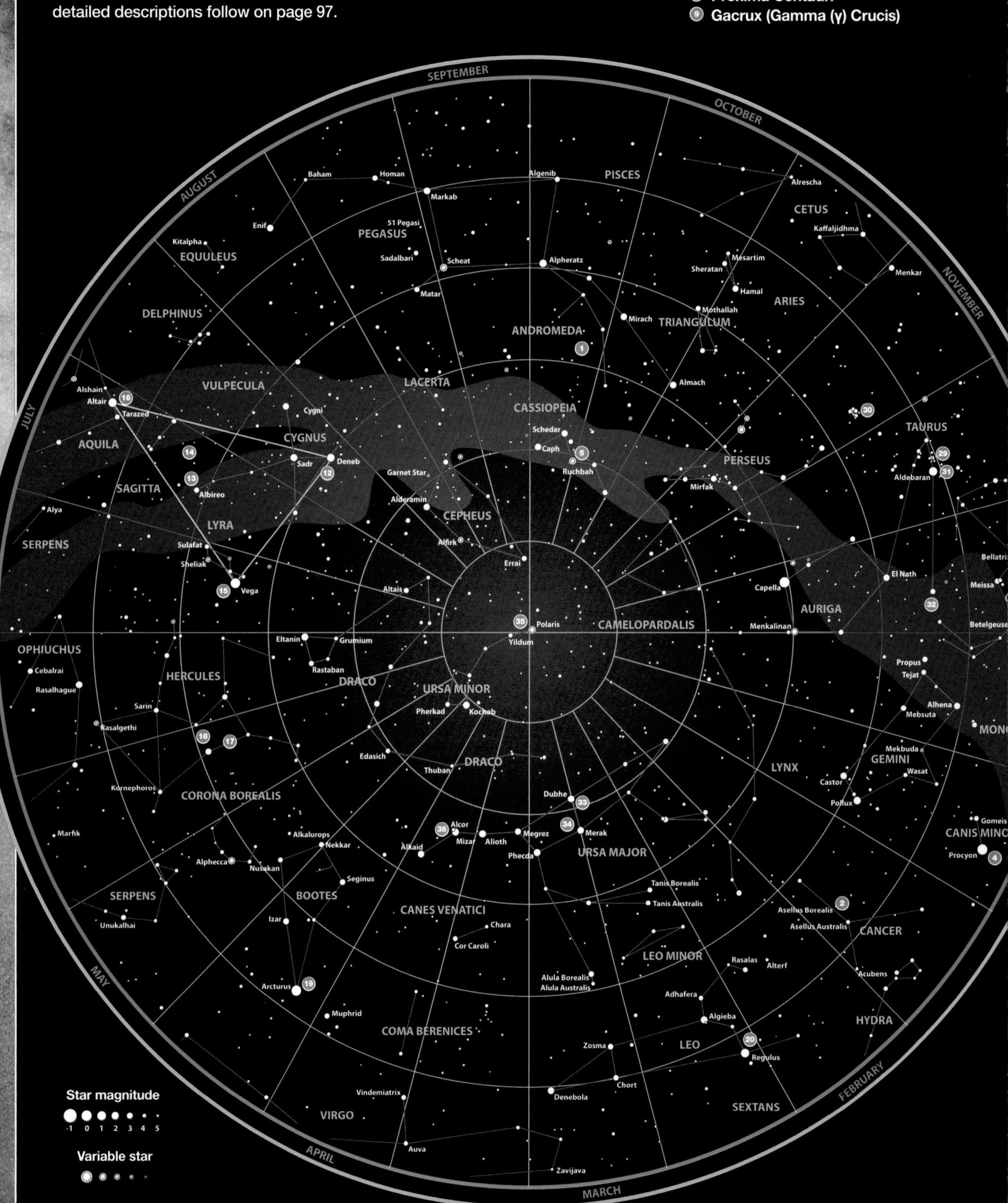

- 10 Coal Sack
- 11 Acrux (Alpha (α) Crucis)
- 12 Deneb (Alpha (α) Cygni)
- 13 Albireo (Beta (β) Cygni)
- 14 Summer Triangle
- 15 Vega (Alpha (α) Lyrae)
- 16 Altair (Alpha (α) Aquilae)
- 17 Hercules Cluster (M13)
- 18 Keystone
- 19 Arcturus (Alpha (α) Bootes)
- 20 Regulus (Alpha (α) Leonis)
- 21 Betelgeuse (Alpha (α) Orionis)
- 22 Rigel (Beta (β) Orionis)
- 23 Great Orion Nebula (M42)
- 24 Sagittarius A*
- 25 Lagoon Nebula (M8)
- 26 Antares (Alpha (α) Scorpii)
- 27 Butterfly Cluster (M6)
- 28 Open Cluster (M7)
- 29 Hyades Open Cluster
- 30 Pleiades Cluster (M45)
- 31 Aldebaran (Alpha (α) Tauri)
- 32 Crab Nebula (M1)
- 33 Dubhe (Alpha (α) Ursae Majoris)
- 34 Merak (Beta (β) Ursae Majoris)
- 35 Polaris (Alpha (α) Ursae Minoris)
- 36 Mizar (Zeta (ζ) Ursae Majoris) & Alcor

■ IDENTIFYING STARS

Around 4,000 stars can be seen with the naked eye on a night of good 'seeing'. Ordinary bird-watching binoculars will reveal maybe 100,000, and an amateur telescope reveals … millions.

Historically, as can be seen from star charts, stars are named and have mainly Arab and Greek origins. The red supergiant star Betelgeuse, marking the left shoulder in the constellation of Orion, the Hunter, is one example. Older texts suggest it should be pronounced 'BEETLE-juice'. It is actually correctly pronounced 'BET-el-jooze', meaning 'house of the twins' in Arabic because the Arabs saw it as being part of the neighbouring constellation of Gemini, the Twins.

After 1604, the German celestial cartographer Johann Bayer established a system for the position and magnitude of the stars. His star atlas, *Uranometria*, labelled many of the stars visible to the naked eye using the Greek alphabet: the brightest star was generally called alpha (α), the next brightest beta (β), and so on. Betelgeuse is, therefore, also known as Alpha (α) Orionis. Early in the 18th century, the Reverend John Flamsteed – appointed by King Charles II to be England's first Astronomer Royal – enlarged the list of 'named' stars by adding Arabic numbers and including fainter stars. In his system, the stars are numbered west to east across a constellation, so Betelgeuse is also 58 Orionis.

Andromeda and Cassiopeia: detail from *Planisphere celeste*, Philippe La Hire, 1705.

Betelgeuse: a red supergiant star, its Arabic name meaning 'house of the twins'.

Orion: Johann Bayer's *Uranometria* (1661) showing the constellation Orion.

Any stars not covered are referred to by letters or numbers allocated in star catalogues, such as Sir Patrick Caldwell-Moore's list of 109 bright star clusters – the Caldwell Catalogue – or the Smithsonian Astrophysical Observatory's vast Catalogue where stars are designated 'SAO' followed by many digits. Another much larger list is the Hubble Space Telescope (HST) Guide Star Catalogue that includes millions of stars. Perhaps the most famous catalogue is the Messier Catalogue (see Appendix 2), published in 1781 and named after the French astronomer Charles Messier, who detailed 103 'fuzzy objects' he encountered whilst searching for comets. His list was revised and expanded over subsequent years. The current Messier list includes 110 objects – mainly nebulae, clusters, and distant galaxies. In 1888, John Louis Emil Dreyer (1852–1926) initiated the New General Catalogue (NGC) of Nebulae and Clusters of Stars – an expanded version of John Herschel's earlier catalogue – and, in 1895 and 1908, added two supplements called the Index Catalogues. Collectively, these objects are designated with NGC or IC numbers and there are over 13,000 of them.

Even with the naked eye, it is clear that stars have different brightness, or 'magnitude'. Star charts generally have a key with variedly sized symbols representing stars from first to sixth magnitude: the larger the symbol, the brighter the star.

THE GREEK ALPHABET

α	alpha	ν	nu
β	beta	ξ	xi
γ	gamma	ο	omicron
δ	delta	π	pi
ε	epsilon	ρ	rho
ζ	zeta	σ ς	sigma
η	eta	τ	tau
θ	theta	υ	upsilon
ι	iota	φ	phi
κ	kappa	χ	chi
λ	lambda	ψ	psi
μ	mu	ω	omega

Once you have located the constellation of Andromeda, which is easy since it is located south of Cassiopeia's 'W' and just off one corner of the Great Square of Pegasus, you will be able to locate the nearest spiral galaxy to us, the Andromeda galaxy (1). This galaxy, also a member of our Local Group, is similar to the Milky Way galaxy, comprising hundreds of billions of suns and with a diameter spanning 20,000ly. It is located some .9 million light years away and, with the exception of M33 in Triangulum which requires excellent eyesight and 'seeing', is the most distant object that can be seen with the unaided eye. The light we see began its journey at the dawn of humanity!

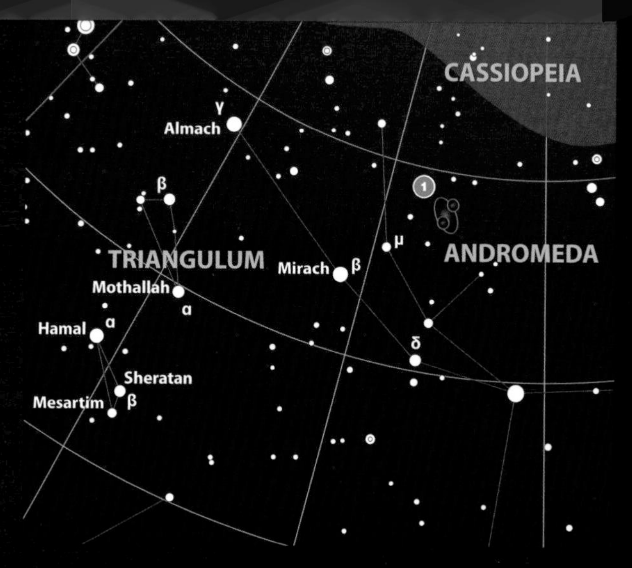

BOOTES – THE HERDSMAN

ollow the Great Bear's curve of the 'saucepan handle' and you will 'arc' round to the distinct golden glow of the K-type giant tar, Arcturus (19), meaning 'guardian of the bear': 'Arctos' is Greek for 'bear'. This is the Alpha (α) star in the constellation of Bootes, the Herdsman, and resembles Dubhe in Ursa Major but, at 37ly distant, it is four times closer. It is a beautiful sight or the naked eye. It is also easy to overlook the fact when scanning the sky that this star is 113 times more luminous than our Sun; its bloated presence could consume 25 million Earths! urthermore, although it appears 'fixed' to us, it is actually whizzing through space at a whopping 87mi/s in the direction of he constellation Virgo. Over the last 1,600 years it has moved wo Moon diameters (1°) against the starry background. This ate of speed is much faster than the other stars in our Sun's eighbourhood, indicating that it could be adopted from a warf galaxy consumed some six or seven billion years ago.

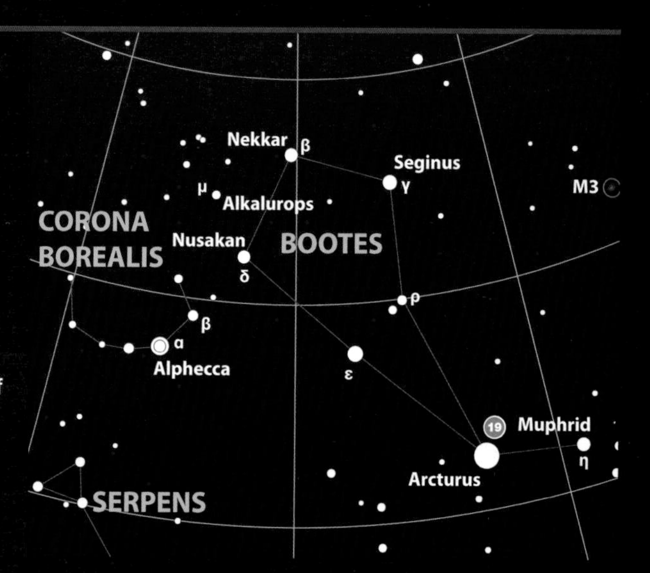

CANCER – THE CRAB

ocated in the constellation of Cancer, Praesepe (2) – atin for 'manger' – also known as the Beehive Cluster M44), is one of the sky's finest open clusters and its warming stars can be seen with the naked eye from dark location. There are, in fact, over 200 stars in the eehive spread over an area three times the diameter the full Moon and at a distance of around 580ly.

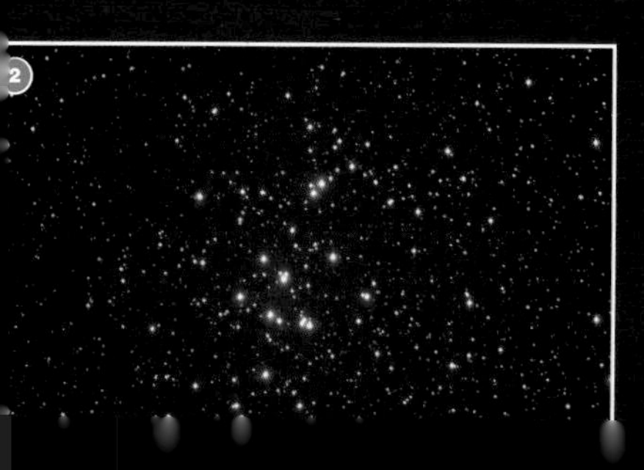

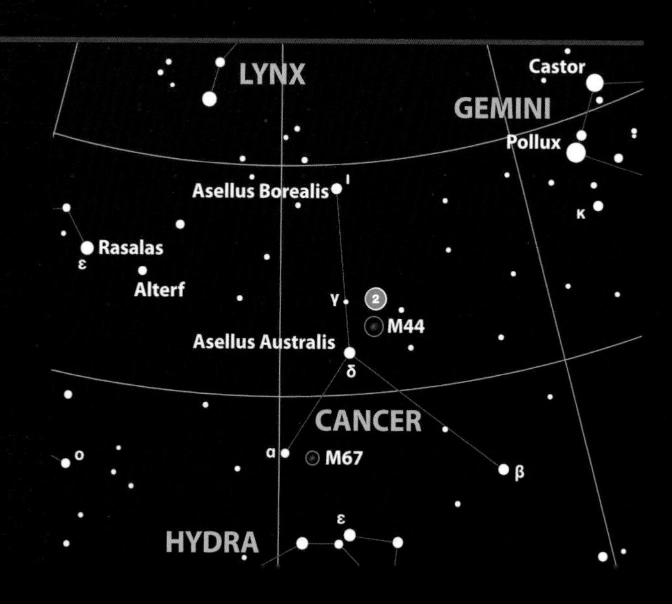

Located at the head of the constellation Canis Major, the Great Dog, dazzling Sirius ③, otherwise known as the 'Dog Star', is the brightest star in the entire sky. Located only 8.7ly from Earth, its enormous brilliance is also due to its being some 40 times more luminous than the Sun. In 1834, Friedrich Bessel noted that Sirius had a strange wobble, indicating a companion, but it wasn't until 1862 that the famous telescope maker Alvan Clark (1804–87) discovered the companion. In its 50.2-year very elliptical orbit, it was found to be a faint white dwarf star – the nearest white dwarf to us – and is now known as Sirius B, or, more appropriately, the Pup star. Canis Major is one of the most striking of all the constellations. Along with its neighbouring constellation, Canis Minor, the Little Dog, they represent hunting dogs leaping at Orion's heels.

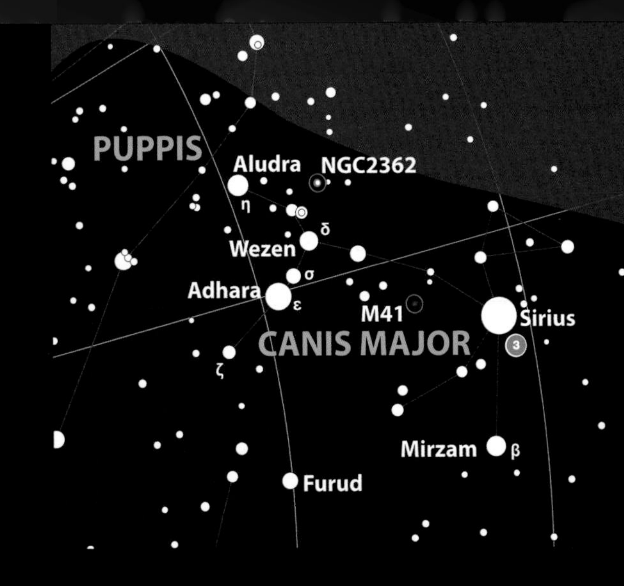

CASSIOPEIA – THE SEATED QUEEN

This constellation, prominent in the northern hemisphere's winter sky, forms a striking W-shape and is visible all year from mid-northern latitudes. It is situated on the other side of Polaris from the Plough/Big Dipper and can easily be seen with the naked eye. In Greek mythology, Cassiopeia was the queen of the ancient kingdom of Ethiopia as well as being the wife of Cepheus and mother of Andromeda. Of note to the naked eye is the star Gamma (γ) Cassiopeiae located at the centre of the 'W'. It is normally the constellation's third brightest star but since it is one of the irregular variables mentioned earlier, its brightness varies as it loses mass on to a surrounding disc, or shell.

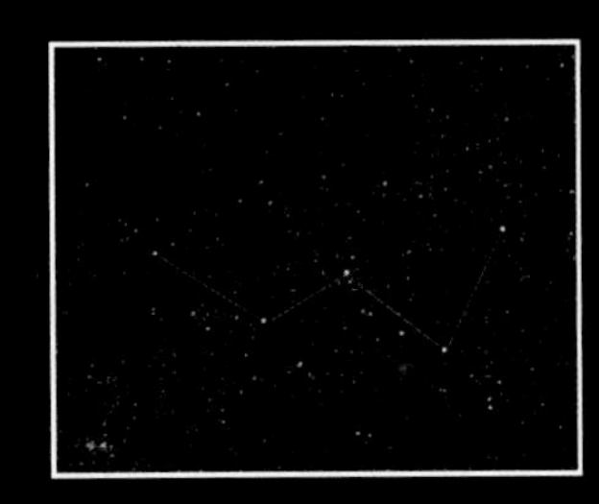

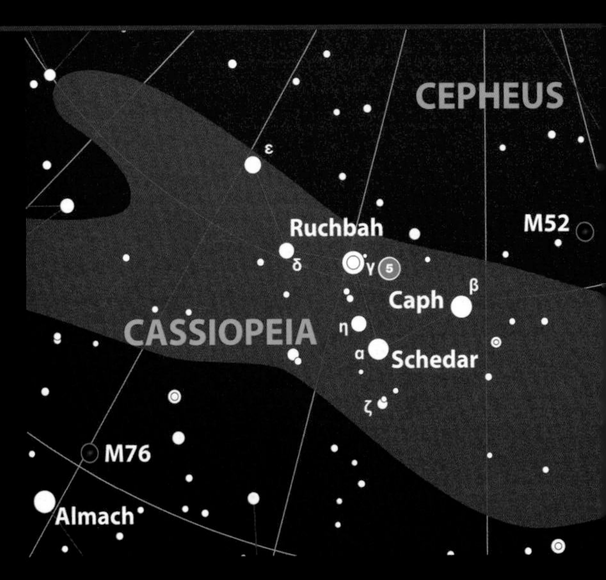

CENTAURUS – THE CENTAUR

For those observing in the southern hemisphere there are the naked eye treats of some of the closest stars to us, such as Alpha (α) Centauri ⑥, and the finest globular cluster in the entire sky, Omega (ω) Centauri ⑦. At the foot of the Centaur, Alpha (α) Centauri is only 4.3ly away. The closest is Proxima Centauri ⑧ at 4.22ly distant. Alpha (α) Centauri is a pretty double star (binary) system with its companion stars revolving around each other every 80 years. Alpha (α) and Beta (β) Centauri form the two bright 'pointers' to the famous Southern Cross constellation. Located some 17,000ly away, Omega (ω) Centauri is simply stunning, even with the naked eye

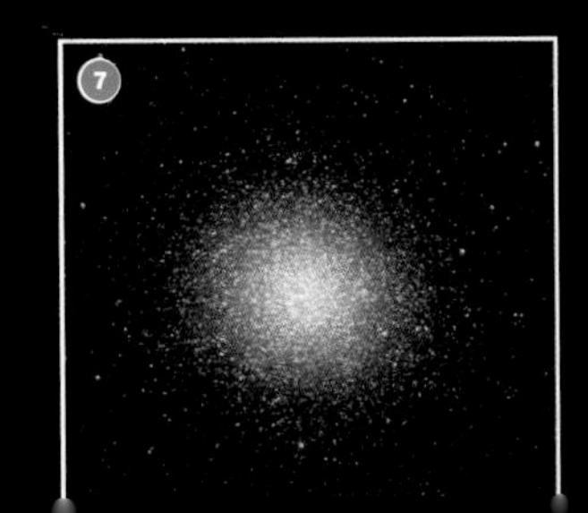

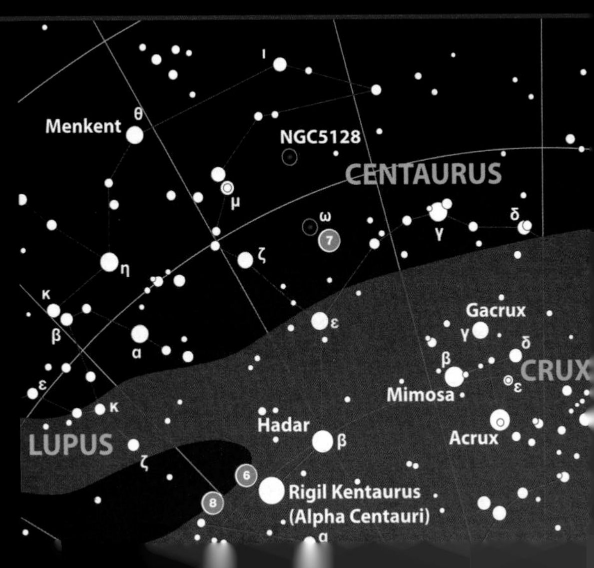

CRUX – THE SOUTHERN CROSS

This is the most famous southern hemisphere constellation. Its distinctive cross shape has guided sailors for centuries since the upright of the cross points to the south celestial pole. Visible to the naked eye is Gamma (γ) Crucis (9), also known as Gacrux, which is a wide double star marking the northern end of the cross. It is an optical double and consists of a 6.4 magnitude star and a bright orange primary. Also look out for the Coal Sack (10): lying just east of the star Acrux (11), this is one of the largest and most dark nebulae in the sky – the very clouds from which future stars will be born.

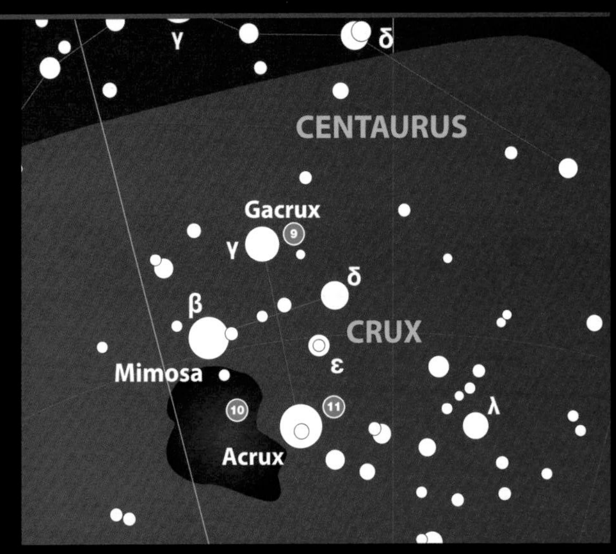

CYGNUS – THE SWAN

This constellation actually resembles its asterism, the 'swan' shape soaring across the rich northern Milky Way. Take a look under a dark sky: here the arm of our Galaxy divides into two streams and the dense, dark nebula between is stunningly visible. With the naked eye, both Deneb (Alpha (α) Cygni) (12) and Albireo (Beta (β) Cygni) (13) are visible, respectively denoting the tail and head of the swan. Deneb is on a par with Rigel in Orion since it is 25 times more massive than our Sun and 60,000 times more luminous. Around 1,500ly away, Deneb is the most distant of the famous 'Summer Triangle' (14) – an asterism formed with Vega (15) in the constellation of Lyra, the Lyre, and Altair (16) in the constellation of Aquila, the Eagle. Albireo, at the tail of the swan, or foot of the cross, when seen through a small telescope, is one of the prettiest sights in the sky: it is really two stars, one appearing blue and the other yellow.

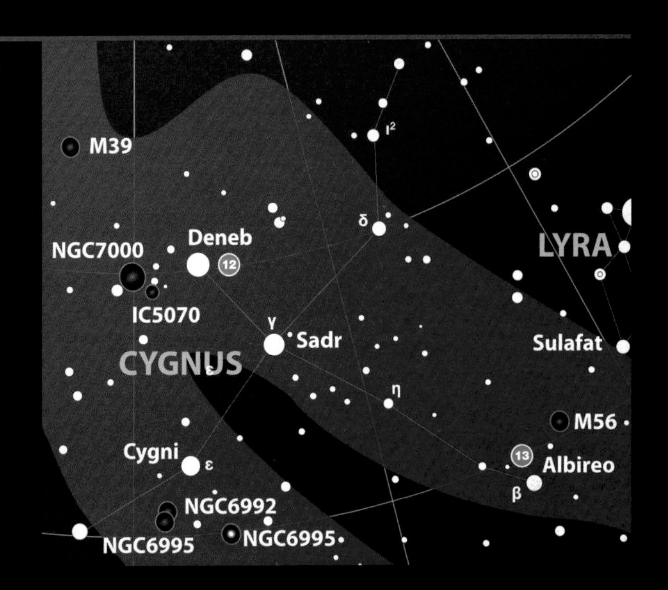

HERCULES – THE HERO

Located in the northern hemisphere, this constellation is home to the famous Hercules Cluster (M13) (17), or the Great Cluster in Hercules as it is known, arguably the most dramatic globular cluster in the northern hemisphere sky. To find the 'Keystone' (18) of this constellation, and thus M13, look for the star Arcturus (19) and then gaze about two hand-spans away to the north-east and you will find Vega: the Keystone is midway between and covers 10° of the sky – your clenched fist held at arm's length. Just look about one-third of the way along the imaginary line between the two stars marking the western corners of the Keystone and, on a dark night, M13 will be visible to the naked eye as a faint fuzzy patch. What you are seeing is a misty sphere of half a million suns spanning some 140ly at a distance of 25,000ly. In binoculars, or even a small telescope, it is a sight to behold.

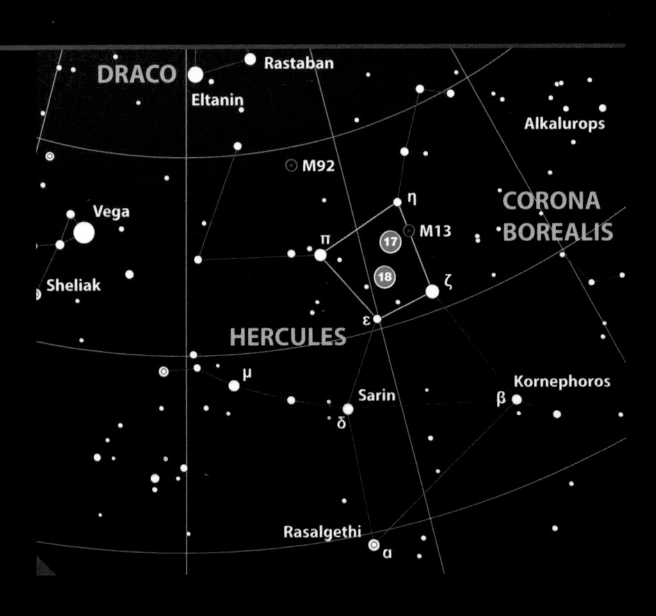

LEO – THE LION

One of the oldest zodiacal constellations, the asterism of Leo, the Lion, was known by the peoples of ancient Egypt and India. It actually resembles its namesake, with its sickle (or backward question mark) tracing out a lion's head and the overall pattern resembling the reclining Egyptian Sphinx. To the naked eye, its radiant first magnitude star, Regulus (20), meaning 'the little king', (Alpha (α) Leonis), is unmistakeable and, at 77ly away, marks the lion's heart.

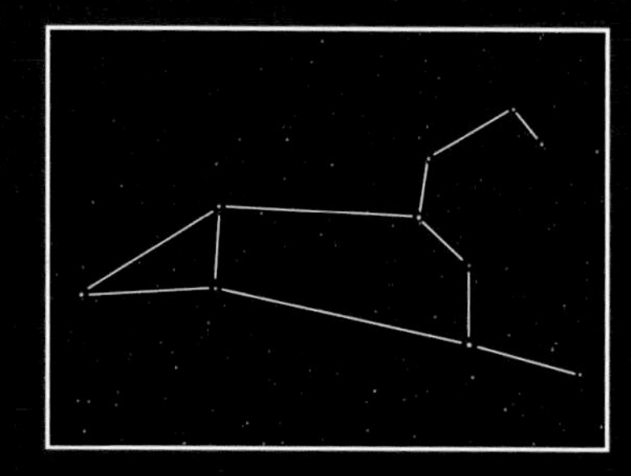

It shines 140 times brighter than our Sun but spins a good deal more rapidly; whereas our Sun spins on its axis once a month, this star only takes 15.9 hours!

ORION – THE HUNTER

Straddling the celestial equator, and visible equally in the northern and southern hemispheres, is the constellation of Orion, the Hunter, a distinctive group of stars recognised for thousands of years. The Syrians called it Al Jabbar, the Giant. The ancient Egyptians called it Sahu, the soul of Osiris. In Greek mythology, Orion was a giant and a great hunter. Artemis, the goddess of the hunt and the Moon, fell in love with him and, in so doing, forgot to light the sky. As a result, she mistakenly killed him with an arrow. Later, after discovering her error and becoming inconsolable, she placed Orion's body in the sky along with his hunting dogs, Canis Major and Minor.

Orion, visible in the winter months, is a treasure trove for the naked eye and becomes even more stunning when viewed with binoculars or a telescope. Betelgeuse (21) is a beautiful, orange-glowing red supergiant star measuring some 600 million miles across. Also known as Alpha (α) Orionis, and at a distance of around 640ly away, it is a first magnitude semi-variable, which, over a period of almost seven years, varies in brightness from 0.3 to 1.2 magnitude. Heading south, and after crossing the three distinct second magnitude stars forming the jewelled 'belt', we reach the diametrically opposite blue supergiant Rigel (22), Arabic for 'foot'. Also known as Beta (β) Orionis, this monster is 50,000 times more luminous than the Sun and has a seventh magnitude companion. At around 770ly away from us, Rigel is still only half the distance to the Great Orion Nebula (M42) (23) : famously, an enormous cloud measuring some 22 billion miles across and home to a stellar nursery – a birthplace for stars. Even though it is 1,500ly away, it is plainly visible to the naked eye under a dark sky and is stunning in binoculars. The enormous nebulosity emanates from its core where four stars, known as the Trapezium, provide the power.

The constellation of Sagittarius is an easy target for those in the southern hemisphere (close to the horizon during summer months in the northern hemisphere) since its asterism clearly profiles a 'teapot' complete with handle (eastern end) and spout (western end). Mythologically, Sagittarius is a centaur – half man, half horse. When gazing at Sagittarius we are looking towards the hub of our Galaxy: its exact centre lies just north of where the borders of Sagittarius, Ophiuchus, and Scorpius meet. We are looking into a band where the Galaxy is broadest and simply littered with dark dust lanes, galactic and globular clusters, bright and dark nebulae, and an ineffable proliferation of billions of stars (24). Binoculars reveal numerous stellar objects, of which one, the pink-glowing, diffuse Lagoon Nebula (M8) (25), situated some 5,000ly away in the Sagittarius spiral arm, can be seen with the naked eye under good conditions. It is another stellar nursery, one half containing newly born stars, the other containing a newly born sixth magnitude blue supergiant, separated by a dark central rift.

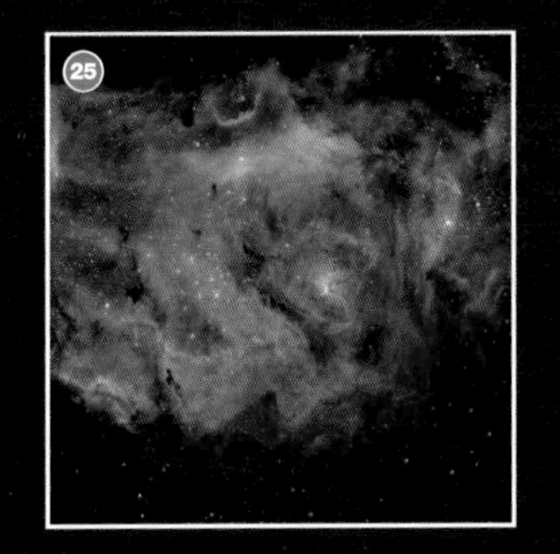

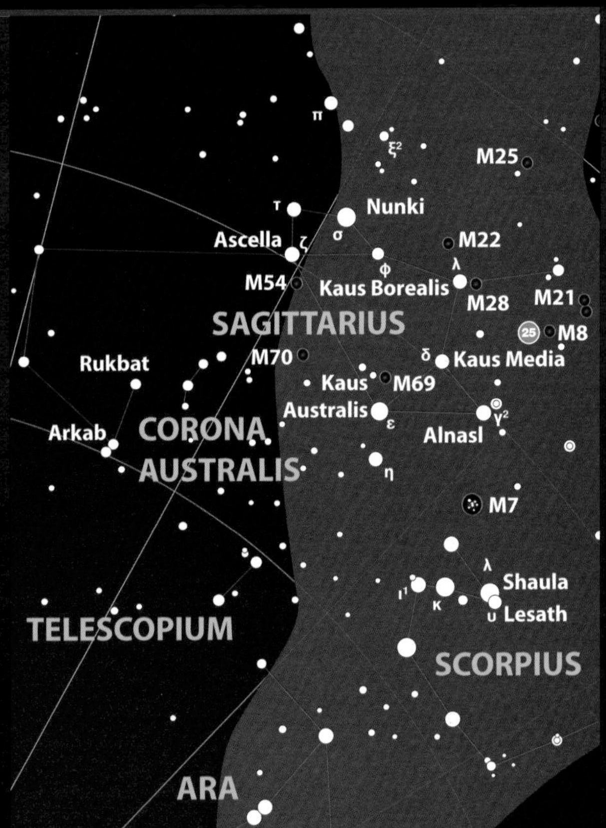

SCORPIUS – THE SCORPION

Greek mythology tells us that Scorpius is the scorpion responsible for killing Orion, hence these two constellations are in opposite sides of the sky! To my mind, Scorpius in the southern hemisphere is the most beautiful of the constellations and clearly conveys a scorpion with its stinger set to kill. It is full of bright stars and rich star fields. Near the northern end is a line of three dazzling stars, with the colossal red supergiant star Antares (Alpha (α) Scorpii) (26) – Greek for 'rival of Mars' – in the centre. Around 5,000 years ago the Persians considered Antares to be one of the Royal Stars and a guardian of heaven. The ancient Chinese saw its distinct red glow as a 'Great Fire' at the heart of the 'Dragon of the East'. The Romans called it Cor Scorpionis – 'heart of the scorpion' – a title also used by the French (Le Coeur de Scorpion). Situated some 520ly away, this star is thought to be some 400 times wider than the diameter of the Sun and 9,000 times more luminous. The star itself would fill the orbit of the planet Mars. It has a much smaller and hotter companion orbiting every 900 years, but a telescope is required to view it. It is a beautiful sight. Also visible with the naked eye are two open clusters: the Butterfly Cluster (M6) (27) near the stinger, and M7 (28), lying south-east of M6. The stars in M6 appear displayed in the shape of a bird or insect with outstretched wings, hence the name, and they are around 1,600ly away. M7 is the most southerly of the Messier objects and not only is it the finest cluster in this constellation but also one of the best in the entire sky.

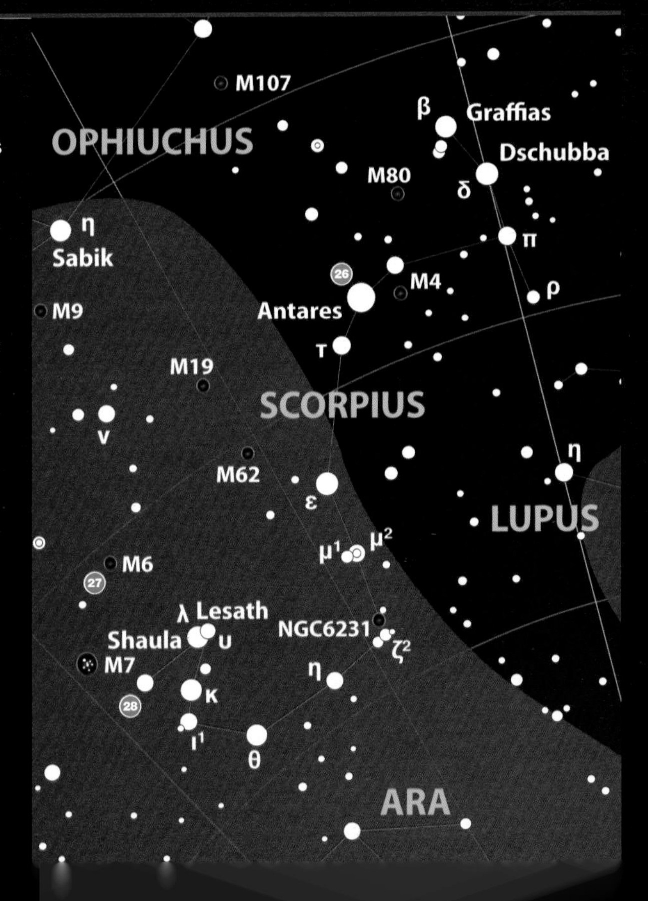

TAURUS – THE BULL

The V-shape marking the head of the bull in the northern hemisphere's constellation of Taurus is also full of treasures for the naked eye: it contains two of the largest star clusters we can see – the Hyades (29) and Pleiades (M45) (30). Since the time of the Chaldeans, some 5,000 years ago, this constellation has been worshipped as a bull, the symbol of fertility and power. Majestic winged bulls (Lamassu) stood guard beside the imposing gates of Assyrian palaces in ancient Nineveh. The ancient Egyptians worshipped a real bull, Apis, thought to be the incarnation of Osiris, and the Israelites worshipped the Golden Calf. The Greeks saw this group of stars as Zeus in disguise.

Only the forequarters of the bull are visible since it is emerging from the waves but, clearly visible to the naked eye, is Aldebaran (Alpha (α) Tauri) (31) – Arabic for 'follower' – an orange giant star and the brightest in the constellation. Only 60ly distant, this glowing giant marks the eye of the bull. Located just above the lower horn is the Crab Nebula (M1) (32), the nebula marking the site of the famous 1054 supernova. Sadly not visible to the naked eye, it is clear in a four-inch telescope or larger.

Forming the bull's shoulder, the 'Seven Sisters', or Pleiades cluster, are unmissable and are the most famous open star cluster in the sky. These young stars are journeying through a giant cloud of gas, illuminating it as they do so.

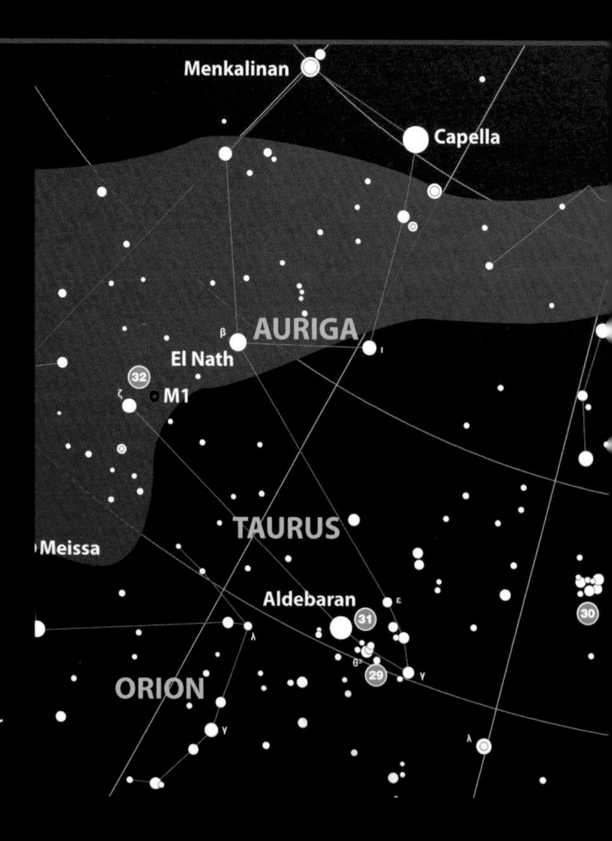

URSA MAJOR – THE GREAT BEAR

Also known as the Big Dipper, or Plough, this is one of the oldest constellations and perhaps the best known. It is the third largest of the constellations – only Virgo and Hydra cover a greater area of the sky. Resembling a saucepan with a bent handle, it comprises seven stars: three in the handle and four in the bowl. The two end stars of the bowl are Alpha (α) (33) and Beta (β) (34) Ursae Majoris, known as Dubhe and Merak respectively and more commonly as 'the pointers'. When observing, if you draw an imaginary line from Merak to Dubhe and then extend this around five times as far again you will reach the pole star, or Polaris (Alpha (α) Ursae Minoris) (35), in the constellation of Ursa Minor, the Little Bear. This seemingly unobtrusive Cepheid variable star, the closest Cepheid variable to us at 430ly distant, is, coincidentally, 0.9° away from the northern celestial pole. It is the star around which the entire sky revolves. The star itself is a yellowish-coloured supergiant nearly 50 times larger than the Sun and giving out some 2,500 times as much light. It is actually part of a triple star system – Polaris A, Ab, and B – bound together by gravity. Star Ab is two billion miles away from A – comparable to the distance from our Sun to the planet Uranus. When gazing skyward, look for the famous apparent double star in the middle of the saucepan's handle – Mizar (Zeta (ζ) Ursae Majoris) and Alcor (80 Ursae Majoris) (36). Separated by just 12 arcminutes it is possible to see the pair with the naked eye. Mizar is a binary system, each component of which is a spectroscopic binary, (not sufficiently separated to be resolved by telescope but detectable by variations in the wavelengths of emitted light) so has four stars. Alcor itself is also a spectroscopic binary. So there are six stars in total.

■ METEOR WATCHING

(See also *Meteors* and *Meteor Showers,* page 49.) Naked eye observing assumes even greater dimension if a meteor shower is in progress, provided it is not hampered by the glare of the Moon. The most spectacular reliable annual showers are the Perseids and Geminids, peaking around 12 August and 14 December respectively. The Perseids are exciting to watch. Mostly no bigger than grains of sand, comet Swift/Tuttle's debris enters Earth's atmosphere at around 60km/s, making the meteors appear fast and bright. They also leave persistent 'trains': glowing trails of ionised atoms and molecules in the upper atmosphere. Any meteors not associated with specific showers are called sporadics. Fortunate observers may well see a fireball – rare but extremely bright, colourful meteors, brighter than the planet Venus. See Appendix 3 for practical websites.

Cascading Perseids: a shower of shooting stars.

■ ARTIFICIAL SATELLITES

Not everything in the sky has been there for millions of years. The launch of the Russian *Sputnik 1* satellite in 1957 signalled the start of the artificial satellite revolution and there are now around 8,000 satellites orbiting our planet, of which around 100 are visible to the naked eye. Ranging from a few miles to thousands of miles above Earth's surface and best seen in summer, they are visible at twilight or in darkness when they catch the Sun's rays. It follows, therefore, that they cannot be seen if they are in Earth's shadow. Look for a moving, unblinking light passing soundlessly overhead.

Orbital debris: a computer-generated image of objects currently being tracked in Earth orbit.

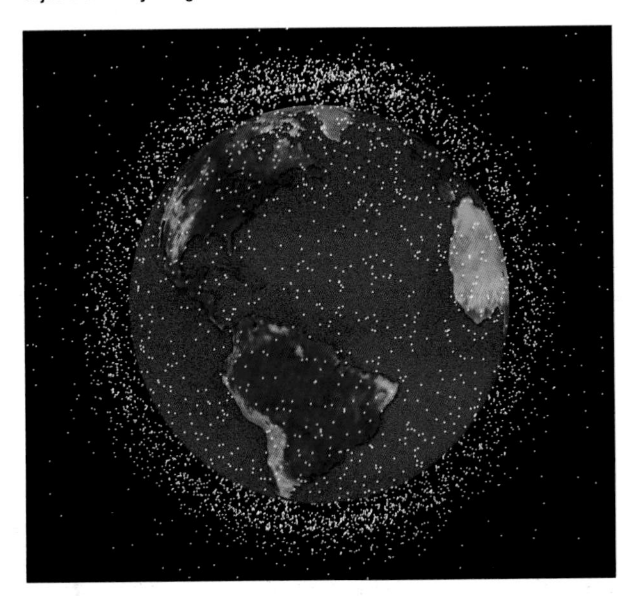

International Space Station: an unmistakeable target for naked eye viewing.

Their differing trajectories make them visible anywhere in the sky:

- Polar: orbiting over Earth's north and south poles.
- Elliptical: orbiting in an oval shape above and below Earth's Equator taking scientific measurements, *eg* monitoring the ozone level.
- Sun-synchronous: inclined a few degrees beyond Earth's poles, permitting an imaging satellite to make pictures with the sunlight always striking the Earth from the same angle.
- Geostationary: communications and TV broadcast satellites orbiting above Earth's equator at altitudes of 30,000 miles and at the same speed as Earth's rotation, which is why they appear stationary. These are too distant to be seen with the unaided eye.

Also included are the 66 Iridium, low polar-orbiting communications satellites used mainly by the US Defense Department. Each has three door-sized highly polished antennae acting like full-length bathroom mirrors turned towards the Sun at a distance of 485 miles; when at exactly the right angle they reflect sunlight in the form of a powerful flare. Visible for up to 20 seconds, but at a peak magnitude of -8 for a second or two (30 times brighter than Venus!), these are stunning for unaided viewing.

The International Space Station is arguably the most exciting man-made object to witness with the naked eye. Periodically, its 90-minute, low Earth orbit makes it visible during twilight. With a telescope it is possible to see its structure or a docked Space Shuttle craft.

Planetarium software reveals when satellites are visible from your location. Alternatively, a quick check of the heavens-above.com website covers all satellite-searching needs (see Appendix 3).

Additionally, there are auroras, solar and lunar halos, the five naked-eye planets, the comets, solar and lunar eclipses, and sunspots, all mentioned earlier. This completes our taste of naked eye viewing. Have you a thirst for more? Let's go deeper with binoculars.

Double vision

Strangely enough, binoculars, not telescopes, are the first piece of astronomical equipment to buy. Why? Simply pick up a pair, gaze through them at a crisp crescent Moon bathed in Earthshine, a first-quarter Moon suspended in inky blackness, or gaze into the arm of our Galaxy arcing overhead and you will have your answer. They are a telescope in miniature and, experts say, reveal 40% more than the naked eye. Certainly, the views of shadowy walled craters along the Moon's terminator, the dazzling rays and the differentiation between dark maria and bright highlands are a revelation ... and these views are only the beginning!

■ WHY?

Binoculars are the perfect 'go-between' for the naked eye and telescopes:

- In the human eye, contraction of the muscles of the iris causes the pupil to dilate, allowing greater light to enter the light-gathering area, the retina. This is our aperture. This aperture can only dilate to a maximum 7mm, and this reduces with age. Consequently, most stars, and other objects, are too dim to be seen by the naked eye. A larger aperture is needed. Binoculars fulfil that need because they gather more light over a wider area, enabling dim objects to be seen more clearly, as any wide-eyed Barn Owl will attest!
- You may already possess a pair for terrestrial pursuits. Use them for astronomy: even seasoned observers find them indispensable.
- They are generally lightweight, inexpensive, and quick and easy to use: the twin barrels are comfortable to look through and they convey stunning three-dimensional images.

Standard binoculars: will enhance any astronomical viewing.

Sweeping for targets: their wide-field view makes finding objects easier.

- Their wide-field view makes it easier to find targets and their low magnification gives steadier images. Greater magnification incurs greater vibration.
- The larger apertures open up the night sky to reveal more than 100,000 stars – significantly more than the approximately 4,000 seen with the naked eye.
- Whereas astronomical telescopes without certain accessories reveal upside-down or mirror-reversed images, or both, binoculars show a correct image, making the transition from naked eye to magnified viewing a little easier. Binoculars reveal objects as they appear in the sky or on star charts.
- In some cases binoculars beat the naked eye and telescope for performance, especially for those big-picture views.

■ WHICH BINOCULARS?

If you already own a pair of binoculars, they may be perfectly suitable. If selecting a pair, the choice is daunting, ranging from virtually useless plastic-lensed toy instruments to six-inch refractors costing as much as a car! Having said that, some binoculars are definitely better than others for stargazing. Here are some tips for purchasing, but always 'try before you buy':

MAGNIFICATION AND APERTURE

Binoculars are characterised by two numbers, combinations being: 10x25, 10x42, 10x50, 8x40, 7x50, or 7x35 ('x' means magnification). However, they can also range from handbag-sized 6x16s to 25x150 monsters. The first number indicates magnification: how much closer objects appear than when seen with the unaided eye. The second number indicates aperture: the diameter, in millimetres, of each of the two front lenses, or 'objective lenses' ('objectives') as they are called. So, 10x50s will magnify by ten times (objects will appear ten times closer) and have objective lenses 50mm (around 2in) in diameter.

This information is invariably printed near their eyepieces. Some binoculars may also be marked with a second set of numbers indicating the width of the field of view (FOV), eg '367ft/1,000yds'. FOV is the diameter of the circle that can be seen through the binoculars. Therefore, if an observer views any object 1,000 yards away, 367ft of height and 367ft of width will be visible.

Naturally, for objects billions of light years away a subsequent conversion to obtain degrees, or angular, distances is necessary, and this is done by simply dividing the first number by 52.4. So 367/52.4 = 7°. If metric, divide by 16 instead. Some manufacturers do this automatically: 'Field 7°', 'Wide Angle 6.5°', or 'FOV 7°' will appear. Generally, most 7x binoculars cover around 7° of sky (14 times the size of the Moon), 10x cover around 5° and those designated wide-angle or ultra-wide-angle cover around 8° to 12° (with distortions around the edge of the FOV).

PERFORMANCE

Magnification and aperture together yield performance: simply multiply one number by the other. So, 10x50 binoculars give a performance rating of 500 whereas 7x50s offer 350. The 10x50 should provide better views. Magnification and objective lens size should be closely related. It is not a matter of selecting the largest binoculars or ones advertised as offering greater magnification.

Which binoculars are best for astronomy? The short answer is 10x50s, since 50mm models collect more light than 35mm ones so images will be brighter and sharper. Why not go for even bigger binoculars then, say 70mm, 100mm, or 150mm? Generally, low-power binoculars reveal the most sky, making targets easier to find, but greater magnification makes objects more conspicuous. Here, however, weight is the main consideration: large, heavy binoculars, although offering higher magnification, are tiring to use and difficult to hold steady – destabilised images result. Binoculars with 56mm aperture are about the limit for hand-holding for any length of time. They also normally have small fields of view and require heavy-duty tripods and mounts (see *Binocular Mounts*, page 108).

EXIT PUPIL

The exit pupil is also another consideration at low magnifications. If you hold a pair of binoculars at arm's length and point them to a bright surface, you will see a small disc of light hovering in front of each eyepiece. This is known as the exit pupil, or the eyepiece's image of the objective lens, and its size is calculated by dividing the aperture by the magnification. So, 10x50 binoculars equal an exit pupil of 50mm/10 = 5mm. This is important for efficient use of the eye: it should be the same size as that of the dilated pupil of the eye. If the exit pupil is too large, an observer has to move their eyes to absorb the entire view. Too small, and the light-gathering area of the eye will be underused. Since, in dark conditions, the human pupil ranges from 7mm to 8mm for anyone under 30, and since a millimetre is lost with every subsequent 20 years of life, an exit pupil of between 2.5mm to 5mm is about ideal.

Moon through binoculars: the lunar craters and dark maria become stunningly clear.

Overall it is a balance, but 10x binoculars are the better suited for gazing at the night sky. Certainly a pair of 10x50s should show all 109 deep-sky objects in the Messier Catalogue (see Appendix 2).

OPTICAL COATINGS

All optical elements of binoculars should have multi-layer anti-reflection coatings, to maximise light transmission and reduce internal reflections after it enters the objective lens. More expensive binoculars generally have better coatings, but the difference between expensive and less-expensive models is virtually negligible. Single-coated lenses (the objectives and eyepieces) can be less effective in terms of reflection reduction, while other surfaces may have none at all; this could mean that much precious light does not make it to your eye – it will be washed out.

Multi-coated lenses have multiple types of coating. Fully coated means that all lens surfaces exposed to air are coated. Fully multi-coated (FMC) means everything is covered. Avoid cheaper binoculars with 'ruby' coatings: these red tinted lenses, although effective for reducing glare in daylight, actually block incoming light when used for astronomy.

PURCHASING: TEST RUN

When comparing various models for purchase, perform the following simple tests:

- Centre a star (or glint of sunlight on a distant telegraph pole) and bring it into sharp focus: the object should be tack-sharp, without fuzziness, double-imaging, or false colour. Move the object slowly to the edge of the FOV: in good binoculars it should remain sharp and point-like to at least 50% of the distance from the centre to the edge. Binoculars that only remain sharp at the centre of the FOV should be rejected.
- Hold the binoculars at a distance and look down the eyepieces. The observed circles of light should be evenly illuminated. Dark squared-off edges indicate poor-quality BK-7 prisms.
- Examine the main lenses: copious white reflections indicate poorly coated optics. Dark lenses with some deep purple or green reflections indicate high-quality multi-coated optics.
- Workmanship: check all moving parts. Look for excessive 'rocking' of the eyepieces and the bridge between – looseness impacts on focusing.
- Both halves of the binoculars must be parallel and optically aligned in order to avoid eyestrain. Set the binoculars on a solid base, aim them at a distant building, and focus. Look in the right barrel and notice where objects are relative to the edge of the field in the left–right (horizontal) direction. Compare this view with the left barrel and everything should be in the same place relative to the edge of the field. Do the same for objects in the up–down (vertical) direction: they, too, should appear in the same places. If there is a significant difference in the position of an image between the right and left barrels, do not purchase the binoculars.
- Make sure the model is comfortable to hold and not too heavy. The popular 7x50s and 10x50s range from around 26 to 50 ounces (0.75–1.4kg): a useful guide weight is 22 to 32 ounces (0.6–0.9kg) for comfortable handholding. It is generally held that 10x is the limit since any juddering caused by shaking arms is magnified ten times in the image.
- If purchasing high-eyepoint binoculars, check the eyecups provide full FOV access, with and without spectacles.
- Whilst looking through the binoculars, cover each of the front lenses in turn. If the image jumps back and forth, the binoculars are 'cross-eyed', or out of collimation, which is a major defect. Do not purchase them.
- Finally, bear in mind that a brand name is not necessarily an indication of superior quality. Often, unknown brands at the same price can be equally impressive.

EYE RELIEF

This is the distance in millimetres that your eyes must be from the eyepieces in order to see the entire field of view. An eye relief of less than 9mm makes for uncomfortable viewing. If you wear spectacles, but not to correct astigmatism, then preferably remove them. If you wear spectacles to correct astigmatism, keep them on but have a long eye relief, also known as high-eyepoint. High-eyepoint binoculars push the exit pupil some 18mm to 26mm away from the surface of the eyepiece lens, compared with around 10mm to 15mm on normal binoculars. These models come with large rubber folding eyecups that, when folded down, give the extra distance needed. When unfolded, they act as a guide for the unaided eye.

CENTRE-FOCUS ADJUSTMENT

There are two types – centre and individual eyepiece focusing. The former has a central knurled dial between the eyepieces to simultaneously adjust the focus of both halves of the binoculars. Individual focus eyepieces are simpler and more robust but focusing one eyepiece at a time can be laborious.

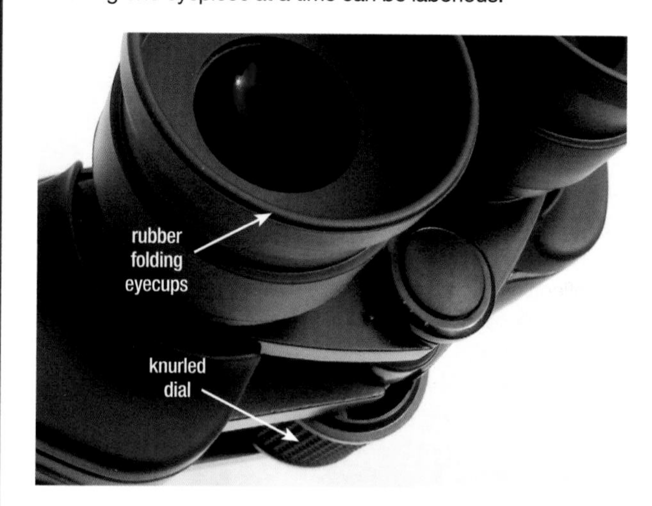

DIOPTRE ADJUSTMENT

The precise focal point for each eye can be different. Therefore some binoculars may have a dioptre setting on, or near, the right eyepiece – the dioptre adjustment – with a minutely adjustable graduated scale and zero point. Using this optimises the instrument's optical capabilities. To do so, focus with the left eye only (close the right eye) using the centre-focus adjustment. Then focus with the right eye (close the left eye) using only the right eyepiece. When the right eye is focused, make a note of the dioptre setting. This is useful when others use your binoculars: to re-personalise them simply reset the scale and then use the centre-focus adjustment to simultaneously focus for both eyes.

L-shaped right-angle adaptor: on a standard tripod.

Roof prism binoculars: with 'straight-through' shape.

TRIPOD SOCKET

This threaded hole is normally concealed beneath a plastic cap at the front of binoculars and is there to connect with an L-shaped right-angle adaptor for a standard camera tripod. When connected, the result is more comfortable viewing and steadier images.

OPTICS: ROOF AND PORRO PRISM

Binoculars use one or two types of prism – roof or porro: without them binoculars would be physically longer and the images would appear upside down. Neither prism is inherently better than the other since both are designed for the same purpose: to provide a correct image and shorten the distance from the objectives to the eyepiece by reflecting light along folded pathways. Roof prisms produce lighter models with a 'straight-through' shape: they are more compact than porro prisms in sizes under 42mm. However, although top of the range, expensive models are excellent, the middle of the road can create light spikes with bright stars and are, therefore, rarely used in astronomy. Porro prisms have a zigzag shape and are the most common. They are bulkier but affordable and produce excellent images.

BK-7 AND BAK-4

These refer to the two chemical types of optical glass used in the internal prisms. BK-7 use borosilicate glass and BAK-4 use higher density barium crown glass. BAK-4 is typically the better choice because it yields bright, crisp images with good contrast, but, again, it is virtually negligible.

PORRO PRISM BINOCULARS

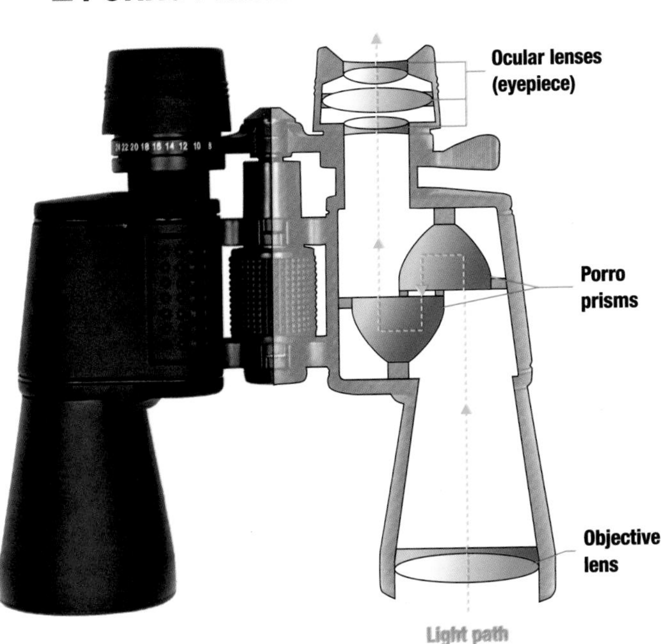

ROOF PRISM BINOCULARS

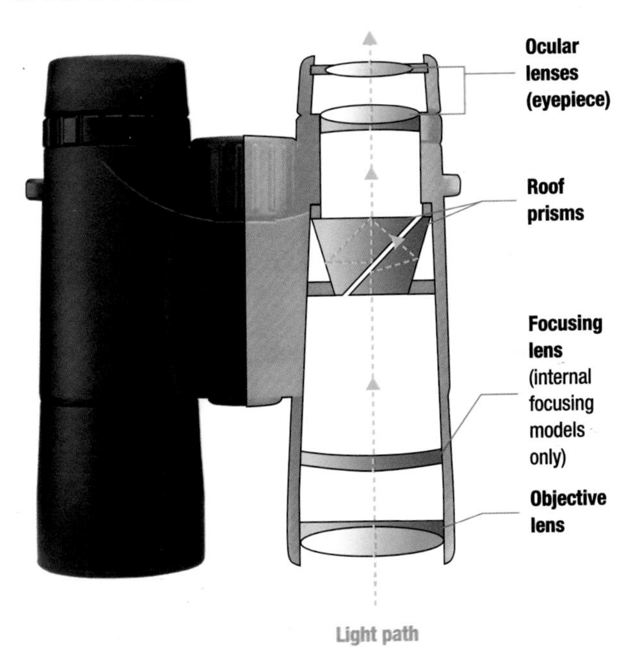

■ BINOCULAR MOUNTS

There is no point having a great pair of binoculars if you cannot hold them steady. Shaking is normally the result of two factors: the model is so heavy muscles shake just holding them, or the magnification is so great that any slight 'wobbling' is intensified. Nowadays, giant binoculars can be anything from 9x63 to 25x100. With anything over 60mm, additional support is needed:

- A camera tripod: this provides a solid platform for steady viewing but creates problems when craning your neck to see objects near the zenith. This can be overcome by purchasing cantilevered swing arms that suspend the binoculars away from the tripod.
- A simple chair or recliner provides comfort when seated and assists you to keep still when viewing.
- A dedicated binocular mount: these provide steady, comfortable viewing but can be expensive and are not always portable.
- A camera monopod: these can be used from a seated position and offer steady comfortable viewing. An additional block of wood placed on the top of the monopod to act as a platform enhances performance.

Heavyweight binoculars: demand the steadiness of a sturdy tripod.

■ IMAGE-STABILISED BINOCULARS (ISBS)

Introduced in 1996, these roof-prism designs are the crème de la crème in the binocular world. With extra optical components and internal AA battery-powered gimbals to detect and compensate for any shaking, they give incredible stabilised views without the need for any additional equipment. As a result, magnifications of up to 18x are possible. These are also available in 10x30, 12x36, 10x42, 15x50 (4.5° FOV) and 18x50 (3.7° FOV) models, the latter two being ideal for astronomy. Viewing is spontaneous and hassle-free … but the larger models come at a price.

Image-stabilised binoculars: steady views, despite shaking, makes them ideal for stargazing.

■ THE BINOCULAR SKY

With binoculars, the night sky shifts up a gear from unaided eye viewing. It is the difference between analogue and hi-definition. In addition to the spectacular views of the Moon already mentioned, the rest of the 'starry bowl' assumes entirely new depth. In 10x50s, the ever-changing aspect of three or four of Jupiter's Galilean moons can be enjoyed, although not in detail. Binoculars reveal the crescent phase of Venus when it is close to the end of evening apparition or rising in the morning. The planets Neptune and Uranus (magnitude +8 and +6 respectively) can be seen as aquamarine 'stars'. The skies can be probed for the dwarf planet Ceres, and the brighter asteroids – Vesta, Iris, Flora, and others – or the stunning sweeping tails of visiting comets. Moving much further away, the mighty Andromeda Galaxy (M31), along with its satellite galaxies M32 and M110, are bright enough to be located on autumn nights even under light-polluted city skies. In dark rural sites, more testing targets are the spiral galaxy M81, and the peculiar galaxy M82, both in the constellation of Ursa Major (the Plough or Big Dipper), and the Triangulum galaxy M33, a face-on spiral.

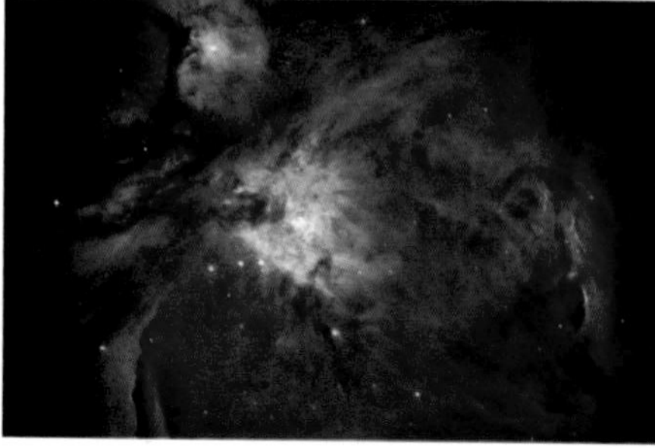

Great Orion Nebula, M42: easy to locate and stunning in binoculars.

SELECTED VARIABLE STARS FOR BINOCULAR VIEWING

Name	Constellation	Type	Period (days)	Magnitude range
Eta (η) Aquilae	Aquila	Cepheid	7.2	3.5–4.4
R Carinae	Carina	Mira	308.7	3.9–10.5
Omicron (ο) Ceti	Cetus	Mira	332	3.4–9.3
Zeta (ζ) Geminorum	Gemini	Cepheid	10.2	3.6–4.2
Delta (δ) Librae	Libra	Eclipsing Binary	2.3	4.9–5.9
Beta (β) Lyrae	Lyra	Beta Lyrae type	12.9	3.3–4.3
R Lyrae	Lyra	M-Type Red Giant	50	4.0–5.0
Algol (β) Persei	Perseus	Eclipsing Binary	2.9	2.1–3.4

SELECTED DOUBLE STARS FOR BINOCULAR VIEWING

Name	Constellation	Magnitudes
α Capricorni	Capricorn	3.6 and 4.3
α Ceti	Cetus	2.5 and 5.6
Albireo (β Cygni)	Cygnus	3.1 and 5.1
ο Cygni	Cygnus	3.8 and 4.8
Regulus	Leo	1.4 and 7.7
α Librae	Libra	2.7 and 5.2
ε Lyrae	Lyra	4.7 and 4.6 (Double Double)

Some double stars can even be split into their separate components: Epsilon (ε) Lyrae, not far from the star Vega in the constellation of Lyra, is a double double, each double having a separation of 2.5 arcseconds.

SELECTED NEBULAE FOR BINOCULAR VIEWING

Constellation	Name	Distance (ly)
Carina	Eta (η) Carina (NGC 3372)	3,700
Cygnus	North America (NGC 7000)	1,800
Orion	Great Orion (M42)	1,500
Orion	Orion (M43)	1,500
Sagittarius	Swan, Omega (M17)	5,500
Sagittarius	Lagoon (M8)	5,000
Sagittarius	Triffid (M20)	5,000
Vulpecula	Dumbbell (M27)	1,000

Survey the constellations of Sagittarius or Scorpius and gaze into beautiful nebulae or the wide dark dust lanes delineating the central Galaxy. Peer overhead and see Cygnus soaring with outstretched wings through the 'fuzz' that has resolved into billions of stars.

Milky Way galaxy arcing overhead: binoculars resolve its myriad dark dust clouds and billions of stars.

SELECTED OPEN CLUSTERS FOR BINOCULAR VIEWING

Name	Constellation	Magnitude (total)	Approx. size (arcmins)	Approx. distance (ly)
M44 (Beehive)	Cancer	3.1	95	600
NGC 4755 (Jewel Box)	Crux	4.2	10	6,800
M39	Cygnus	4.6	30	7,300
M35	Gemini	5.5	30	2,800
NGC 869 (Double Cluster)	Perseus	4.4	35	7,000
NGC 884 (Double Cluster)	Perseus	4.7	35	8,100
M6 (Butterfly)	Scorpius	4.6	26	1,500
M7	Scorpius	3.3	50	800
M45 (Pleiades)	Taurus	1.2	110	440
Hyades	Taurus	0.5	330	150

Open clusters, groups of stars relatively close to us in the disc of the Galaxy, resolve into beautiful individual stars.

SELECTED GLOBULAR CLUSTERS FOR BINOCULAR VIEWING

Name	Constellation	Magnitude (total)	Approx. size (arcmins)	Approx. distance (ly)
Omega (ω) Centauri (NGC 5139)	Centaurus	3.7	36	17,000
M13	Hercules	5.9	17	23,000
M92	Hercules	6	11	26,000
M15	Pegasus	6.4	12	34,000
M22	Sagittarius	6.0	18	10,000
NGC 104 (47 Tucanae)	Tucana	4.5	31	16,000

Globular clusters, the gigantic aggregations of stars that are billions of years old and scattered throughout the Galactic halo, are no longer invisible, but become fuzzy balls in binoculars.

THE DEVIL'S IN THE DETAIL

Any one of these images answers the question of whether binoculars are required. If further affirmation is needed, simply revisit the sample constellations outlined for unaided eye viewing earlier in this chapter.

Since you will see more with binoculars, you will also – rather like graduating from a national road map to a more detailed urban guide – need a more detailed star atlas; a simple star chart and planisphere (star wheel) are not deep enough. Circular charts in current monthly magazines are good starters, as are books and Internet websites where you can print customised star-dome maps, but an 'opened up' night sky demands at least a detailed fifth, or sixth, magnitude star atlas (see Appendix 3).

Having appreciated the constellations, and other wonders, of the naked-eye sky and having scanned deeper with binoculars, it is inevitable that our 'urban guide' will lead us to a desire for Google Earth detail! Welcome to the daunting world of amateur telescopes … and a 'street guide' sky that is truly universal.

Amateur telescopes

There is an extraordinary array of telescopes available to amateur astronomers. From straightforward 'push and shove' to more expensive high-tech computerised models, the choice is as bewildering as the cosmos itself. For anyone starting out, it is a minefield. For anyone already engaged in amateur astronomy … it is still a minefield, one impossible to completely navigate within the confines of this manual. However, the following may be helpful for distinguishing polar axis from polar bears when making your initial purchase and when starting out.

■ TELESCOPE TYPES

The main optical designs for amateur telescopes are:

- The achromatic and apochromatic refractor.
- The Newtonian reflector.
- The catadioptric.

All these telescopes will reveal, to a lesser or greater extent, details on the Moon as small as a mile across, the rings of Saturn, the clouds of Jupiter and its four major Galilean satellites, nebulae, star clusters, the brightest galaxies, and the deep sky.

Craters on the Moon: spectacular in binoculars, the lunar craters are stunning in a small telescope.

■ REFRACTOR TELESCOPES

Refractors refract, or bend, light through lenses, which is then focused to an eyepiece at the bottom of the telescope tube – the optical tube assembly. Devised in the late 16th and early 17th centuries, the simple single lens meant that early telescopes suffered from chromatic aberration – although the light was bent and magnified, the lens also acted like a prism splitting it into its component colours and resulting in a 'blue halo' around objects. Refracting telescopes today have an objective lens that is a combination of different types of glass with a very short space between – known as the air-spaced achromat (achromat means 'without colour'). The apochromatic refractors (apos) eliminate false colour by using double, triple, or even quadruple lens elements with one element made of fluorite or ED (extra-low-dispersion) or Super ED glass: these bring the colours to the same focus. Others use small corrector lenses near the focuser.

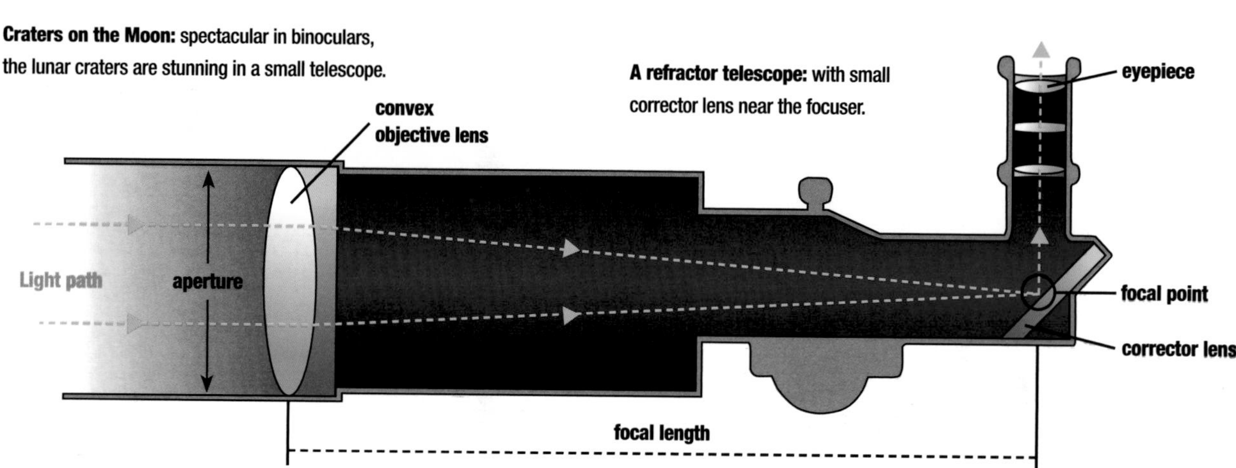

A refractor telescope: with small corrector lens near the focuser.

NEWTONIAN REFLECTOR TELESCOPES

First developed by Sir Isaac Newton in 1668, these classic designed telescopes use mirrors to collect light. Light enters the top of the telescope and travels to a concave (curving inwards) primary mirror – preferably with a parabolic curve – at the bottom. From here, it is reflected back up the tube to a central flat secondary mirror sited at a 45° angle near the top. The light then exits the scope through the side of the tube where the focuser – the mechanism for focusing the eyepiece – is placed.

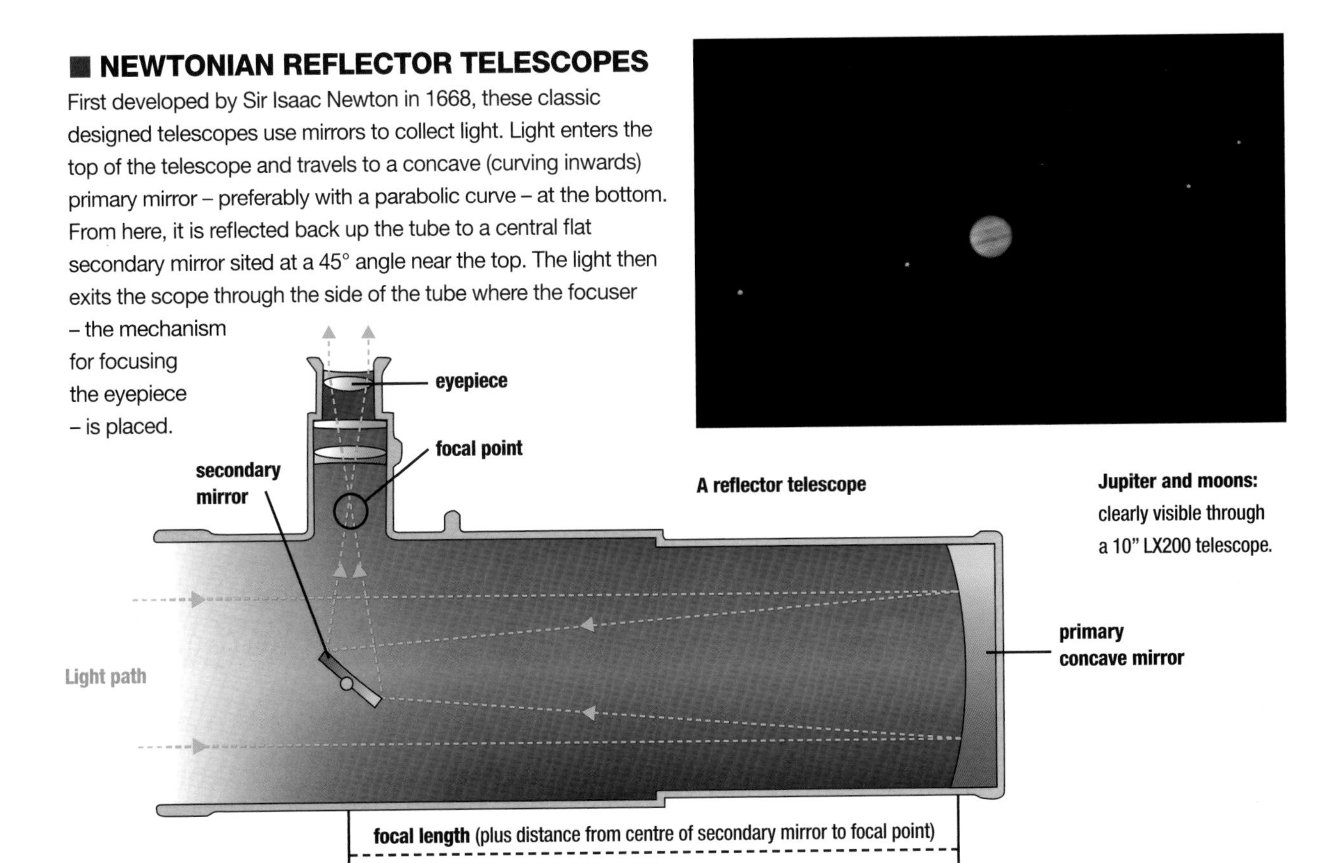

A reflector telescope

Jupiter and moons: clearly visible through a 10" LX200 telescope.

CATADIOPTRIC TELESCOPES

Examples are Maksutov-Cassegrain (invented independently by Bouwers and Dmitri Maksutov in the Soviet Union in the 1940s and also known as 'Maks') and Schmidt-Cassegrain (SCT) telescopes (using optical technology invented by Bernard Schmidt and Guillaume Cassegrain in the 1930s). These compact hybrid models use curved mirrors and an aspherical corrector lens – more commonly called a corrector plate – to bounce light back and forth in the tube. Light enters through the corrector plate at the top of the tube. On an SCT, light enters through the corrector plate, bounces off the curved primary mirror at the back, travels back up to the convex (curved outward) secondary mirror, and then down again to exit through a central hole in the primary mirror to an eyepiece sited at the back of the telescope.

On a Mak telescope the steeply curved corrector plate is known as a meniscus (crescent-shaped) lens. It is convex on one side and concave on the other. Like the SCT, light enters through the corrector plate, bounces off the curved primary mirror at the back, and travels back up to the curved central secondary mirror – here really just a reflective coating on the back of the corrector lens – and then down again to exit, like the SCT, through a central hole in the primary mirror to an eyepiece sited at the back. The incoming light is effectively 'manoeuvred' four times.

Other types of optics are the Ritchey-Chretien telescope (RCT) – a specialized Cassegrain design that has a hyperbolic primary mirror and hyperbolic secondary mirror to eliminate optical errors (coma); the Schmidt-Newtonian, which combines a Schmidt corrector with Newtonian optics to minimise the off-axis coma found in fast Newtonians, and the Maksutov-Newtonian, which offers an aberration-free view across a wide field at low power as well as images such as those seen in refractors at high power.

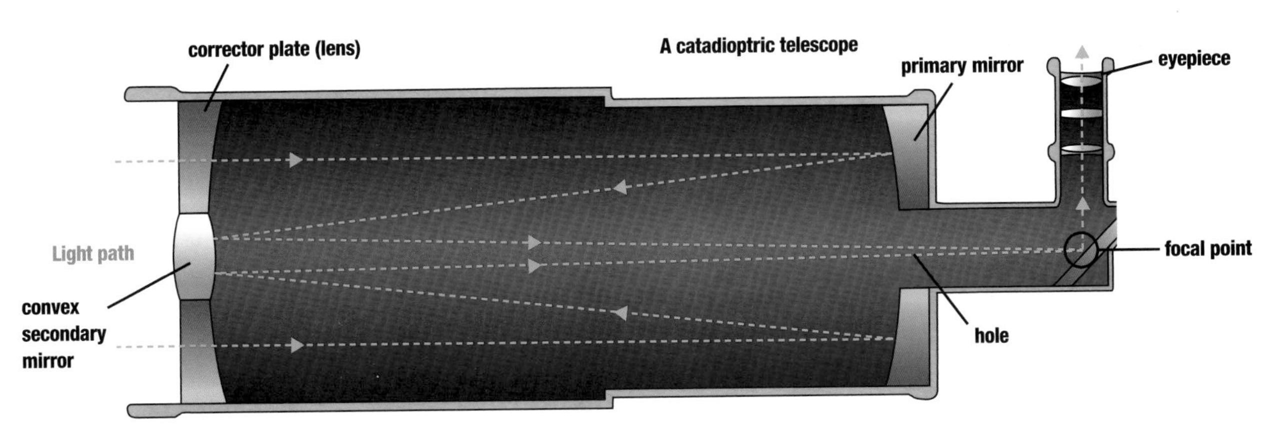

A catadioptric telescope

■ TELESCOPE FUNDAMENTALS

Always refer to the instruction manual when assembling a new telescope (although these are not always entirely helpful). Many scopes arrive 'ready to go' but others require some minimal assembly: tripods need extending and bolting together, the optical tube assembly (OTA) needs affixing, fittings need attaching.

The three most commonly used telescopes are the achromatic and apochromatic refractor, the Newtonian reflector and the catadioptric. What do they look like?

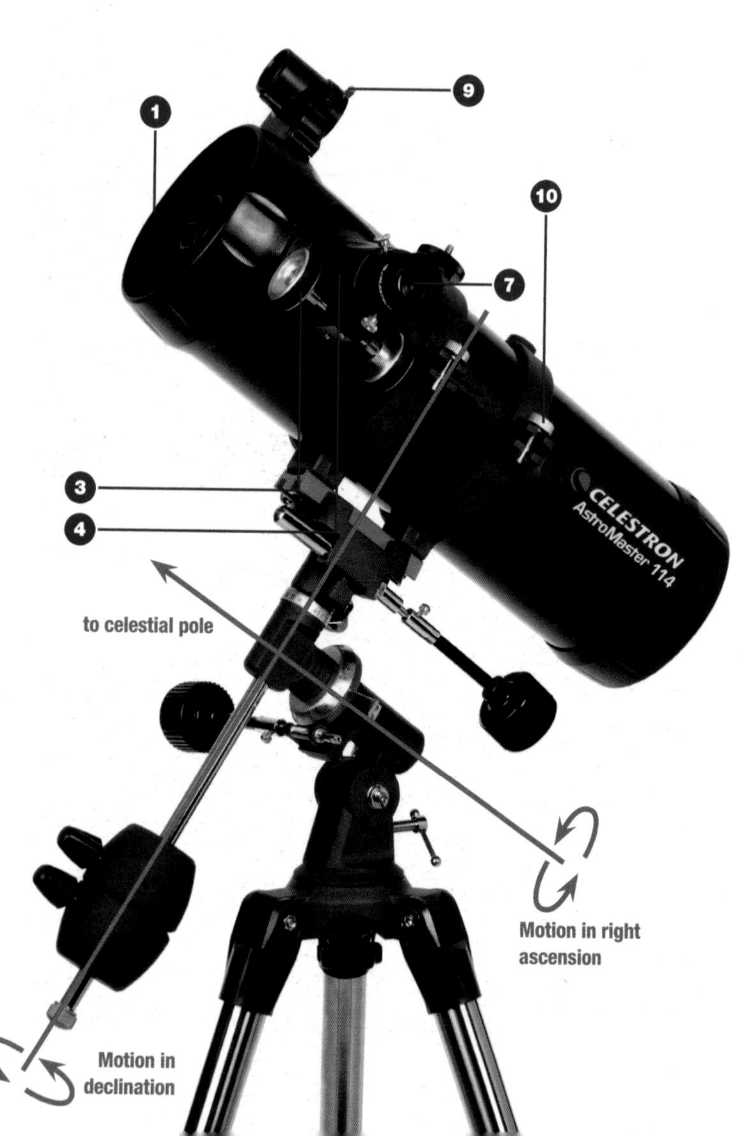

A TYPICAL APOCHROMATIC REFRACTOR OTA (TOP IMAGE) **AND A TYPICAL NEWTONIAN REFLECTOR OTA** (AT LEFT)

1. (Top image) Lens: some apochromatics use double or triplet lenses with one element made of ED glass. Others use small corrector lenses near the focuser. (Left image) Newtonian reflectors do not have objective lenses but, instead, utilise mirrors.
2. Dewcap: helps prevent dew forming on main lens.
3. Focuser: a rack-and-pinion type that slides the eyepiece to enable focusing. Should have a Focuser lock: when tightened, it prevents the focuser moving.
4. Adapter tube: enables use of 1.25-inch eyepieces and can be removed and replaced with a camera adapter for photography.
5. Diagonal: a 90° prism for more comfortable right-angle viewing position.
6. Thumbscrews: thumbscrews tighten the diagonal and eyepiece in place. Very easily dropped and lost, so be careful!
7. Eyepiece.
8. Finderscope (viewfinder): a low-power scope for locating and centring objects. Some require attachment to the OTA, others come attached.
9. Finderscope adjustment screws: two screws set at right angles – used to align the 'finder' with the OTA's main optics.
10. Tube rings: the side bolts on these rings can be loosened to unclamp the tube and rebalance it in declination (altitude). The whole tube can be rotated for a more comfortable viewing angle.

A TYPICAL CATADIOPTRIC TELESCOPE OTA

1. Eyepiece.
2. Eyepiece holder and thumbscrew: tightens eyepiece in place.
3. Diagonal prism: for more comfortable right-angle viewing position.
4. Finderscope: a low-power, wide-field sighting scope with cross hairs to enable easy centring of objects in the telescope's eyepiece.
5. Finderscope alignment screws: use these to adjust the alignment of the finderscope with the OTA's main optics.
6. Smartfinder not visible - see page 130.
7. Focus knob: moves primary mirror in finely controlled motion for precise focus. Rotate anticlockwise for distant objects, clockwise for nearby objects.
8. Declination lock: controls vertical movement of scope. Turn anticlockwise to unlock and obtain free manual rotation around the vertical axis. Be careful to support the OTA when unlocking.
9. Fork arms: hold OTA in place.
10. Declination setting circle/declination pointer: graduated scale for celestial coordinates and with moulded triangular pointer underneath the circle. Line up the desired declination with this pointer.
11. Handbox: enabling handheld automated control for nearly all the functions of the telescope.
12. Handbox holder: holds the handbox in a convenient position on the fork arm handles.
13. Dustcover: gently remove before use. Replace after each session but first ensure any collected dew has evaporated.

ALIGNING A FINDERSCOPE

The finderscope, also known as a viewfinder, must first be aligned with the scope. Below is a generic method for doing so:

1. Insert the finder into its supplied bracket (a above). If there is a spring-loaded bolt, pull this back. Some finders have a rubber O-ring that secures into a recessed niche on the finder tube and keeps the finder steady. On others, after insertion tighten the thumbscrews (b above) to a firm feel only.
2. Slide the finderscope track (c above) at the base of the assembly into the mounting slot. (On some scopes the mounting slot has to be screwed onto the OTA first.)
3. Insert your telescope's eyepiece. (1)
4. Unlock the right ascension (RA) (14) and declination (Dec) (8) locks to move the telescope freely on both axes.
5. Point the scope at a well-defined object, such as a telegraph pole, at least 200 yards away. Centre the object in the eyepiece and tighten the RA and Dec locks.
6. Look through the finderscope and loosen or tighten the thumbscrews until the cross hairs are precisely centred on the object sighted in the scope's eyepiece. When first observing at night, select a sharp bright object and make any adjustments with the thumbscrews as necessary.

SMARTFINDER

The above is also a variation on a theme for aligning a Smartfinder (red dot finder), except when looking through a Smartfinder the bottom or side alignment screws require turning until the red dot points precisely at the object in the scope's eyepiece.

■ TELESCOPE SIZE AND POWER

Two features determine a telescope's size: focal length and aperture.

FOCAL LENGTH

Focal length is the distance, in millimetres, between the light-gathering surface and the focal point – the area where the light, whether bent by mirrors or lenses, comes to a point (also known as a 'small plane' or 'spot'). With refractors and reflectors it is normally equal to the length of the telescope tube. However, with Maks and Schmidt-Cassegrains the focal length, although long, is folded twice, allowing for a shorter tube. The focal length measurement is normally indicated on the tube or in the accompanying instruction manual. When combined with a given eyepiece, it determines the power of the telescope.

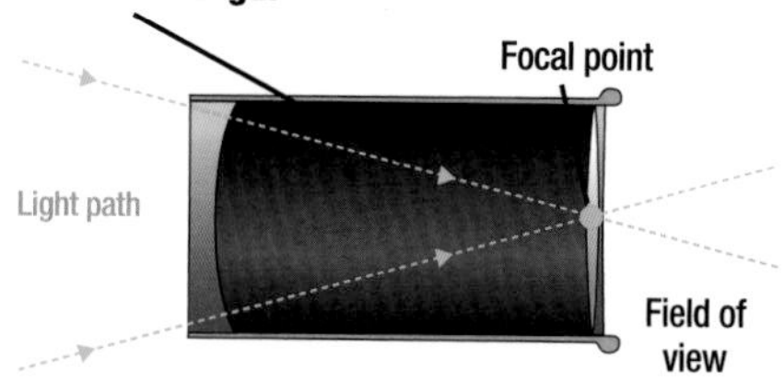

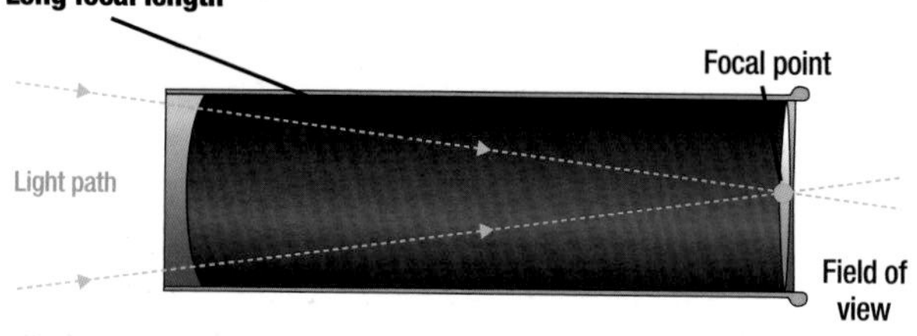

APERTURE

This is the diameter of the telescope's main light-collecting surface, whether it's a lens or a mirror. It goes without saying, therefore, that aperture is key. However, no matter how small the aperture, if the optics and focus are correct, an object will appear sharp. For every doubling in aperture size, the light-gathering power increases by a factor of four: a six-inch (150mm) mirror has four times the surface area of a three-inch (75mm) mirror and, therefore, collects four times the light and produces images that are four times brighter. A telescope's ability to clarify objects is known as its 'resolving power' or 'resolution'. Again, size of aperture is key when it comes to resolving power: doubling the aperture doubles the resolving power.

A four-inch (100mm) aperture resolves stars one arcsecond apart but an eight-inch (200mm) resolves stars just 0.5 arcseconds apart. The human eye can resolve only a 60 arcseconds separation. Note that Focal Reducers can be used to apparently reduce focal length, thereby achieving a more concentrated, brighter and flatter view but also a wider field of view.

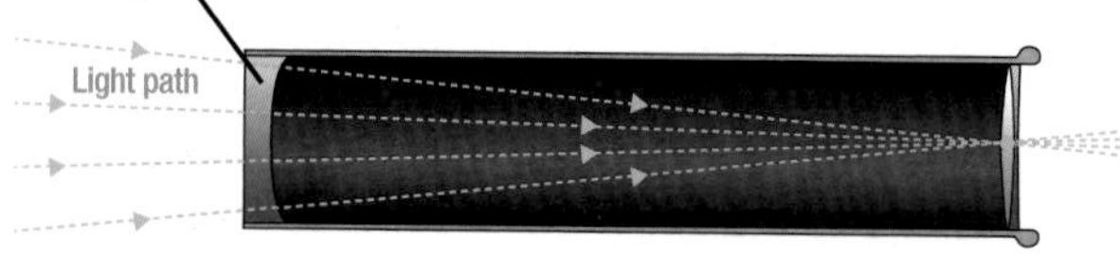

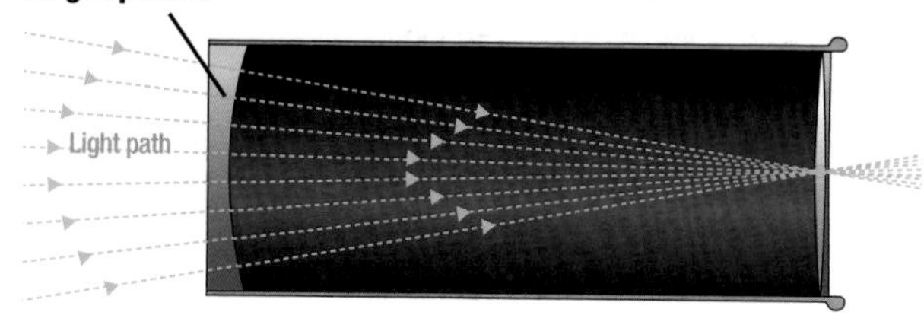

■ MAGNIFICATION

Eyepieces also have focal lengths (see also *Eyepieces*, page 124). Dividing a telescope's focal length by an eyepiece's focal length gives the telescope's magnification – how much an object is enlarged. As an example, a telescope with a focal length of 1,000mm and an eyepiece of 25mm focal length magnifies the image 40 times (1,000/25 = 40). Magnification is actually quite unimportant; swapping eyepieces enables an observer to magnify a telescope's power by any amount. But the maximum power deliverable by that telescope will be equal to around 50 times its aperture in inches: for example, a 2.4-inch (60mm) telescope aperture cannot operate much higher than 120 power. A 4.5-inch (110mm) aperture offers 225 power. If these limits are exceeded, yes, the image appears larger but it will be dim and blurred.

■ FOCAL RATIO

Dividing a telescope's focal length by the aperture gives the all-important focal ratio, or f-ratio. A telescope with a focal length of 1,000mm and an aperture of 200mm (8in) is known as an f/5 model and would be classed as 'fast'. Contrastingly, the 1,500mm focal length model with the 100mm aperture would be an f/15 model and classed as 'slow'. These speeds originate in the world of photography. When using a telescope for deep-sky astrophotography (see Chapter 5), a 'faster' f/5 model will record a nebula in a quarter of the exposure time needed by an f/10 model. But when it comes to visual observing there is little significance between the f-ratios. The different speeds are not an indication of light-gathering ability or the brightness of the objects viewed. For example:

1,500mm focal length divided by 100mm aperture
= f/15 focal ratio.
1,500mm focal length divided by a 25mm eyepiece
= 60 x magnification.

500mm focal length divided by 100mm aperture
= f/5 focal ratio.
500mm focal length divided by a 25mm eyepiece
= 20 x magnification.

Although the f-ratios are different and, with the same eyepiece, the magnification is different, the actual brightness of the observed object still remains the same.

■ TELESCOPE MOUNTS

Telescope mounts are as important as the telescope itself: they are the workhorses, and, as the adage goes, 'no foot, no horse' – so no mount, no scope. Primarily they are required to move smoothly, easily, and quietly, but also to provide a rock-steady base. With today's rapidly increasing technology, mounts offer a good deal more besides and just as there are many optical types of telescopes, so there are many types of mounts, generating numerous mix-and-match possibilities. The following is intended as a helpful overview:

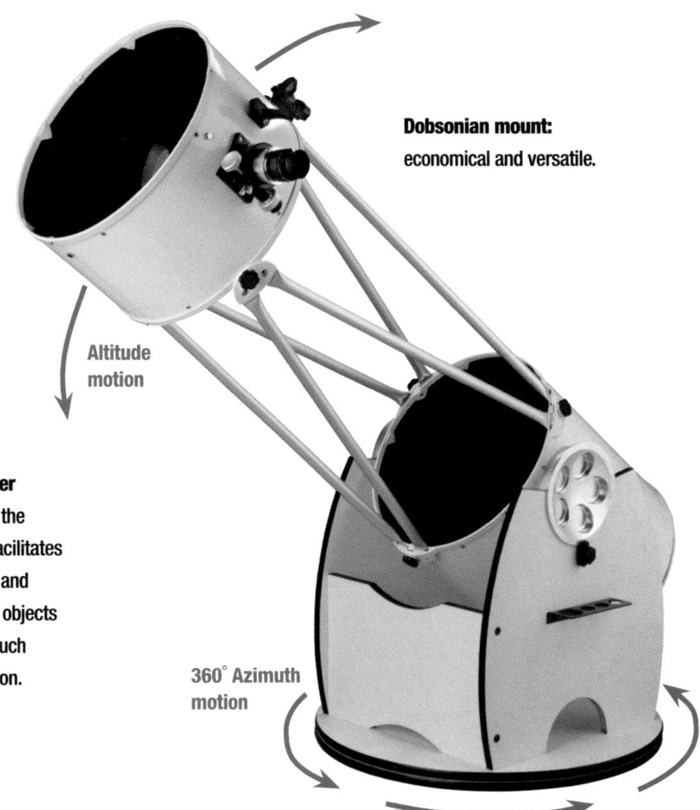

Dobsonian mount: economical and versatile.

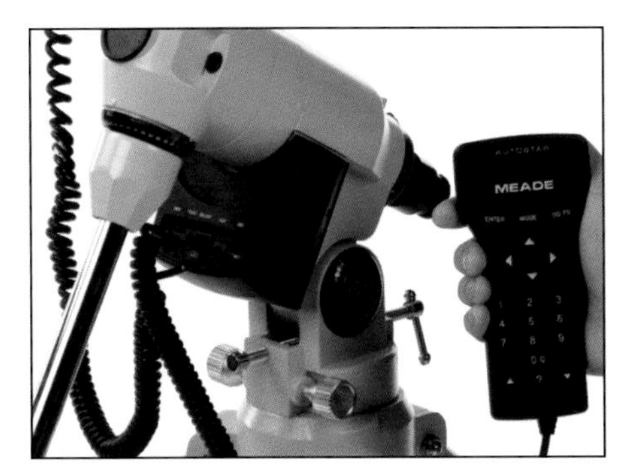

Computer guided: the mount facilitates locating and tracking objects at the touch of a button.

ALT-AZIMUTH (ALT-AZ) MOUNT

This term is short for altitude (vertical) azimuth (horizontal) mounts, and conveys how this simple telescope mount moves: up and down and side to side, to follow a star across the sky.

DOBSONIAN (OR DOB) MOUNT

Designed by amateur astronomer John Dobson in the 1970s, this melamine-covered wooden or aluminium alt-az type mount, with Teflon pads to provide smooth action, bears his name, and it has undergone an evolution. Renowned for being straightforward and sturdy, they used to require manual pushing for observing. Today, although still normally used with Newtonian reflectors, these mounts have evolved, with lightweight collapsible open truss tubes that can be assembled or disassembled with the turn of a simple Allen key. It, too, offers 'grab and go' potential. This, coupled with the Dobsonian's circular base design delivering 360° rotation, makes them extremely versatile and durable in terms of use, portability, and storage: they can easily be carried in a family car or stored in a garage.

They do not offer tracking, at least not without the optional, but expensive, tracking platform or computerised motor system. As such they are not really suitable for serious, long-exposure astrophotography. But some Dobs offer computerised finding: an observer pushes the scope around the sky until a display confirms a captured target. All in all, however, they make great value for money starter-combinations, especially in the 6- to 10-inch range, although whopping 30-inch models are also available. And at the premium end, there are those with handcrafted optics and furniture-type finishes too.

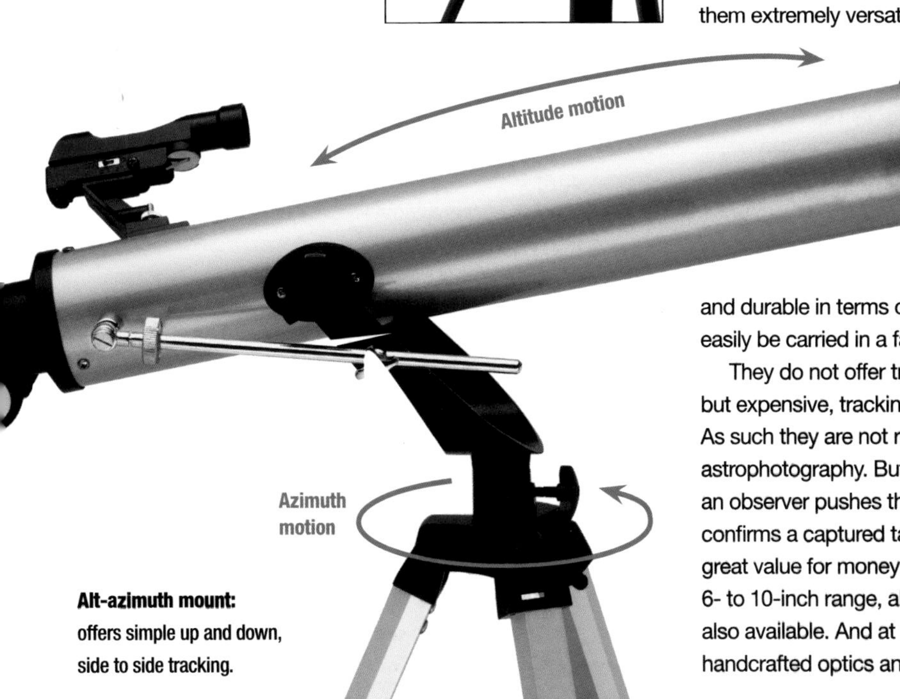

Alt-azimuth mount: offers simple up and down, side to side tracking.

FORK MOUNT

This mount is formed from a large metal fork attached to each side of the optical tube assembly (OTA) on either side of the telescope tube. They are essentially alt-az mounts used with short tubes – the Schmidt-Cassegrain models – and allow the scope to move up and down. The fork itself sits atop a rounded, or rectangular, drive base that swivels 360°.

The popular Celestron NexStar 8SE with fork arm

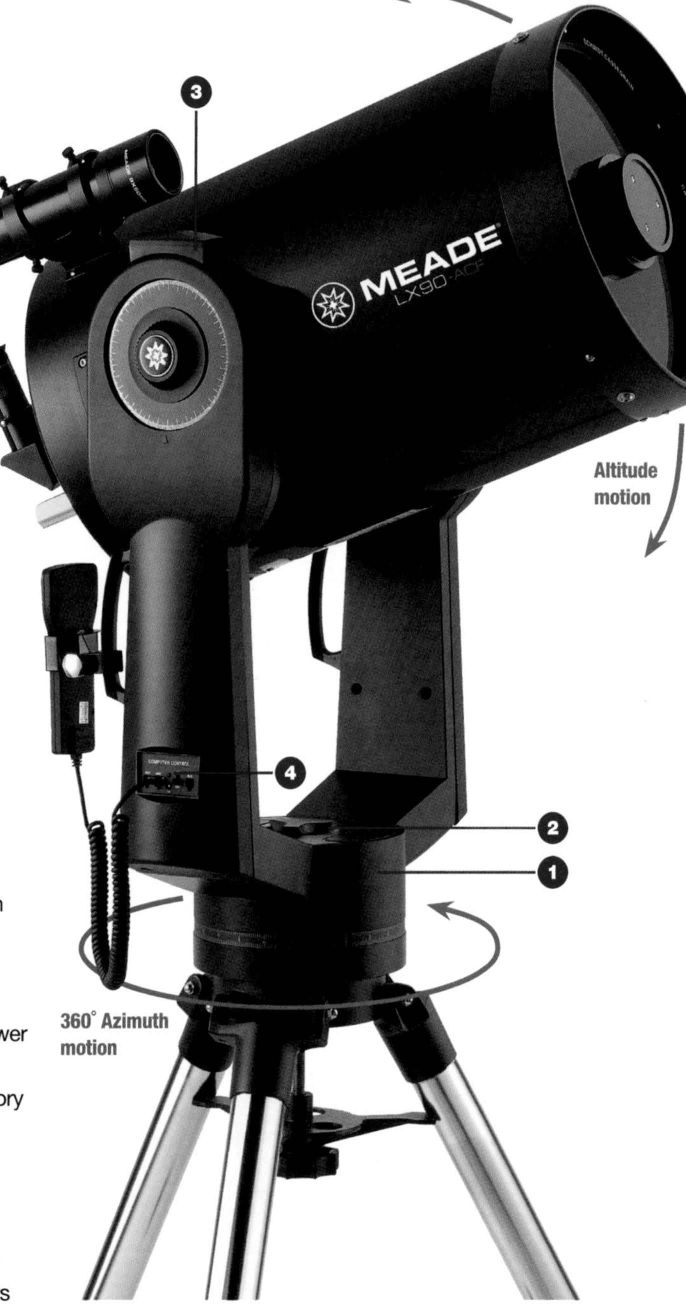

A FORK-MOUNT CATADIOPTRIC TELESCOPE

1. Drive base.
2. RA lock: controls manual horizontal rotation of telescope: turn anticlockwise to unlock and have free manual rotation about the horizontal axis.
3. GPS receiver: receives information transmitted from Global Positioning System (GPS) satellites.
4. Computer control panel: comprises handbox port, red power light indicator showing power supplied to motor drive, on/off switch for control panel and handbox, auxiliary accessory port, and 12V power connector for connecting to vehicle cigarette light plug or standard 115V AC home outlet.

FORK MOUNT WITH WEDGE (EQUATORIAL)

Some fork mounts utilise another component – the wedge – that tilts the fork in the direction of the north (or, for observers in the southern hemisphere, the south) celestial pole. With the wedge, the fork mount is effectively angled into equatorial mode (see *Equatorial mount*, opposite). Once the telescope has been initially polar-aligned with the north (or south) celestial pole (see *Polar-alignment*, page 121), it will track celestial objects for as long as required.

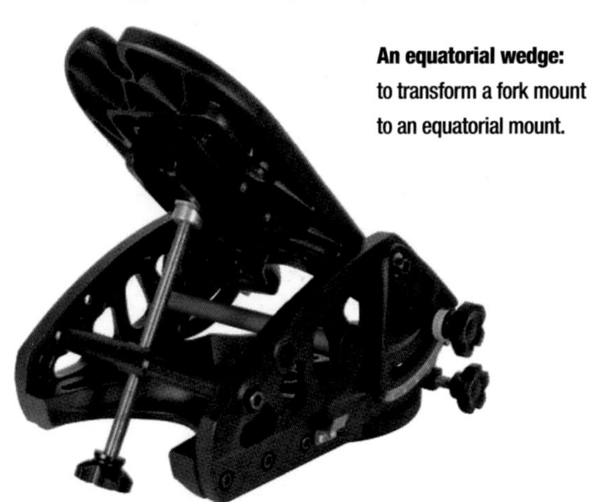

An equatorial wedge: to transform a fork mount to an equatorial mount.

EQUATORIAL MOUNT

These were invented by Joseph von Fraunhofer in the early 1800s. The most common equatorial mount is the German equatorial mount (GEM). As with the fork mount using a wedge to angle it into equatorial mode (see left), GEMs are designed to tilt to an angle at which the azimuth (horizontal) circle of motion of the scope can be matched to the equator and thus enable tracking of celestial objects throughout the night. Again, the mount must first be polar-aligned with the north (or south) celestial pole (see *Polar-alignment*).

There are also even more advanced equatorial mounts, in the form of polar-alignment scopes that offer greater accuracy when aligning with the celestial poles: essential for astrophotography (see *Polar-alignment*).

When shopping for these types of German equatorial mounts, it becomes clear that there are many designs, their names synonymous with various telescope manufacturers.

GERMAN-STYLE EQUATORIAL MOUNT

There are a plethora of different models of German-style equatorial mounts. The latest EQ-8 model from Sky-Watcher is but one of many and is for the serious amateur observer/imager. However, Sky-Watcher offer an entire series of mounts:

The EQ-1: these are often provided with small, lightweight telescopes and can often be inadequate for more serious night sky viewing.

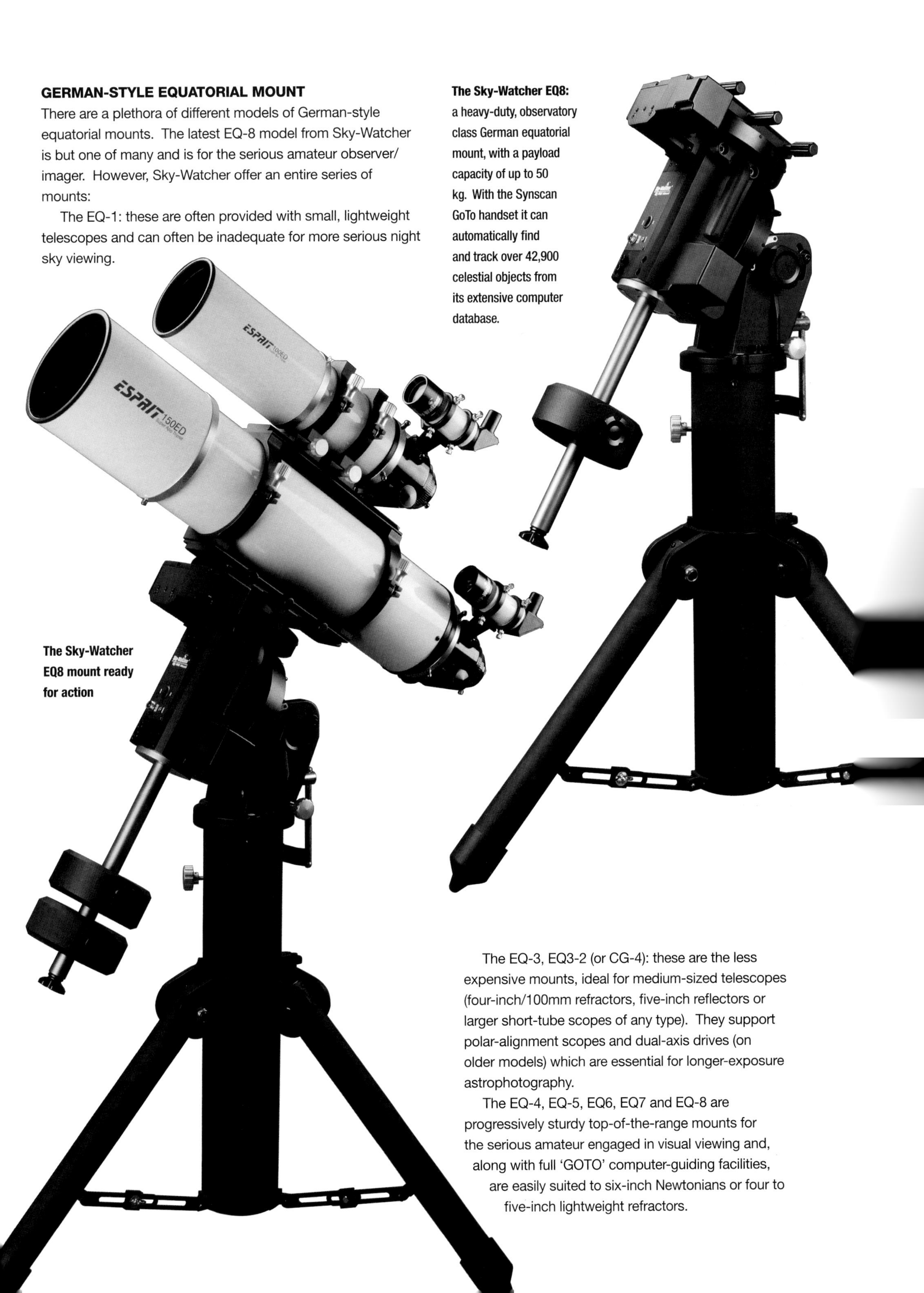

The Sky-Watcher EQ8: a heavy-duty, observatory class German equatorial mount, with a payload capacity of up to 50 kg. With the Synscan GoTo handset it can automatically find and track over 42,900 celestial objects from its extensive computer database.

The Sky-Watcher EQ8 mount ready for action

The EQ-3, EQ3-2 (or CG-4): these are the less expensive mounts, ideal for medium-sized telescopes (four-inch/100mm refractors, five-inch reflectors or larger short-tube scopes of any type). They support polar-alignment scopes and dual-axis drives (on older models) which are essential for longer-exposure astrophotography.

The EQ-4, EQ-5, EQ6, EQ7 and EQ-8 are progressively sturdy top-of-the-range mounts for the serious amateur engaged in visual viewing and, along with full 'GOTO' computer-guiding facilities, are easily suited to six-inch Newtonians or four to five-inch lightweight refractors.

COMBINED ALT-AZ AND EQUATORIAL MOUNTS

'GOTO' fully computer-guided mounts are the norm these days for amateur astronomy and do exactly what they say on the proverbial tin: they find, go to and track objects at the touch of a button. They typically come in the two formats already mentioned - equatorial or altazimuth – each serving a different purpose. Alt-az mounts are better suited for visual observing but equatorial mounts are the preferred option for imaging. The example shown below – Star-Watcher's popular AZ-EQ6 GT model – is a mount that can be assembled and operated in either equatorial or altazimuth mode.

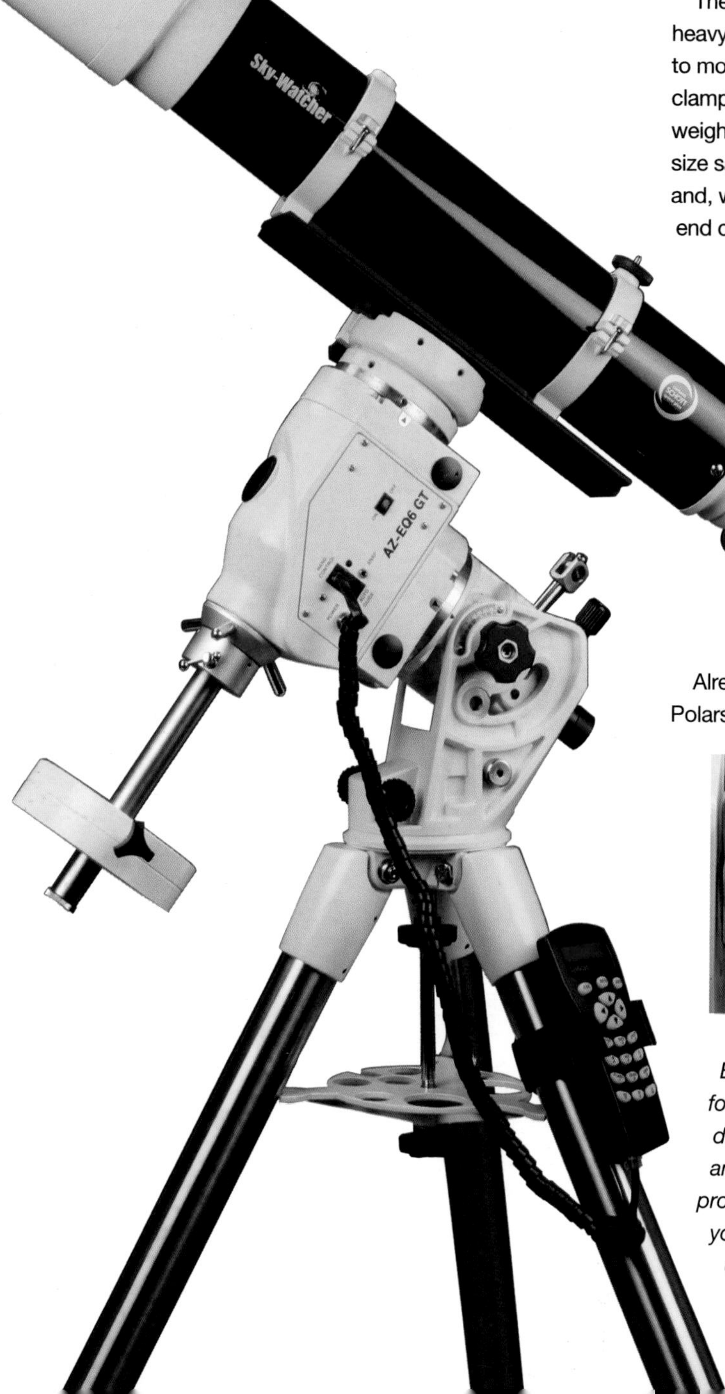

The Sky-Watcher AZ-EQ6 GT in equatorial configuration.

In order for astro-photographers to acquire the best quality images, their first prerequisite is for a mount to be in an equatorial configuration. An equatorial mount follows the apparent movement of the stars across the sky in a smooth continuous arc whereas an altazimuth mount moves in a series of discrete horizontal and vertical steps. An altazimuth mount will keep an object centred in the field of view. However, over a long period of time, the field of view appears to rotate resulting in what is known as 'field rotation'. This field rotation produces elongated stars around the periphery of an image and blurs the central celestial object.

The AZ-EQ6 GT is versatile in another way too. With its heavy duty stainless steel tripod it is a straightforward process to mount a single telescope in the standard dual-size saddle clamp (see page 119) and counterbalance this with a suitable weight on the counterbalance bar. However, a second dual-size saddle clamp can be affixed to the counterbalance bar and, when tightened securely onto a machined flat on the end of the counterbalance bar, allows two telescopes to be installed at the same time in altazimuth configuration. Both telescopes balance one another by adjusting the amount of counterbalance bar extension. The second saddle clamp, with its neat altitude adjustment, allows an observer to point both telescopes at the same elevation.

Already installed in the base of the RA axis is an illuminated Polarscope with, for use in the northern hemisphere, etchings for the Polaris circle, The Great Dipper and Cassiopeia and, for the southern hemisphere, Octans. *See page 122 for polar alignment.*

A polarscope already installed at the base of the RA axis.

Each telescope mount, whether alt-azimuth or equatorial, fork or Dobsonian, GOTO or with external RA and Dec drives, offers its own challenges. Setting up such mounts and their accompanying telescope/s can be a complex process, regardless of level of knowledge. Always ensure you take the time to carefully read the supplied manuals, especially the small print!

■ TELESCOPE COMPARISONS

Having graduated from unaided eye, or binocular, viewing to telescope viewing (or perhaps not: many enjoy unaided observation of the night sky), how do you choose your first telescope? Don't become the proverbial 'rabbit in the headlights' or suffer 'paralysis by analysis'. Page 120 outlines some of the advantages and disadvantages of telescope types but, ultimately, any decision is dependent upon:

- Ability: is simplicity or sophistication required? Are you simply enjoying the night sky or thinking about astrophotography?
- Interest: is the telescope for viewing the Moon and planets or for deep-sky observing?
- Funds: although prices are generally dropping, the greater the sophistication, the greater the cost. Pick a scope within the price range that suits you and remember that relatively inexpensive scopes can also do a really great job.
- Portability: is the scope to be used at home or elsewhere? The bigger the aperture and the longer the tube, the less transportable and user-friendly it is.

In conclusion, the best telescope for anyone is one that is frequently used. It should, therefore, have good optics, a sturdy mount, be portable and user-friendly, and come at a cost appropriate to its class. There are over 1,000 models out there! These considerations are for guidance only. For further reading and guidance refer to Appendix 3. Most importantly, your telescope should be about having fun and enjoying astronomy.

A saddle clamp.

Sky-Watcher AZ-EQ6 GT with 2 scopes in alt-az configuration.

A selection of Sky-Watcher AZ-EQ5 mounts with telescopes.

TELESCOPE TYPES: PROS AND CONS

REFRACTORS: ADVANTAGES

- Normally maintenance-free. Treat the objective lens well and it will last a lifetime.
- Easy to aim: the focuser is located at the bottom, or back, of the telescope tube and is, therefore, in line with the target. Just point!
- Free from central tube obstructions: in some telescopes reflecting mirrors obscure light to the focuser, compromising image contrast.
- Less troubled by cooling-down time. Air temperature inside OTAs should equal that outside: if not, eddying air currents disturb incoming light and distort images. In reflectors and catadioptrics light bounces back and forth in the tube, increasing distortion: 'equalising' the interior with the exterior temperature is vital. In refractors light enters only once, exiting via the rear focuser: the light disturbance is minimal despite temperature variation.
- Available in many focal ratios.

REFRACTORS: DISADVANTAGES

- Generally more expensive than reflectors per unit of aperture, but prices are dropping. Commercial competition has resulted in good optics at lower prices.
- Tricky for viewing objects near the zenith due to the position of the rear focuser.
- Troubled by false colour: the shorter focal length achromatic scopes can suffer from chromatic aberration.

A Newtonian 'spider': these hold mirrors in place but may cause diffraction spikes – small rays – around brighter stars.

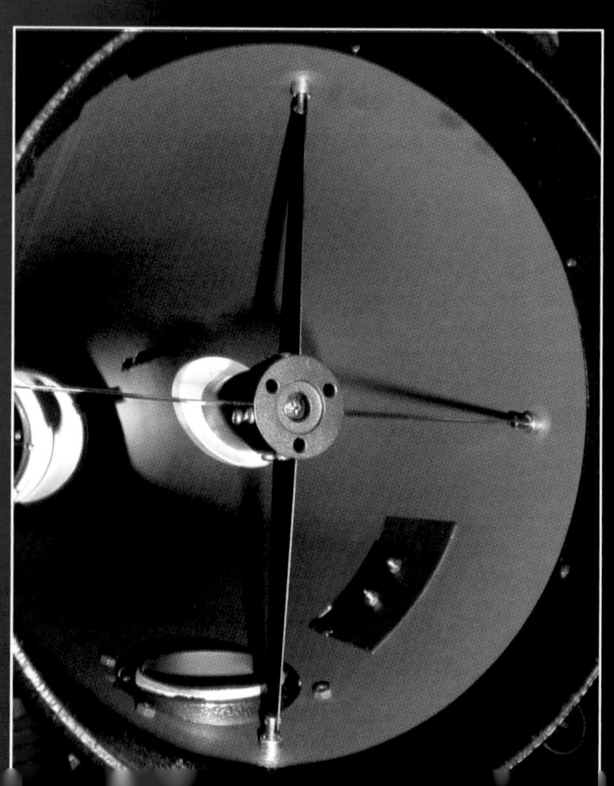

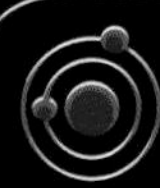

COLLIMATION

This is the process of aligning the centre of the telescope's mirrors with the centre of the focuser and is vital for sharp focusing. Invariably, especially with new scopes, mirrors will be collimated, but occasionally knocks suffered in transit or temperature changes will compromise alignment. To test for inaccurate collimation, slowly defocus a bright star, such as Polaris. If the expanding round disc is not symmetrical there is a problem. If the scope is a refractor or Maksutov, you will need to return the scope to the manufacturer: access for making adjustments is generally restricted. Collimation can be done on SCTs by adjusting the three screws on the secondary mirror cell: this alters its tilt enabling the light to fire straight down the centre of the scope. With reflectors both mirrors require adjusting and it is a trickier process. In all instances, collimating can be an intimidating and protracted procedure for anyone starting out. I would recommend utilising 'hands-on' advice from an experienced observer or referring to DIY sections on astronomy magazine websites (see page 132 and Appendix 3).

NEWTONIAN REFLECTORS: ADVANTAGES

- Generally less expensive than other telescopes per inch of aperture: a much larger aperture reflector can be had for th[e] same price as a very small aperture refractor.
- Free from false colour: light is not refracted (bent) through a lens. Recent models have a high-quality parabolic primary mirror to eliminate spherical aberration.
- Since one end of the tube is open, the inside air temperatu[re] takes less time to equalise with the outside temperature.
- Generally more comfortable for viewing since the focuser is at the top of the tube.

NEWTONIAN REFLECTORS: DISADVANTAGES

- Troubled by secondary mirrors that can block incoming ligh[t] and lower contrast: mirrors held in place by a 'spider' desig[n] that causes small rays – diffraction spikes – to radiate from brighter stars. Newer models now incorporate ultra-thin secondary mirror supports to counteract diffraction spikes and light loss.
- Troubled by a longer cooling-down time as light travels insi[de] the tube more than once. The bigger the telescope, the thic[ker] the primary mirror, and thus the longer it takes to cool.
- Require collimation (see boxout): the mirrors and focuser ha[ve] to be periodically realigned with each other to avoid distorti[on].
- Difficult to aim since the focuser is at a 90° angle to the tube.
- Disadvantaged by exposed and delicate mirrors: although often multi-coated with silicon dioxide as standard, they re[quire] occasional cleaning and possible eventual replacement.

The Meade LX850 14" on a German Equatorial Mount. Credit: Meade Instruments Corp

CATADIOPTRICS: ADVANTAGES

- Highly portable since the focal length is contained in a short tube: an eight-inch (200mm) model has a tube only one-third as long as an eight-inch Newtonian reflector. Even large-aperture ten-inchers are very manageable.
- Solidly constructed, with interior mirrors protected by the closed tube.
- Recent models offer excellent advanced coma-free optics.

CATADIOPTRICS: DISADVANTAGES

- Hampered by lengthy cooling-down periods since the tube is closed. As light bounces up and down the tube several times, any air currents intensify disturbance and increase image distortion: cooling-down is more essential than in refractors or reflectors, where there is less light travel.
- Troubled by secondary mirrors that can cause a central obstruction: this affects contrast, but the lack of 'spider' vanes means there are no diffraction spikes when viewing bright stars.
- Not always well suited to wide-field viewing, although their 'slow' focal ratios (mostly around f/13 or higher) yield higher magnification.

POLAR-ALIGNMENT

Stars appear to rise in the east and set in the west as a result of Earth's 24-hour rotation around its axis. The axis is centred on the north celestial pole located 0.9° from Polaris, the North Star. The south celestial pole lies one degree from the star Sigma (σ) Octantis. The result of this spin is that objects very quickly drift from view. The higher the magnification, the faster the drift: objects can slide away in less than a minute. Since drifting objects move diagonally – the angle being dependent on the target and observing location – the vertical and horizontal action of the alt-az mount in particular requires constant adjustments in order to keep track. To keep track, all telescopes must first be aligned with the north, or south, celestial pole – polar-alignment.

QUICK POLAR-ALIGNMENT

To polar-align fork-mounted telescopes with a wedge, or a German equatorial mount (GEM), for general observing, or for straightforward wide-angle astrophotography, a quick alignment to within one or two degrees of the celestial pole is fine:

GERMAN EQUATORIAL MOUNTS

Simply make sure the polar axis of the mount is set at an angle equal to your latitude on Earth and then aim it at Polaris, or for southern hemisphere observers Sigma (σ) Octantis.

FORK MOUNTS WITH WEDGE

Look up one of the fork tines and adjust the tripod legs until the scope is approximately aligned with Polaris or Sigma (σ) Octantis.

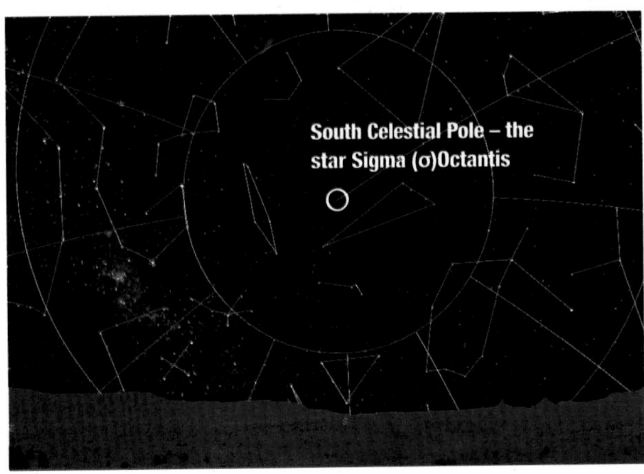

■ ACCURATE POLAR-ALIGNMENT

For advanced applications, especially astrophotography, the telescope's polar axis needs to be within five arcminutes of the true celestial pole. This is 0.9° from Polaris, in the direction of Alkaid – the end star of the Plough's (Ursa Major) handle. For those in the southern hemisphere, it is 1° from faint +5.47 Sigma (σ) Octantis (see previous page). Alternatively, another way to locate the true south celestial pole is to draw a line through the stars Gamma (γ) and Alpha (α) Crucis and extend it for 27°.

FORK-MOUNTED WEDGE AND GERMAN EQUATORIAL MOUNTS

The polar axis on a fork mount is the one around which the forks revolve. The other is the declination axis whereby the OTA swings up and down between the arms. On German equatorial mounts the declination axis has the scope at one end and the counterweights at the other: the polar axis, with the declination axis attached, must be aimed at the pole.

For a more accurate polar-alignment for both these mounts, follow the guidelines below. With practice, it takes just a few minutes:

1 Level the tripod and adjust the wedge's altitude setting for your location, ie 50° latitude.
2 Site the telescope with the forks aimed northward.
3 Carefully unlock the Dec lock and swing the OTA to indicate 90° on the declination circle on the side. The OTA should be parallel with the forks. Relock. (Alternatively, use the hand controller to point the scope to 90° Dec.)
4 Swivel the telescope (move the tripod or make adjustment on the wedge) left to right to centre Polaris in the finderscope. Do *not* adjust the right ascension or declination at this point.
5 Move the scope up and down to centre Polaris in the finderscope. Again, raise or lower the tripod or adjust the wedge but do *not* adjust the declination axis.
6 To aim at the true north celestial pole, move the scope so that the finderscope cross hairs are 0.9° from Polaris following a line in the direction of Alkaid, the last star of the Plough's (Ursa Major) handle. In a finderscope with a 6° FOV, when the true celestial pole is centred in the cross hairs, Polaris, or Sigma (σ) Octantis, is around one-third the distance from the centre to the edge of the field.

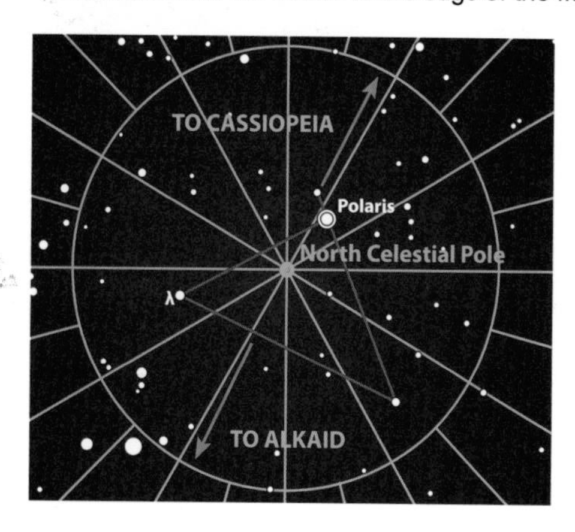

The precise location of the true North Celestial Pole.

POLAR-ALIGNMENT SIGHTING SCOPES

There are also equatorial mounts that are even more advanced, in the form of polar-alignment sighting scopes that offer the more accurate alignment essential for astrophotography. These are small, wide-field telescopes mounted along the RA axis with reticles (rather like stencils) that give a naked-eye view and indicate how far to offset from Polaris to centre on the true celestial pole. There are points for 'guide' stars or a cross hair linking a line from the Plough through the true pole and on to Epsilon (ε) Cassiopeiae in Cassiopeia. It's simply a matter of rotating the polar axis so that this line, or a guide star, matches the real sky.

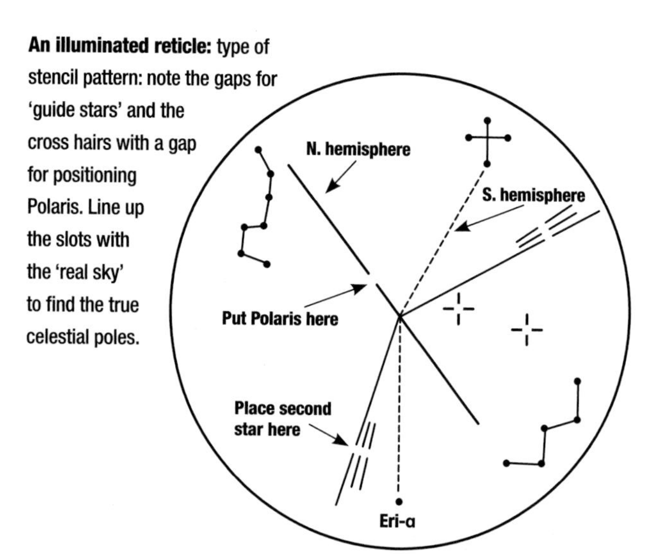

An illuminated reticle: type of stencil pattern: note the gaps for 'guide stars' and the cross hairs with a gap for positioning Polaris. Line up the slots with the 'real sky' to find the true celestial poles.

■ VERY ACCURATE POLAR-ALIGNMENT

This section is really for the advanced astrophotographer, where shooting has to keep within just a few arcseconds of a target for longer periods of time. For casual observing, feel free to ignore!

ONE-STAR POLAR-ALIGNMENT

1 Polar-align the telescope following the steps in *Accurate polar-alignment* left.
2 Select a bright star on the celestial equator for which you have up-to-date known RA coordinates, and centre this in the eyepiece.
3 Keeping the star centred, turn the RA setting circle so that it agrees with the known RA coordinates.
4 Turn the telescope back: move the RA setting circle first, then the Dec circle (do not exceed the first 89° mark) until they indicate the current RA and Dec coordinates for Polaris, the Pole Star.
5 Relock both RA and Dec locks.
6 By releasing the scope's fine adjustment levers, fine-tune the mount's altitude (latitude) and azimuth (longitude) settings to carefully manoeuvre the mount until Polaris is centred in the eyepiece. Do not move the RA or Dec settings.
7 With Polaris centred, unlock the scope and return it to the initial 'bright star'.

8 If necessary, readjust the RA circle again and repeat the whole procedure until the positions are spot-on precise. (Subsequent procedures should require fewer adjustments.)
9 If the scope is not permanently sited, make sure you record the RA and Dec coordinates ready for the next precise polar-alignment and note where you placed the instrument.

TWO-STAR POLAR-ALIGNMENT

This method takes longer than the one-star alignment but gives greater accuracy.

1 Polar-align the telescope following the steps in *Accurate polar-alignment,* page 122.
2 Point the telescope at a bright star due south on the celestial equator. Use a reticle eyepiece if you have one: align the cross hairs so that they are parallel to the lines of RA and Dec movement.
3 Make sure the motor drive is working.
4 Make sure you know which direction is north in your eyepiece and then observe the star for a few minutes. Ignore any drift east or west but look out for the star drifting north or south. If it drifts northward, the polar axis is pointed too far west, ie it is looking to the left of the true celestial pole. If the star drifts southward, the polar axis is pointed too far east, ie right of the pole.
5 Turn the mount in azimuth (longitudinally) in the necessary direction and watch the star again: there should be no drift even after observing for some 20–25 minutes. If happy, move on to step 6.
6 Aim the scope at another bright star, rising in the east this time. Again, observe for a few minutes, ignoring any RA drift. If the star drifts northward, the polar axis is aimed too high, ie it is aiming above the pole. If it drifts southward, it is aimed too low, ie it is aiming below the pole. Make the necessary altitude adjustment to the polar axis, return to the star and observe again: look for any further drifting and repeat the steps to finally eliminate it.

FULL 'GOTO' COMPUTER-GUIDED MOUNTS

These offer tracking at the touch of a button without the need for initial polar-alignment and are invariably used with advanced applications or astrophotography. Some scopes may, however, require an observer to initially enter the time and viewing location; but in other models, after tightening the RA and Dec locks the observer simply has to power up. The telescope then automatically:

- detects its base is level, using three geometric compass points to define a level 'plane' – like ensuring a table is level and solid, three legs are required. Once the three points are calculated the necessary compensation is made.
- locates magnetic north and calculates true north. Once done, it will go to two (or three) alignment stars in turn to orientate itself with the sky. Each time, the observer has only to press the arrow keys on the hand controller to centre the brightest star in the eyepiece and confirm by pressing another key. 'Alignment Successful' appears in the red liquid crystal display (LCD).
- attempts a 'GPS fix' – the inbuilt Global Positioning System receiver on the OTA attempts to acquire and synch up with signals from the GPS satellite system.

Once the telescope knows its limiting positions, its level, the location of true north, the observer's site location, and the date and time, it can automatically move at the touch of a button to any of over 100,000 astronomical objects in its database, including artificial satellites, the Hubble Space Telescope, the International Space Station and more.

Procedures vary from one model to another: some models require that the OTA is manually set in the alt-az 'home' position first.

1 Release the Dec lock, set the OTA to 0° on the Dec setting circle and relock to a firm feel.
2 Level the tripod base.
3 Move the base of the scope so that the computer control panel faces south.
4 Unlock the RA lock, move the OTA until it points north, and relock.

After these initial steps, the observer can manually select one or two stars for the one- or two-star alignment procedure (outlined earlier). Always follow the manual. Once a target is acquired, quiet motors pulse to keep it centred in the eyepiece all night long, or until another object is requested. At the end of a viewing session, simply instruct the scope to 'park' before powering off: when powering up next time, it will know its position and no-fuss viewing can resume.

Some 'GoTo' scopes on German equatorial mounts may also require a quick polar-alignment procedure first, but software is provided to assist.

Handbox controllers: with wireless models also available, these offer push-button remote-control operation.

Telescope accessories

You have the basic outfit, but how can it be accessorised? Your telescope may be functional, so let's customise to help it work faster, view further and with refined resolution. Can we fine-tune it to obtain its optimum performance? In short, how does a telescope become your telescope? It does so... with a little help from some 'friends'. Welcome to the arena of astronomical accessories...

■ EYEPIECES

Together with filters, eyepieces are the most important accessories every telescope owner must consider. Typically at least two are supplied as standard but, invariably, only one – the lowest magnification – will be worth having. Just as a photographer acquires extra, high-quality lenses, so all amateur astronomers benefit from sets of top-quality eyepieces and filters. The information below may help when faced with, yet again, a bewildering array of choice.

FOCAL LENGTH

A telescope's mirror or lens collects and focuses incoming light but it is the eyepiece (also known as an 'ocular') that magnifies the image: to change the magnification, simply change the eyepiece in the focuser. Eyepieces are not sold by magnification but by their focal length, marked in millimetres on the barrel. This figure indicates the focal length of the miniature optics (anything from three to eight lens elements) inside the eyepiece and typical numbers are 25mm, 12mm, or 9mm. The power produced by any particular eyepiece is calculated by dividing the focal length of the telescope by the focal length of the eyepiece: a telescope with a 2,000 focal length and a 20mm eyepiece yields 100 power. The shorter the focal length of the eyepiece, the higher the power produced.

A selection of low, medium and high magnification telescope eyepieces. Two Barlow lenses (back row, far right) increase the focal length of a scope and magnify the eyepiece power. A focal reducer (middle row, far right) is a lens placed near the focus of a scope to reduce focal length by a fixed amount. Note the illuminated reticle eyepiece (front row, second from left) for through-the-telescope monitoring.

THREE TYPICAL EYEPIECES ARE:

- 35x to 50x (low power): gives wide fields for finding targets and offers panoramic views of star fields.
- 80x to 120x (medium power): resolves clusters and double stars.
- 50x to 180x (high power): yields detail on planets.

Magnification greater than 180x enlarges images but creates blurriness. With the odd exception, avoid zoom eyepieces: they may combine a range of powers but they produce poor images.

Eyepieces: a selection of the Televue DELOS Series of eyepieces.

Inserting an eyepiece: carefully release setscrew, slide eyepiece into the diagonal (or adapter/focuser) and tighten setscrew to firm feel only.

APPARENT FIELD OF VIEW (AFOV)

How much is seen through an eyepiece is dependent on its magnification and its apparent field of view (AFOV), the latter also being dependent on its optical design. If you hold an eyepiece up to the light and glance through it, you will see a circle of light: the apparent diameter of this circle is the apparent field of view, measured in degrees. Standard ones have AFOVs of 45° to 55°. This does not mean that an AFOV of 50° reveals an area of sky some 50° across. In fact, a 50°, 25mm eyepiece will show less than 1° of sky. Using a large AFOV eyepiece is like looking through a ship's porthole, then squinting at the same target through the cabin keyhole! Wide-angle eyepieces have 60° to 70° AFOVs and premium wide fields offer 82° to 84° fields: these are generally preferred for deep-sky observing since they show a larger area of sky, although they can distort stars into lines or appear V-shaped around their edges. They are not suitable for planetary viewing where high contrast and sharp images are essential.

TRUE OR ACTUAL FIELD OF VIEW (FOV)

True or actual field of view is the expanse of sky visible through the eyepiece expressed in angular degrees; so if the Moon, appearing to measure 0.5° across, fills the eyepiece, then the eyepiece has a true FOV of 0.5° (30 arcminutes). This field will vary, dependent on the focal length of the telescope used.

BARREL DIAMETERS

As well as focal length, eyepieces come in three standard sizes based on the physical diameter of the tube that slips into the focuser, or diagonal: 0.965in, 1.25in, and 2in. Avoid low-cost starter telescopes with 0.965 focusers since the available eyepieces for this size are few and generally inferior. The 1.25in eyepieces are most common: they offer better quality and are available in a wide selection of models. Some eyepieces have even larger barrels – the 2in models. These are used on premium telescopes for lower magnification, wide-angle viewing. When purchasing, consider a 2in focuser: there will be an inexpensive, screw-in adaptor for a 1.25in eyepiece included in the package.

Three types of barrel diameter: the 2-inch (left) and 1.25-inch (middle) are preferred.

FIELD OF VIEW COMPARISON

NARROW APPARENT FIELD OF VIEW
A view through an eyepiece with a narrow apparent field of view.

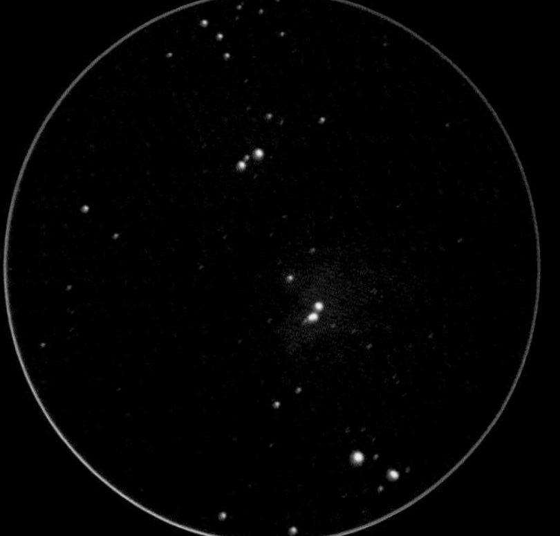

WIDER APPARENT FIELD OF VIEW
A view through an eyepiece of similar focal length but wider apparent field of view. A greater area is revealed.

SHORTER FOCAL LENGTH
The same FOV as image on left but a shorter focal length gives greater magnification. The true FOV is the same as the far left image but at greater magnification.

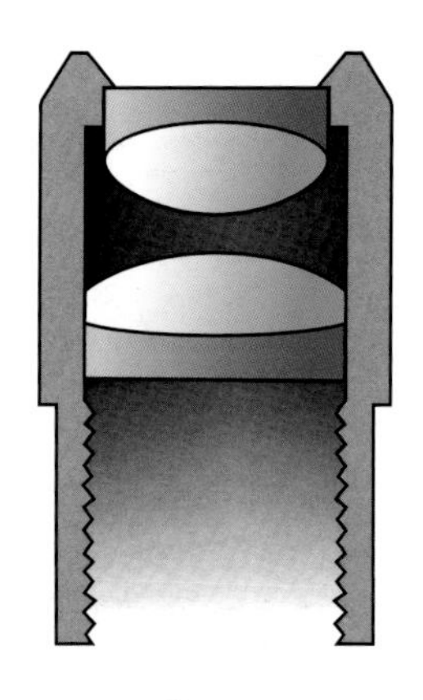

Plössl lens
4 lens elements

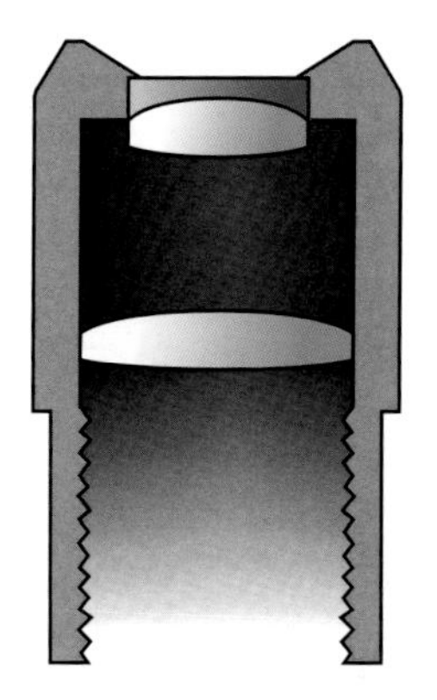

Kellner lens
3 lens elements

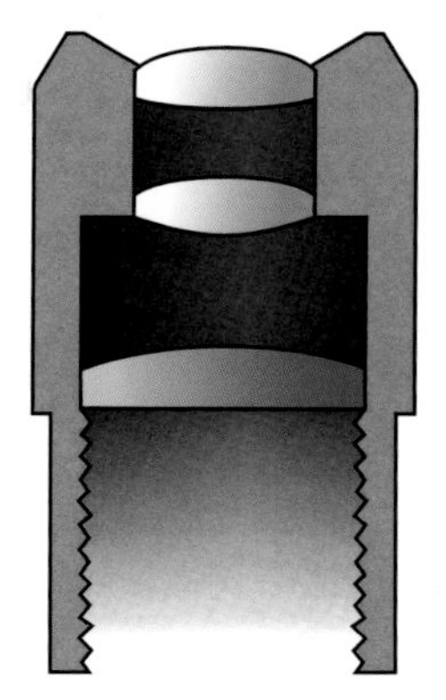

Orthoscopic lens
3 lens elements

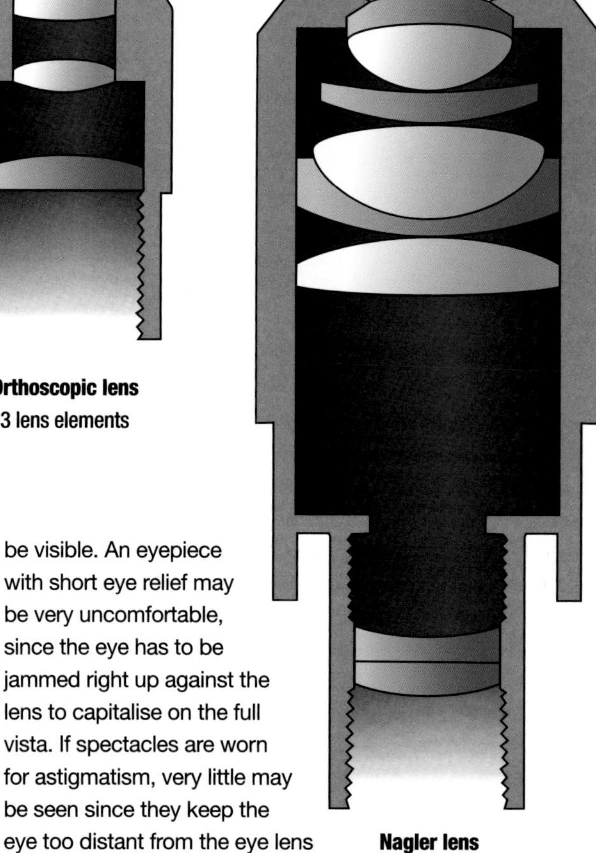

Nagler lens
8 lens elements

DESIGNS

Most telescopes are sold with one or two general-purpose eyepieces, but greater use will lead to a need to upgrade, and there is a phenomenal choice of eyepieces out there. Quality and price varies:

- Kellner, Edmund Scientific's RKE, Modified Achromats eyepieces (AFOV 35° to 45°): these use three lens elements to produce a good image at low cost, especially for long-focal-length scopes.
- Orthoscopic (AFOV 45°): these offer good eye relief with few ghost images and aberrations, although their narrow field is not ideal for deep-sky viewing.
- Plössl (AFOV 50°): these, like Orthoscopic, are sometimes known as 'symmetrical' eyepieces. Both types come in sets and use four lens elements. Although Plössl offer a wider field than Orthoscopic, both yield superb sharpness and contrast and are the best buy in the current marketplace.
- Wide Fields (AFOV 65° to 70°): these offer wider fields of view with minimal edge aberrations and are available at a moderate price: Tele Vue Panoptics (derived from 'panorama optic'), Vixen LVW, Meade Super Wide Angle, Lanthanam Superwides, William Optics SWAN, and Pentax XW are a few examples.
- Tele Vue Radians (AFOV 60°): again, these offer wide fields with long eye relief and good contrast, but at a cost!
- Tele Vue Ethos, Nagler T4, T5, Meade Ultra Wide Angle (AFOV 82° to 84°): although somewhat heavier, these are the premium-priced eyepieces, containing five to eight lens elements. Available in many focal lengths, they truly do reveal a much wider 'picture window' view of the universe. The Ethos in particular delivers a stunning 100° AFOV with pinpoint images right to the edge.

EYE RELIEF

As with binoculars, eye relief is important when using eyepieces. Every eyepiece has an eye lens at one end of the barrel and a field lens at the other. Eye relief is simply a measure of how close an observer's eye has to be to the eyepiece's eye lens in order for the entire field of view to be visible. An eyepiece with short eye relief may be very uncomfortable, since the eye has to be jammed right up against the lens to capitalise on the full vista. If spectacles are worn for astigmatism, very little may be seen since they keep the eye too distant from the eye lens and only a tiny central portion of the field becomes visible.

BARLOW LENSES

Eyepieces with a short focal length (4 to 8mm) give high power. However, the higher the power, the shorter the eye relief, and the shorter the eye relief the closer an observer's eye has to be to the eyepiece, which makes for uncomfortable viewing. Enter the Barlow lens. These insert between the eyepiece and the telescope, or between the eyepiece and diagonal. They effectively increase the focal length of the scope and multiply the power of any eyepiece. When used with comfortable longer-focal-length eyepieces, which already offer better eye relief, they give more power. For 50% increased amplification they can even insert between the diagonal and focuser on some refractors so that a 3x becomes a 4.5x. Available in 1.8x to 5x magnifications, the most common is the 2x Barlow, which transforms a 20mm eyepiece into a more powerful 10mm one. They come in various tube lengths, ranging from less than 3in to the generally greater amplifying 6in.

Lens coating: evidenced by the bluish tint.

COATINGS

All modern eyepieces are coated to enhance light transmission, and reduce flare and ghost images. Minimal coating is a layer of magnesium fluoride on both exterior lens surfaces that gives them a bluish tint. A single-layer coating on all surfaces is termed 'fully coated'. In premium eyepieces, all lens surfaces are multi-coated and called 'fully multi-coated'.

EYEPIECE CASES

Many manufacturers sell special cases to protect eyepieces from bumps, dust, and dew, but any foam-clad protective box will do. The hard-sided, briefcase type containers normally sold for photographic equipment are ideal.

Eyepiece and filter protection: a hard-sided, foam-lined case is ideal.

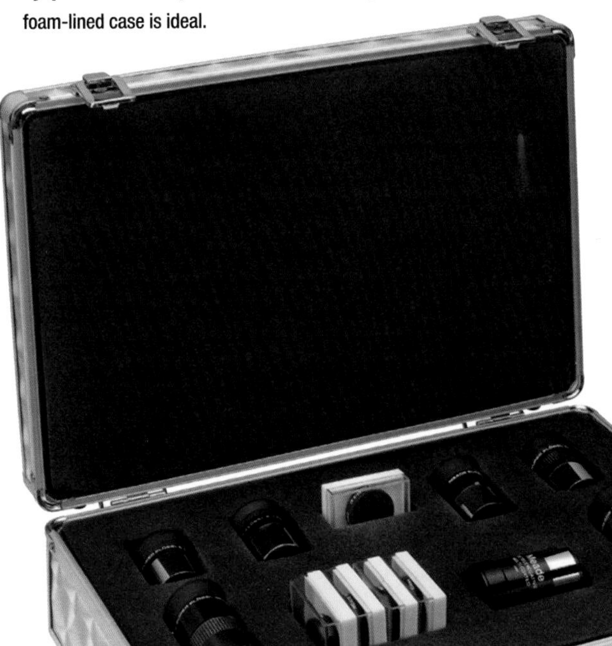

CLEANING

EYEPIECE AND LENSES

The first rule with any optics, whether eyepieces or lenses, is to keep them covered when not in use. Given time, however, they undoubtedly become coated with dust, fingerprints, or eyelash oil. To remove, blow away the loose dust with a lens blower or a can of compressed air. Any loose specks can then be lifted with light strokes of a camel-hair brush. When it comes to cleaning, never use cleaning solutions or cloths sold for spectacles. Instead, for small surface areas like an eyepiece, moisten a Q-tip cotton swab with camera-lens cleaning fluid and, after gently cleaning, follow up with a dry one to remove streaks. For larger surface areas, use a lens cleaning tissue or a micro-fibre lens cleaning cloth.

The recommended cleaning fluid formula advised in telescope manuals is a 50:50 radio of pure isopropyl alcohol and distilled water and one drop of biodegradable unscented liquid dish-washing (not dishwasher) soap to make a pint of solution. This is safe on any anti-reflection coating and dries clean.

Never disassemble any eyepiece and never pour fluid directly on to the lens: it seeps inside, causes staining, and feeds fungal growth. With some refractor lenses you can remove the front eyepiece lens assembly to get at the rear lens, but never disassemble or remove double or triple lenses. With Schmidt-Cassegrains, the usual advice for cleaning lenses is *don't* – it can do more harm than good. A few specks of dust on the corrector plate are innocuous and certainly less harmful than scratches caused by over-zealous or inappropriate cleaning.

MIRRORS

Newtonian aluminised mirrors really only require the odd blast of canned air or minimal strokes of a camel-hair brush to keep them clean, and this is done by carefully removing them from the end of the telescope tube and unclipping them from the cell. Always refer to the instruction manual, or specialised Internet help sites, for guidance on disassembling. Washing is only necessary if they become grimy or thick with dust. This involves initially running the mirror under cold water to remove any dirt, then placing it carefully on a towel and submerging it in a sink beneath half an inch of warm water with a few drops of gentle liquid soap added. Use cotton balls to gently dab it clean, using a linear direction – do not rub or use circular motions. Repeat if necessary in the other direction. Drain the sink and rinse with cool water. To avoid water stains, finish off with a distilled water rinse and allow the mirror to dry naturally.

FILTERS

If a good set of eyepieces is the making of a telescope, then a good set of optical glass filters is the making of an eyepiece (although this is not always necessary for those just starting out). Cheap plastic filters abound and they are nearly always useless. All eyepieces, albeit with different construction, have a common function: to enhance an image by, paradoxically, reducing (filtering) the amount of light entering the eye. Available for 0.965in, 1.25in, and 2in diameters, they are mounted in cells with a standard filter thread: these then screw into the base of an eyepiece. They come in three types:

1. Solar.
2. Lunar and planetary.
3. Deep-sky or nebula.

SOLAR VIEWING

WARNING: NEVER LOOK AT THE SUN WITH THE NAKED EYE. NEVER LOOK AT THE SUN WITH ANY OPTICAL DEVICE UNLESS YOU ARE CERTAIN IT IS SAFELY FILTERED.

Even a fraction of a second of unfiltered light can cause partial or permanent blindness. Do not take the risk.

So what does 'safely filtered' mean? It means you do not gaze through the bottom of glass beer bottles! It means you do not use copious layers of black-and-white or coloured photographic film, or any form of polarising or neutral-density filters. It means you do not use sunglasses, no matter what design or 'UV protection'. And you do not use smoked glass, no matter what tint. Be especially aware of cheap solar 'filters' provided free with 'bargain basement/Christmas gift' telescopes. These invariably comprise dark green glass: when the scope is aimed at the Sun, the glass heats to hundreds of degrees, cracks, and sunlight scorches in.

Astronomical and photographic outlets do sell 'solar eclipse viewers'. These are safe 'sunglasses' with an optical density (OD) – a measure of the light being transmitted by the filter – greater than 5.2. An OD of 5.2 means light intensity is less than 1/100,000th of the light from the Sun. The filter absorbs harmful UV radiation and reduces infrared radiation to less than 1/2,000th of the level that could damage your eyes. But they are not to be used with binoculars, telescopes, or cameras.

Another safe, and easily accessible, filter is a No 14 welders' filter, especially designed for the welding industry and, therefore, of the correct density. These filters can be safely fixed to binoculars.

For more detailed viewing using an unfiltered small telescope, the safe 'solar projection' method outlined right is necessary.

SAFE SOLAR VIEWING – THE UNFILTERED 'SOLAR PROJECTION' METHOD

To view the Sun safely with an unfiltered amateur telescope take the following steps:

1. Firstly, never be tempted to peer at the Sun at any stage: observe a projected image only.
2. Use a telescope with not more than an 80mm aperture to avoid damaging the eyepiece. Never use a Newtonian reflector: heat reflected from the primary mirror will shatter the secondary. Plastic parts will also melt. Small refractors work well. With larger refractors use cardboard with a hole in it placed over the aperture. *Never* use a SCT or Maksutov scope for solar projection – the heat build-up is too intense and potentially damaging to the instrument.
3. Cover the finderscope.
4. Observe the shadow of your telescope on the ground and move the scope until the shadow is minimal. This means the Sun is now shining directly down through the OTA and out through the eyepiece.
5. Use a 30x low-cost eyepiece – heat can damage coatings.
6. Project the image on to a piece of paper placed a foot from the eyepiece.
7. Protect the projected image from direct residual sunlight by shielding it with another sheet of card or paper around the focuser/eyepiece: this accentuates the image.
8. Focus the eyepiece to sharpen the image.

To prevent heat build-up when not observing, remember to cover the telescope aperture. This is a good method for general solar observing, especially for sketching sunspots.

SAFE SOLAR VIEWING – USING SOLAR FILTERS

The alternative, and safe, way to directly view the Sun through a telescope is with a specialised glass solar filter (also known as a full-aperture or pre-filter). This entirely covers the front aperture of the scope and reduces sunlight intensity before it enters the tube. Metal-coated (nickel-chromium alloy) glass filters reveal a natural-looking Sun but cost substantially more. However, a less expensive way is either to make, or purchase, metal-coated thin plastic Mylar film filters. These are available in snug-fit sizes for most telescopes. A word of warning, though: not just any old Mylar will do. Some Mylar does not block harmful ultraviolet or infrared light, so make sure you purchase filters, or filter material, only from astronomical manufacturers. Mylar produces a blue-tinted image of the Sun but a yellowish hue can be acquired if used with a No 23A eyepiece filter (see below). Both glass and Mylar filters reduce the amount of light across the entire spectrum and yield 'white light' views of sunspots and brighter regions near the limb – faculae.

To eliminate all light from the Sun, except for the single wavelength of light emitted by Hydrogen atoms, a special Hydrogen-alpha filter is needed. Whereas spectacular solar prominences are normally only seen during a total solar eclipse, these filters reveal them on any cloudless day … for a price! For the more dedicated solar observers, there are also dedicated H-alpha solar telescopes.

LUNAR AND PLANETARY FILTERS

These are inexpensive filters, available in many colours, that elicit subtle differences. Lunar filters reduce glare and colour aberration and can ease eyestrain when studying the Moon. Planetary filters – when used on the planet Mars, for example – really enhance an image: they can increase contrast, accentuate surface markings, and even highlight the polar caps. Other filters penetrate planetary atmospheres to reveal gas belts, storms, such as the Great Red Spot on Jupiter, or cloud bands on Saturn.

Moon viewed through a yellow filter: craters and maria have greater definition.

Filters are labelled with the same Kodak Wratten numbers used in photography. Of the wide range available, the most useful are outlined in the following table:

No	Colour	Object
8	Light yellow	Moon
11	Yellow-green	Moon
12/15	Deep yellow	Moon
21	Orange	Mars
23A	Light red	Venus, Mars
56	Light green	Jupiter, Saturn
58	Green	Mars
80A	Light blue	Mars, Jupiter

Also useful for planetary observing when using an achromatic refractor is a minus-violet filter: this suppresses chromatic aberration (false colour), especially the purple halo symptomatic of unfocused light often seen around the planets Venus, Mars, and Jupiter.

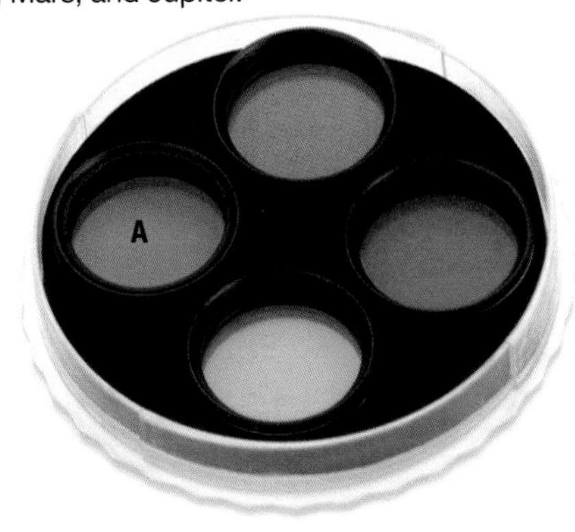

Filter set: which includes a 15, 21, 80A and polarizing filter (A); useful for daylight observation of the first or last quarter Moon since it darkens the background sky and heightens contrast.

Eyepiece filters: used to enhance telescopic viewing.

DEEP-SKY OR NEBULA FILTERS

For anyone observing the deep sky, whether under dark or light-polluted skies, high-tech silvery nebula filters soon become essential, especially for targets such as planetary nebulae, emission nebulae, and supernova remnants (see *Interstellar medium and nebulae*, page 70). Unlike stars that emit light across a broad spectrum of colours, nebulae emit light at specific wavelengths, mostly red and green from hydrogen and oxygen atoms. Light pollution (mercury and sodium), for example, emits yellow and green light in another area of the spectrum. Filters are used to isolate light emitted from targets: they selectively block the unwanted light and thereby enhance the object's image. Here we move into the realms of Broadband filters, Narrowband filters, Oxygen-III filters, and H-beta filters:

- Broadband filters are intended for mild light-pollution reduction (LPR). They transmit the widest range of colours across a broader spectrum and are used for viewing all types of deep-sky objects.
- Narrowband (nebula) filters block unwanted light more effectively by narrowing down the light transmission, and are mostly used on emission nebulae.
- Oxygen-III and H-beta ultra-narrow filters, also known as line filters, are far more discriminating: they allow the transmission of single colours from certain objects at a specific wavelength. The result is high contrast and maximum eradication of light pollution: these are great for supernovae and planetary nebulae.

Broadband filters: for mild light-pollution reduction.

Oxygen filters: for maximum eradiction of light pollution.

Do not use nebula filters for observing star clusters, galaxies, or reflection nebulae: they will make these targets appear more faint. For a more rewarding read about wavelengths and the electromagnetic spectrum, see Chapter 4.

■ OTHER ACCESSORIES

Like many hobbies, a quick thumb through the equipment advertisement pages in magazines or a search across the Internet reveals that accessories abound, whether for naked-eye, binocular, or telescope viewing. Below are some items that are useful but not essential:

FINDERSCOPES (VIEWFINDERS)

As we saw earlier, finderscopes (also known as viewfinders) are small, but important, wide-field secondary telescopes mounted on the main telescope, the prime purpose of which is to aid in the initial pointing of the telescope to its target before viewing through the telescope's eyepiece: it is easier to locate an object in a wider view than become lost in the magnified but much narrower view of the scope's eyepiece. Some even have a special reticle (see earlier) indicating true celestial north in relation to Polaris. Clearly, finderscopes are supplied as standard. However, some are better than others: the better a finderscope, the easier it is to find an object and, in this instance, size does matter so bigger does mean better! Most entry-level telescopes are supplied with inadequate 5x24 finderscopes (5x magnification and 24mm aperture) but others have better 6x30s. For eight-inch, or larger, telescopes, 7x to 9x are essential since they show lower magnitude stars whilst maintaining a wide FOV. All have cross hairs for centring the object and usually yield a correct image to facilitate the use of star charts. Before use, they must be aligned with the main scope (see *Aligning a finderscope (viewfinder) and Smartfinder*, page 113).

Finderscope.

To further enhance navigation, finderscopes known as reflex, or unit-power, finders (red-dot finders), can be used. These are small battery-operated units that easily attach to the main telescope with screws or double-sided tape. Once affixed and aligned with the main scope's eyepiece, they can be used instead of, or as well as, the finderscope to pinpoint targets: when looking through them, they reveal an LED-illuminated gunsight-style red dot, or bullseye, projected on to glass. As an observer moves around, this dot appears fixed in the background naked-eye sky.

Red-dot finder.

Telrad finder.

DIAGONALS

Since focusers on telescopes can end up in awkward, uncomfortable positions for viewing, and often observers of differing heights are using the same model, it is handy to have a diagonal. Available in 1.25in and 2in diameters, these slip between the focuser and eyepiece and can easily be turned to a more accommodating position. They use a mirror, or prism, to change the angle of view by 45° or 90°. All telescope designs produce upside-down images but these diagonals can switch the image the right way up. Furthermore, some even invert it left to right so that a mirror image is seen. Serious observers sometimes use 1.25in or 2in premium prism diagonals: these have dielectric coatings that enhance reflectivity and produce extremely bright images.

Series 5000 2-inch Enhanced Diagonal.

45° Erect-Image Diagonal Prism.

DEW SHIELDS

As the air temperature drops at night so moisture condenses to form dew on lenses, evidenced by halos around bright stars. In addition to curtailing a night's viewing, the dew, since it is an offspring of industrialised acid rain, damages lens coatings and silvered mirrors, especially on refractors, Schmidt-Cassegrains, and Maksutovs. The best solution is a dew shield. These are protective devices that extend beyond the front lens or corrector plates. They can be purchased

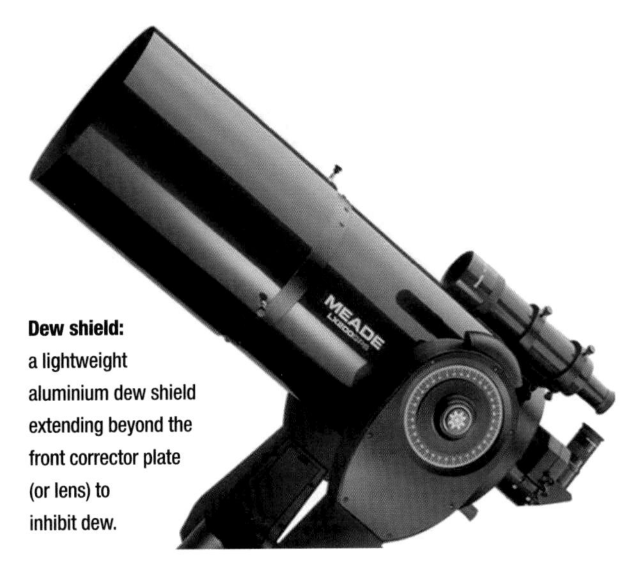

Dew shield: a lightweight aluminium dew shield extending beyond the front corrector plate (or lens) to inhibit dew.

separately or made at home and usually enable an hour extra dew reprieve. After this, as a stopgap measure, a handheld 12V low-heat hairdryer blowing warm air for a minute or two across the lens helps. Make sure you stand about 2ft away and the heat is low, otherwise it will distort images. Alternatively, a heat gun or dew removal gun – normally used to remove ice from windscreens – can be plugged into a car's cigarette lighter outlet. Both are temporary and neither is ideal.

Dew fighting is an uphill battle. Another more permanent fix are low-voltage dew heaters. These are cloth bands containing heat elements that wrap around exposed optical surfaces, including eyepieces, finders, and other accessories, and are fastened with Velcro. They connect to a DC-powered adjustable control box enabling control of the heat supply. Modern digital versions are temperature regulated: they sense the ambient temperature and make automatic heat adjustments as necessary. The bands and heaters can be purchased separately and customised. All provide sufficient heat intensity to prevent dew from forming.

Dew shield: the black interior absorbs stray light for higher contrast.

WEATHERPROOF COVERS

Waterproof or Sun-proof reflective Mylar telescope covers protect equipment when parked for long observing sessions.

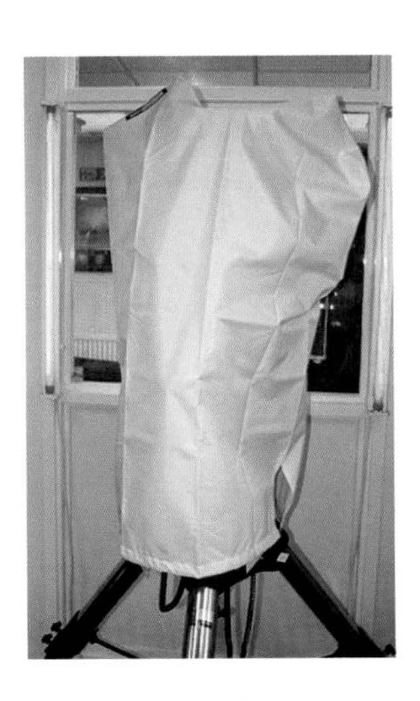

ACCESSORY TRAYS

Accessory trays are useful to keep eyepieces, filters, charts et al., clean, accessible, and secure when observing. Yes, you have pockets, but it's amazing how they fill with all sorts of things and how easily gravity takes over!

ADDED COMFORT

Adjustable chairs, stools, tables, and even ladders (for large Dobsonians!) sound superfluous, but they become essential for comfort, and less irritation, when spending hours at the eyepiece.

MANOEUVRABILITY

Consider using a wheeled trolley if you purchase a larger scope, or have heavy equipment. They take the strain out of viewing at more distant locations.

Binoviewers.

BINOCULAR VIEWERS

These insert into a telescope's eyepiece on most SCTs and top-grade apochromatic refractors and separate the light beam in two. Both beams are then collimated for each eye. When combined with a Barlow lens they sufficiently extend the focal point to permit focus – and the result is virtually 3D vision, pretty spectacular when viewing Solar System objects or the Moon!

COLLIMATION TOOLS

For optimal performance of your reflector telescope, as well as Ritchey Chretien and Schmidt-Cassegrain models, accurate alignment of the optics (or collimation as it is known) is vital, especially if the scope has a focal ratio of f/6 or less. Various collimation tools are available:

STAR TEST: You can collimate your reflector, without tools, by doing a simple 'star test' – centre a bright star in the eyepiece, make it out of focus and look to see where the shadow of the secondary mirror is within the expanded disc of light. Ideally, it should be centred.

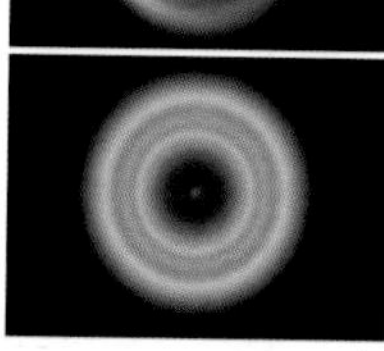

The Star Test method: an uncollimated scope (at top) and a collimated scope (bottom).

COLLIMATION CAP: this is a plastic cap, reflective underneath, and with a small central hole.

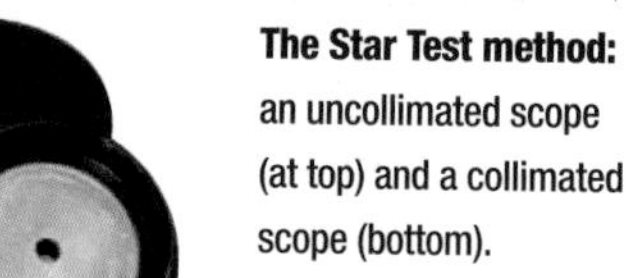

A collimation cap.

CHESHIRE EYEPIECE: this is a sight tube with a peephole at the top end for looking through. It has a polished internal surface tilted at 45° which is aimed at a larger hole in the side of the tube. Some tubes have cross-hairs at the base for aligning the secondary mirror.

A Cheshire Collimating Eyepiece for collimating Newtonians and Schmidt-Cassegrains.

LASER COLLIMATOR: these can, usefully, be used in the dark and offer accuracy when collimating. They are placed in focusers and project a laser beam down the OTA; an observer has to adjust the tilt of the mirrors until the beam returns precisely to itself. Its accuracy relies on careful adjustment of a scope's secondary mirror.

Laser collimator.

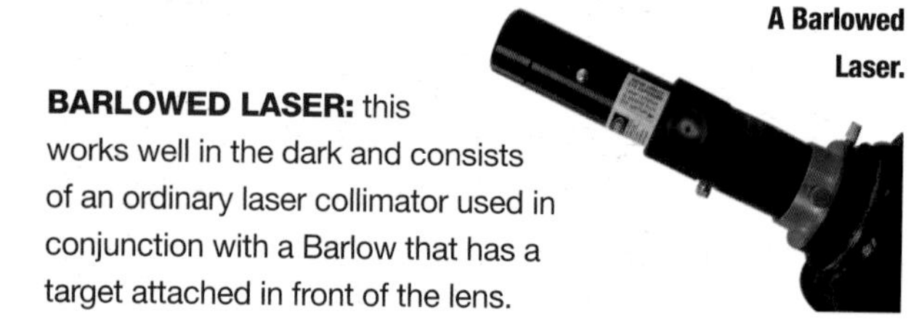

A Barlowed Laser.

BARLOWED LASER: this works well in the dark and consists of an ordinary laser collimator used in conjunction with a Barlow that has a target attached in front of the lens.

COMA CORRECTORS

These are coma-correcting lenses that insert, like Barlows, into the light path of Newtonian reflectors to eliminate aberrations caused by the parabolic primary mirror.

Coma-correcting lens.

DIGITAL SETTING CIRCLES

With the advent of falling prices for 'GoTo' scopes, the technology of digital setting circles is less popular nowadays. However, they still hold good for many types of telescope. They advise, via digital readout, your scope's exact position: like 'GoTos', you simply punch in your desired target's catalogue ID, move your scope whilst watching the display, and when the coordinates display zeros your object is acquired.

POLAR-ALIGNMENT FINDERSCOPES

These are fixed 'inside' and aimed up the polar axis on German equatorial mounts. Each has a reticle – a type of stencil – that can be used to sight 'guide stars' in slots or place cross hairs to align with Polaris and find the true north celestial pole (see *Polar-alignment sighting scopes*, page 122). By aligning the reticle with the naked-eye view a more accurate polar-alignment can be achieved. These scopes are definitely not essential.

Polar-alignment finderscope.

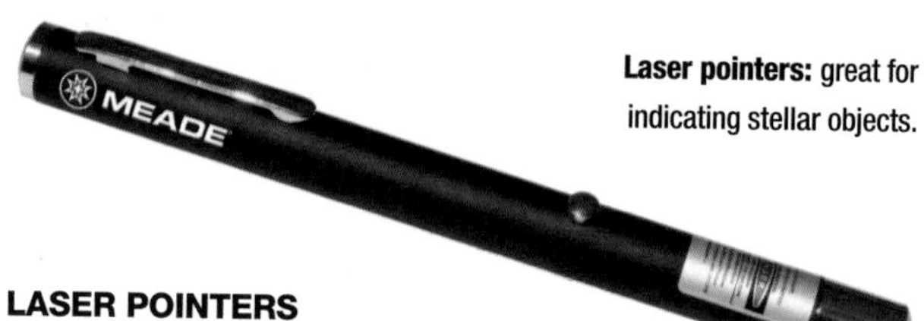

Laser pointers: great for indicating stellar objects.

LASER POINTERS

Laser beams are great for targeting constellations and certainly have a 'wow factor' with friends! They can also double-up as finders – but use discretion. Be warned: laser pointers are banned at star parties. Exercise caution too: do not fire them at people, animals, passing satellites, or aircraft.

ASTRONOMY APPLICATIONS (APPS)

These days pretty much everything can be done using an iPhone or iPad, and this includes real-time navigating, learning and enjoying the night sky. A selection of suggested Astronomy Apps available from an App Store, such as iTunes, are as follows:

- StarMap: (for iPhone and iPad) – for navigating around the night sky.
- StarWalk: (for iPhone and iPad) – aim your phone at a celestial object and this app will identify it and give a detailed explanation. It also offers a raft of other features.
- Sky Gazer: this is for beginners and gives familiarity with the most obvious stars in the sky.
- SkySafari: (for iPhone and iPad) – offers everything you need to explore the stars and other objects.
- Distant Suns 3: (for iPhone and iPad) – similar to StarWalk
- Star Chart: (for iPhone and iPad) puts a virtual star chart in your pocket.
- Moon Atlas: (for iPhone and iPad) – offers a 3D globe of the Moon with over 1800 names features and Apollo Moon Missions data.
- Moon Globe: (for iPhone and iPad) – perfect for students and those interested in our nearest neighbour.
- DeLuxe Moon: provides precise times of Moon phases, animated Zodiac circles, and more.
- Moon Phase for iPhone: offers a host of useful information about our satellite.
- VisiMoon for iPhone: a Moon calendar with lunar phases.
- GoSkyWatch Planetarium: (for iPhone and iPad) – helps to locate and identify planets.
- Solar Walk: an orrery which allows you to play with solar system objects.
- Star Search: an astronomical tour for beginners.
- Grand Tour 3D Pocket Solar System: transit the solar system in 3D.
- APOD Viewer: enables a Grand Tour of the solar system and galaxies.
- Planetarium: enables exploration of the solar system.
- Pocket Universe: offers an introduction to astronomy.
- Google Sky: an online sky/outer space viewer, offering images from NASA satellites, the Sloan Digital Sky Survey, and the Hubble Space Telescope, and more.

Also for iPad:

- Luminos: offers 3D views of planets and moons and covers 2.5 million stars and thousands of other celestial objects.
- Red Shift: covers over 100K stars and 500+ deep sky objects.
- Planets for iPad: a free app that offers planetary exploration

Other useful astronomy iOS apps for smartphones or tablets cover: space education; exoplanets; space weather trackers; meteor shower calendars; polarfinders; satellite tracking; telescope control; planetarium apps for Android or iOS. The list is exhaustive.

The Universe at your fingertips.

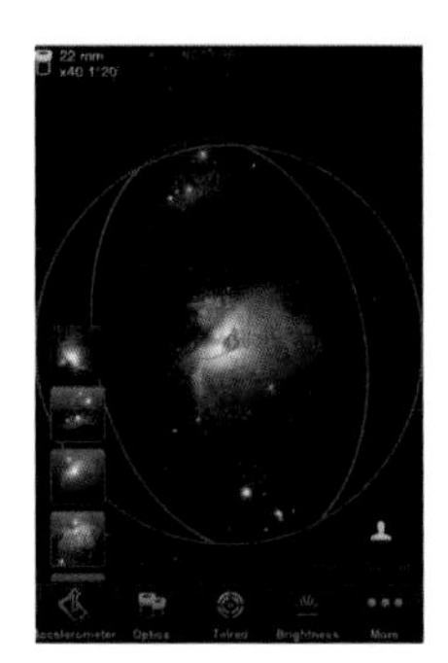

StarMap: one of many Astronomy Apps for iPhone or iPad used for navigating the night sky.

CONCLUSION

Before purchasing any telescope, or accessories, at whatever level, ensure you tick the boxes on the following simple essentials:

- Is it a metal or wood construction with minimal use of plastic, especially on moving parts?
- Is the mount sturdy and steady?
- Does the focuser slide back and forth smoothly (geared rack-and-pinion or roller-style) and does it have a 1.25in or 2in diameter, not 0.965in?
- Is there a good finderscope (viewfinder) – preferably 6x30mm, or greater for eight-inch scopes?
- Are the eyepieces interchangeable – not zoom nor fixed – and are they at least 25mm and 10mm with a 1.25in diameter barrel?
- If not purchasing a motor-driven or computerised mount, are there slow-motion controls on both axes of the mount, whether equatorial or altazimuth, to enable fine-tuned navigation?

And finally ... don't be in a hurry. Take time to shop around. Some companies sell telescopes solely through dealer networks. Others have select dealers and take orders via mail or the web. Still others sell scopes direct from the factory. Smaller manufacturers sometimes only begin construction when an order is placed and require deposits of up to half the cost of the telescope up front. Delivery time can be lengthy.

Read astronomy magazines, where advertisements abound. Scour websites and watch out for reviews, but beware of biased opinions!

Always read the small print, especially since you may decide to return your purchase: shipping, crating and re-stocking charges can be incurred, not to mention all the aggravation.

Where possible, try to purchase locally from a reputable outlet. This is often the safest and easiest option: there will be expert advice, a product can be inspected up close, and it's easier to return it if necessary.

Join a local astronomy club too. Ask questions of members. Participate in their observing nights or star parties to get a feel of the instruments in action.

High-tech astro ware

The world of astrophotography is explored in Chapter 5. Anyone starting out can, understandably, be reluctant to venture into this field but the fast-changing computerised world of astronomy is inescapable. We have already witnessed the wonders of 'GoTo' telescopes and computerised tracking. However, there is still *no* substitute – no substitute – for getting to know the night sky. Much can be said for simply stepping outside, looking up and absorbing the real thing, or rummaging around with a simple paper star chart and red flashlight to 'star hop' from one constellation to another. It's the difference between digging out a road map or flipping on the satnav: the former induces imagination, perspective, and 'big picture' knowledge, the latter makes life easier, but at the expense of orientation and the development of skills. Beware of distancing yourself!

However, as the adage goes, 'all roads take you there', and it is impossible not to connect, at some stage, with the sometimes confusing, but amazing, realms of Internet resource and night-sky software. A tiny sample is outlined below:

■ SOFTWARE

As we have seen, a telescope armed with the latest CCD imaging technology delivers not only automatic alignment at the flip of a switch, but dazzling and informative multi-media guided tours (audio, video, text, and still images) of the universe as well. This is available for anyone starting out. Other telescopes can be linked, via an interface, to laptops operating astronomy software. What is this software and how is it used? Quite simply, astronomy software packages come in two styles: sky-simulation (planetarium programs) and star-charting (planning programs):

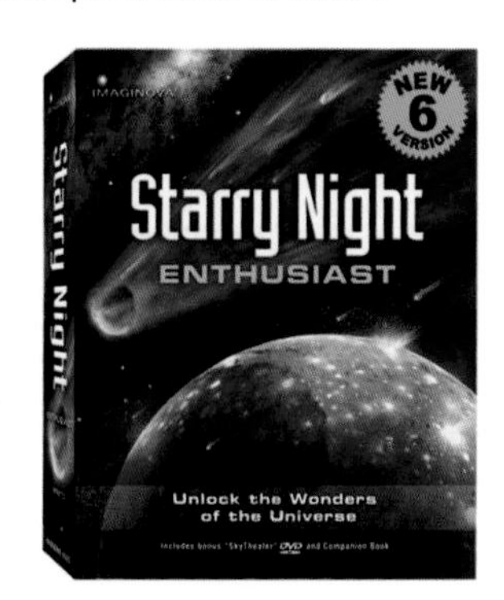

Starry Night: unlocks the Universe within the comfort of your own home.

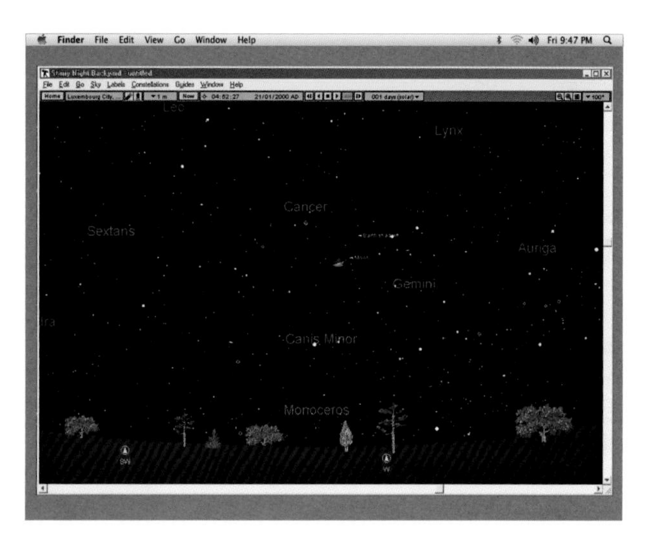

Starry Night: a simulated 3D planetarium with a large deep sky objects database, artificial satellite, flare and eclipse data, and more…

The Sky: a simulated planetarium.

SKY-SIMULATION

These realistic planetarium packages offer graphic representations of the night sky. They are electronic star maps and are ideal for most observers' needs, especially those starting out. Ranging from dots, lines, and symbols – depicting stars, constellations, and deep-sky objects (DSOs) respectively – to almost photographic representations of stars in true spectral colour, DSO images, and other informative data, they enable an observer to have the sky at their fingertips. When interfaced, the screen shows exactly where the telescope is pointed. Slew the scope with a 'GoTo' facility and an on-screen bullseye moves with it. Stop, click, and the object's data – far more in-depth than a handheld controller – is revealed. Current, most popular top-end programs are: *C2A Planetarium Software; Cartes du Ciel; Celestia; Digital Universe Atlas; Equinox6; Google Sky; HNSKY; KStars; Nightshade; Redshift; Sciss; paceEngine; Starry Night Pro; Stellarium; The SkyX; WinStars; Worldwide Telescope; XEphem.*

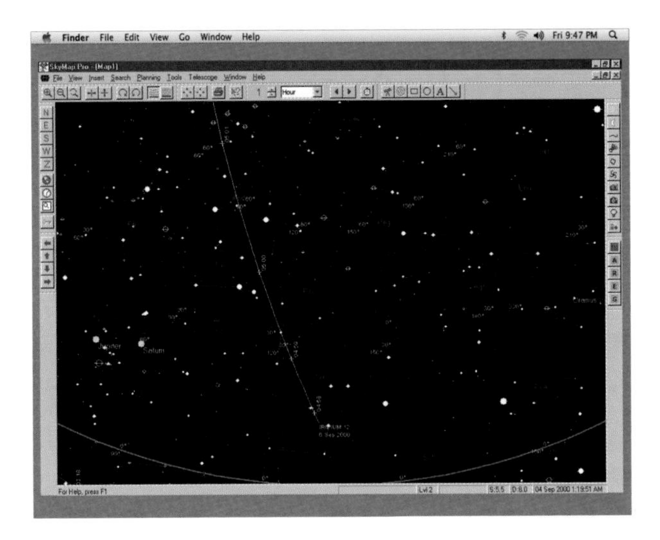

SkyMap Pro sky-charting software: an electronic star atlas.

STAR-CHARTING

These packages offer less 'prettiness' in terms of their planetarium package and are really electronic star atlases for the more advanced, serious deep-sky observer. The best are *Megastar* and *SkyMap Pro*, but there are many others.

Additionally, other programs offer a more database style with star-charting facilities available to a level. Their mainstay is their extensive list of DSOs that can be sorted and filtered for planning a night's deep-sky observing. Popular programs are *Deep Sky Planner*, *Deepsky Astronomy Software*, and *AstroPlanner*.

LUNAR ATLASES

Several stunning lunar atlas programs have also been developed to assist with lunar observing. Recommended ones are *Lunar Phase Pro* and the freeware *Virtual Moon Atlas*, the latter being an intricate, interactive map with a 'GoTo' scope-guiding facility, from Patrick Chevalley, the author of *Cartes du Ciel*. Both are downloadable but, unless using a minimal graphics version, it does take time. CDs are available as an alternative.

Lunar atlas: offering 3D virtual tours.

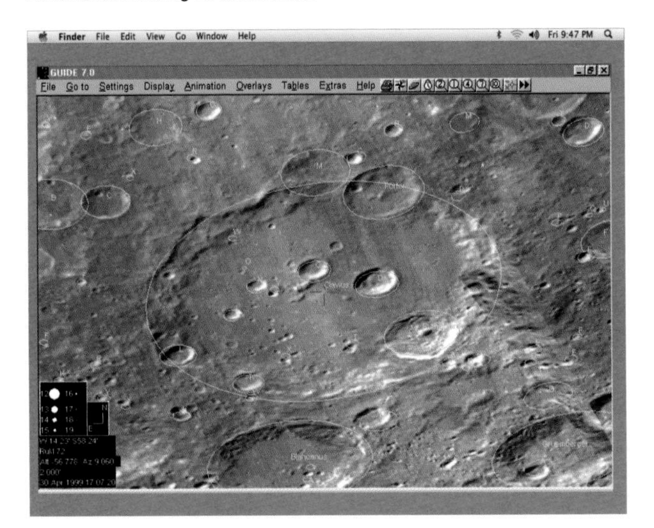

CONNECTING

Getting a software-loaded laptop physically connected to a fully aligned telescope requires a USB-to-serial adaptor: all computers use a standard USB (Universal Serial Bus) to connect to peripherals, whereas telescopes mostly have RS232 connections that only accept telephone-style RJ11 jacks or D-shaped DB9 connectors – help is needed to get them talking to each other. These adaptors, along with their driver software, are the go-between, and are available from computer or office suppliers. Before tackling the software side, make sure you also have the correct cable from the serial adapter to the telescope – normally supplied by the telescope manufacturer – and that this is correctly connected to the scope. Always read the instruction manual, since some cables connect to hand controllers instead. Once connected, take care to select the correct port and telescope model in the 'Set up' section of the astronomy software. When done, the laptop will recognise the 'new' serial port and telescope and will enable both to talk to each other.

Sounds easy? Of course, for the laptop and telescope software language to talk there has to be a translator, and the current popular one is ASCOM (Astronomy Common Object Model). This free program, downloadable from its website, drives the communication between sky-simulation or sky-charting software and many types of telescopes. It is the general interface between computers and a whole range of astronomy equipment including mounts, focusers, cameras, filter wheels, and domes. In some instances, for certain scopes, an add-on driver might be necessary but these, too, are freely downloadable.

With rapidly advancing technology, another hook-up option is a wireless connection enabling cable-free communication for laptops with a Bluetooth transmitting and receiving tool (hardwired in Macs but necessitating a plug-in 'dongle' into a PC's USB port). One telescope manufacturer even supplies the remote program CD as standard. All that is required is the connection cable. Some mounts also offer a similar function for free. No doubt other companies will follow. Even multi-functional smartphones, with sophisticated planetarium software, can now connect to telescopes, with cables or, with Bluetooth and the right receiver, without. No doubt, this technology will soon be superseded too! Always refer to the instruction manual, however, because, depending on the hardware/software, installation can be somewhat tricky.

And for those not concerned with physically peering through the eyepiece and who wish to operate a distant telescope (see *Amateur observatories*, page 138) or, in some cases, hugely distant scope, whilst keeping snug indoors, there is the 'remote' option. Here, a distant telescope is manoeuvred via lengthy cabling or a WiFi connection. In the case of 'hugely distant' telescopes, specialised hardware and software is required, for which companies are out there – but it comes at a cost. An easier, and less expensive, remote option can be to buy Internet telescope time at renowned worldwide observatory sites (see *Remote astrophotography* in Chapter 5). This is available to schools and amateur astronomical societies and is done by registering at the respective Internet site.

Telescope observing

With a telescope, additional to the objects selected for naked-eye and binocular viewing, the sky is a treasure trove. Grab any of the following: planisphere; magazine current circular star chart; paper or computer-generated customised star chart; deep-sky atlas (available from sixth magnitude for 'starters', on up to 11th magnitude for serious advanced observers – see Appendix 3); a favourite observers' guidebook; your LED flashlight ... wait around 30 minutes for your eyes to 'dark adapt' ... and start to enjoy the journey shared in Chapters 1 and 2.

For many, the planets are favourite targets when starting out with a telescope, so below are just a few planetary observing considerations:

MERCURY

- Observe in the evening sky with naked eye or binoculars.
- It is difficult to observe with a telescope since it is normally less than 15° above the horizon and often lost in bad seeing.
- It suffers from false colour (red fringe on lower side of planet and green [blue] on upper) caused by atmospheric refraction (light wavelength dispersal), and also chromatic smearing.
- Since it orbits inside Earth, its phases are visible: with a three-week viewing window, the planet changes from full globe to a slim crescent.
- The best views are during the day: locate it in morning twilight with naked eye or binoculars (August to November is the best time in the northern hemisphere). Use a motor-driven, polar-aligned equatorial scope with a dew cap attached. Ninety minutes after sunrise the planet will be around 30° above the eastern horizon, giving sharp views if good seeing.
- Mercury next transits the Sun on 9 May 2016.

VENUS

- Although stunning with the naked eye, with the exception of its phases, Venus is featureless through a telescope because of its thick atmosphere.
- Avoid observing when the planet is in a dark sky at low altitude: atmospheric dispersion and poor optical effects will ruin the image. Daytime viewing is best, preferably on an autumn morning shortly before or after sunrise. It can be located with the naked eye in a clear sky after 3:00pm (near eastern elongation) or before 10:00am (near western elongation).
- It can be seen much higher above the horizon than Mercury.
- It appears smallest during its 'full' phase when on the opposite side of the Sun from Earth. When a slim crescent, it appears larger since it is much closer – around 25,000,000 miles (40,000,000km).

MARS

- Look for a reddish 'star'.
- Since it has a thin atmosphere, Mars' changeable surface features are enticing, even in beginner telescopes. Most visible are the polar caps. Other features, such as the dark continents and deserts, may be possible, but difficult. Good optics and high magnification are required.
- Larger scopes may reveal dust storms and occasional clouds but will be restricted by our planet's atmospheric turbulence. The typical good-seeing threshold for resolution is 1.0 arcsecond for a four-inch telescope, 0.5 arcsecond for an eight-inch, and 0.3 arcsecond for a 15-inch (80km on Mars at its closest approach). However, the best detail is seen for only two to four months every two years when the planet is at its closest approach – within 35,000,000 miles (56,000,000km) of Earth – when the disc is sufficiently large. Global dust storms permitting, any decent 70mm aperture instrument should reveal surface features.
- Use appropriate filters (see page 129).

JUPITER

- This dynamic planet is always worth locating in any telescope.
- An 80mm refractor (100x to 130x) will reveal its main belts, the Great Red Spot, and the shadows of the four large Galilean moons. Consult astronomy magazines and websites for corkscrew diagrams to track the four main Galilean satellites as they orbit and transit the planet (see *Occultations and transits* opposite).
- Good seeing may also reveal changes in the equatorial and temperate belts. Remember to use filters. Typically, a four-inch refractor or eight-inch SCT should ensure that anywhere between three and ten of the belts are visible, depending on atmospheric circulation.
- The planet is very bright: to keep contrast and resolution, 25x to 35x per inch of aperture should offer the best views.
- Consult astronomy magazines and websites for details of when the Great Red Spot is well placed for viewing. Remember, it is not visible at the edge of the planet's disc (due to limb darkening) but becomes clear when one-quarter of the way across. Most of the planet's features are visible for up to 2½ hours.

SATURN

- The first sight of this planet through a telescope of whatever aperture is never forgotten.
- Enjoy observing the various butterscotch-coloured equatorial belts and the famous diaphanous rings (30x to 60x): their shadow, cast above or below the rings, is narrow but visible on the planet. The rings, though fixed, appear to slowly open and close over a period of 14 to15 years (half its 30-year orbital period). This is caused by our changing viewpoint as we orbit the Sun: by 2016 they will be at their widest.
- Occasionally white spots appear in the upper atmosphere and will be within reach of a four-inch instrument.
- An 80mm refractor will reveal the Cassini Division but a decent six-inch telescope will enhance the view, and for experienced observers there is the much smaller Encke gap (after Johann Encke in 1837).
- Saturn's largest moon, Titan (around five times the diameter of the rings when at its furthest from the planet) is visible in four-inch telescopes and larger, but its shadow on the planet is only seen when the rings are edge-on. An eight-inch scope will reveal seven moons. A 70mm refractor may reveal Rhea. Since Iapetus has opposing dark and bright sides it varies in magnitude from 10th to 12th so will require some patient observing: look around 11 ring diameters distance from the centre of the planet. A six-inch will reveal Dione and Tethys, but the dimmer Mimas and ice-geysering Enceladus demand intensive observation. Consult astronomy magazines and websites for corkscrew diagrams of satellite orbits and rare Titan transits.

OPPOSITION DATES (when closest to Earth and brightest in sky)

Friday	June 03	2016
Thursday	June 15	2017
Wednesday	June 27	2018
Tuesday	July 09	2019
Monday	July 20	2020
Monday	August 02	2021
Sunday	August 14	2022

SUPERIOR CONJUNCTION (passes behind the Sun as seen from Earth)

Saturday	December 10	2016
Thursday	December 21	2017
Wednesday	January 02	2019
Monday	January 13	2020
Sunday	January 24	2021
Friday	February 04	2022
Thursday	February 16	2023

URANUS

- This planet, visible as a 'star', is almost impossible to see with the naked eye. It is an easier target for binoculars, however, and with a decent six-inch scope its featureless aquamarine disc will emerge (light and dark markings do appear, but they are beyond the reach of amateur instruments).
- Uranus takes 84 years to orbit the Sun. Refer to annual astronomy amanacs for its exact location. It is generally not a must-see for observers.

NEPTUNE

- With its small angular size, this planet, like Uranus, is a challenge to find: again, refer to current astronomy magazines and websites for its location. It takes 165 years to orbit our Star, is currently in Capricorn and about to move into western Aquarius for many years. A six-inch or larger instrument will show a 2.5 arcsecond blue disc.
- The largest of its moons, Triton, is 13th magnitude and will only be visible in larger scopes of ten inches or more.

OCCULTATIONS AND TRANSITS

Occultations occur when a celestial object of larger angular size passes in front of a smaller object, such as the Moon in front of a distant star or planet as seen from Earth. This is different from an eclipse whereby one object passes into the shadow of another, although, technically, a solar eclipse is an occultation of the Sun by the Moon! If the background object is not completely hidden from an observer, it is known as a grazing occultation: within a couple of miles of a lunar occultation's predicted path, the background star will wink off and on as it passes behind the Moon's hills and valleys. If a smaller object is in the foreground, the event is called a transit, such as a planet transiting the Sun.

On average, Mercury transits our Star 13 times per century. The next Venus transits will be 10-11 December 2117 and 8 December 2125. On almost any night, one transit can be observed in the Jupiter system where the Galilean satellites pass in front of the planet's disc. A Titan transit of the planet Saturn happens rarely.

Lunar occultations can be thrilling with the naked eye and binoculars, but it is especially exciting with a low-power, widest-field amateur telescope, or captured with a camcorder aimed down the scope's eyepiece (see Chapter 5). Check monthly astronomy magazines or the Internet for exact timings for your location and use these in conjunction with accurate timing sources such as shortwave radio or the International Occultation Timing Association (IOA) website.

Don't forget to sketch your observations too, and keep them in a logbook. Some astronomy magazines provide observing forms, or templates, but simple sharp pencils and paper will suffice. Select an appropriately sized eyepiece – don't overpower – and sketch the main features first, for example Jupiter's belts, before adding finer features. These can be kept as a record of how features change, or sent to official organisations where they can be incorporated into scientific data. To capture a photographic image for sketching later, see Chapter 5.

Amateur observatories

Every telescope needs a home. If you enjoy the night sky with binoculars, or are interested in meteors and aurorae, then wherever you place your 'bins' is your home, and an observatory is unnecessary. But for anyone keen to observe the Solar System's brighter objects, and beyond, then a semi- or more permanent site is needed.

Why? Well, whether tent, portable or permanent dome, run-off shed or run-off roof observatory, the following still applies:

- They offer wind protection and, if well situated, better thermal stability for equipment.
- Bright Solar System objects require at least a six-inch aperture telescope: the larger it is, the heavier and less portable it becomes and the more time is needed to set up and dismantle afterwards.
- Some telescopes are highly accessorised and computerised, especially for deep-sky imaging: assembling and disassembling is a chore that ultimately becomes untenable.
- Large-aperture heavy telescopes, with sturdy steady mounts for deep-sky imaging, demand a permanent site with a fixed, correctly built telescope support.
- Equatorially mounted telescopes can require accurate polar alignment: accuracy is compromised by non-permanent set-ups. A fixed observatory promotes one-off alignment that can be perfected over time.
- Large reflectors, and catadioptrics in general, lose precise collimation when moved and consequently require frequent resetting.

A POD: available in various colours, portable or permanent, these tough designs offer a wider view of the sky.

Robotic observatory: height 1.3m, includes fibreglass dome with base and one shutter, all hardware and motors for rotation and shutter movement.

A portable, semi-permanent, and useful facility comes in the form of nylon tents encasing strong plastic or aluminium poles. These are easily purchased and can be erected or dismantled whenever and wherever required. When staked they offer shelter and good wind protection. For a semi-permanent or more permanent solution, commercial companies offer portable (PODS) or permanent pre-fabricated domes. Other customised domes can be aligned with existing structures or built from scratch.

For the even more serious and adventurous, there is the run-off shed option. This is a shed-like construction, either in one or two sections, that moves on wheels, tracks, or rails, to reveal the sky. The telescope and mounting are permanent.

For anyone with derelict outbuildings, for example, crying out for purpose, there is the run-off roof observatory (see opposite). Here, the walls are fixed but the roof slides open on wheels in tracks or on rails. There can also be a hybrid – the domed observatory, with fixed walls but an upper section (normally a dome, although it can be cylindrical or otherwise) that rotates. Again, this subject demands a book in itself! What follows is a brief overview of a personal case study that hopefully highlights generic considerations when contemplating any observatory or, indeed, the subject of amateur astronomy itself.

■ THE RUN-OFF OBSERVATORY: A PERSONAL CASE STUDY

This observatory is a run-off roof DIY design and was the result of owning a 50sq ft area of neglected brick outbuildings and a new heavyweight 12in Meade LX90 ACF that needed a permanent home, and fast! Its construction involved book and Internet research, work, and expense (although not a great deal more than the cost of a large commercially produced dome). Ultimately, it involved vision, an understanding of an amateur astronomer's needs and common sense, and took into account the following:

Run-off roof observatory: Guard dog is optional.

- Avoid using, or extending, a dwelling, especially an exposed roof area: this suffers from increased wind and light pollution. Telescopes need equalising with the outside air temperature (see *Telescope types: pros and cons*, page 120). Heated houses create thermal instability and subsequent light distortion. Furthermore, buildings transmit vibration. Roof floors can be reinforced but this incurs planning permission and expense.
- It is pointless searching for celestial targets in the region 25° above the horizon: low-level atmospheric turbulence distorts light. Street, house, and security lights worsen the problem. My observatory is surrounded by hills, some urban glow, and a few houses. The lower brick (not thermally ideal – see below – but it adhered to the original construction so avoided planning restrictions) and upper timber-clad walls blocked low-level artificial lighting without diminishing higher altitude views. Coniferous trees, wooden screens, or fast-growing plants on low-level trellises can help with light pollution, as well as providing a good windbreak and security camouflage. If possible, site your observatory where there is most shelter from the elements. For further advice on light pollution refer to the International Dark-Sky Association or the UK's Campaign for Dark Skies (CfDS).
- Northern hemisphere observers generally gaze overhead or towards the south for the most interesting views and southern hemisphere observers look towards the north, so keep this in mind when settling on a site, especially with regard to the celestial poles.
- Ensure an observatory is away from heated buildings if possible. Moreover, try to site it near grass or foliage: man-made brick and concrete constructions and bare earth generally absorb more sunlight during the day and then re-emit it as localised infrared radiation at night, distorting images.
- Garages can convert to observatories but concrete foundations are often too shallow and subject to vibration: deeper concrete might be necessary if the site is near a road or railway. Similarly, as a cheaper and quicker option, other shed-type outbuildings can be considered but they must have solid foundations and reinforced walls. Mass-produced rectangular and square apex, and pent roof, summer houses and sheds are convertible but they invariably require metal bracings or strengthening with timber, especially if the roof is moveable. Even plastic storage sheds can be customised. In my case, I was fortunate. There already existed a 19x8ft outhouse with an adjoining 10x10ft concrete base, which was raised by 2ft for advantageous viewing and minimal vibration. The lower brick wall-surround was rebuilt to 2½ft (0.76m) high and completed to its former height with 3x2in CLS studding clad with 0.25in WBP ply.

The viewing area: 10ft raised viewing area with lower brick wall-surround and 3x2-inch CLS studding clad with 0.25-inch WBP ply.

- The walls are sufficiently low to stand up straight when the roof is closed but not too high to compromise viewing angles. I preferred structural integrity and shelter from wind, but it is possible to have the upper section of a wall fold down once the roof is opened. Furthermore, my telescope can be parked in polar-aligned mode at a 35° angle but the roof can still be comfortably opened and closed. The walls also allow for a user-friendly exterior door customised for accommodating the two-foot rise and safe access steps.
- Size is a major consideration, especially if there is a limited budget, limited space, and planning restrictions (see below). Commercial fibreglass and plastic domes start with a 6ft (2m) diameter but these are too small to be of any real use. Since it is a one-off investment, go for at least a 10ft clear area, whether purchasing or building – you won't regret it. You may also become obsessed with astronomy in the future and purchase a larger scope with more accessories. In my situation, the 10x10ft viewing platform proved just right. It allowed for one centrally located, tripod-mounted telescope and the all-important room to move around it. Remember, telescopes can be offset from their mounts and require a wide clearance for 360° operation at low altitudes.
- Building regulations vary from one country to another. In the UK it is essential you contact your local planning authority to find out if planning permission is required. It is certainly required for new buildings over three metres high, and for those over four metres high if they have a ridged roof. Whether permanent, or semi-permanent, permission for an observatory is dependent on design and any application has to be sent to the planning department of the local authority, along with a fee for submission, a plan, and a description of what is proposed. As a guideline, the 'three metres' limit normally means unridged domes are fine, but there are other stipulations that must be checked. When an application is received, it will be verified to ensure building regulations are met and a planning officer may well inspect the completed construction.

Run-off roof opened: enabling a 360° vista.

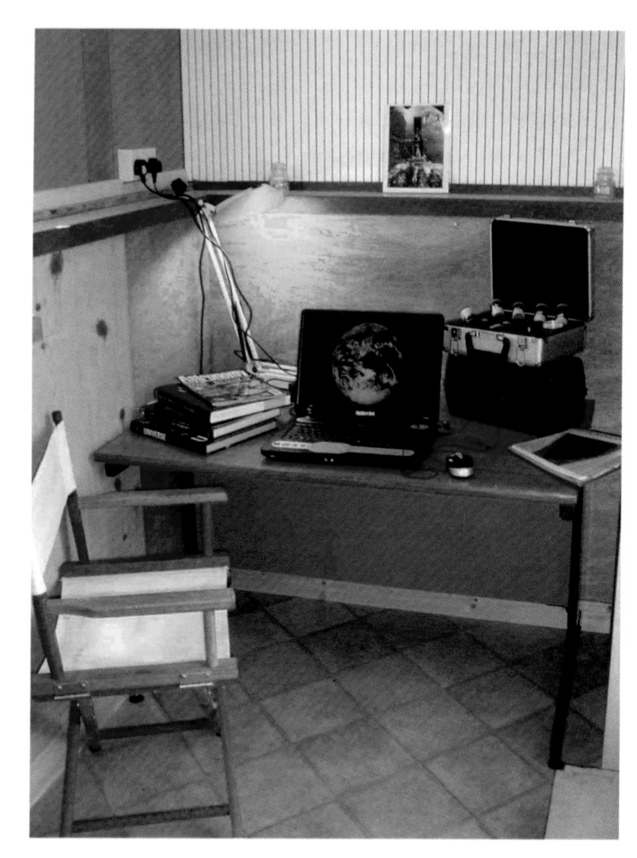

Observatory office: ideal for laptop use, remote telescope operation, and comfort.

- Another size consideration for run-off sheds, and run-off roof observatories, is the area needed to enable the roof, or shed(s), to move: they can require twice as much space again as the building itself. All designs are a variation on a theme but in my situation, because of the building's original L-shape, we found the roll-off roof could only slide one way – eastward over the once-covered patio area. This worked out well, as the whole operation was contained and the roof could slide back fully without need for extra operational space.
- Any dome, shed, or building should also allow space for a desk, adjustable rotating chair, shelving for books and charts, accessories, et al… and for friends to visit! Rectangular or square designs can often be more accommodating in terms of useable area. Again, in my case I was lucky. After gutting and repairing a massive hole in the roof, the adjoining 19x8ft brick and timber outhouse made for a perfect office.
- We covered the original rotten roof with a second one but pitched in the opposite direction to enable the sliding-roof design. This second roof gave added insulation and had sufficient overhang for water run-off. The rotting dividing wall was replaced with an insulated stud partition and an interior solid door meant the viewing area and office could be separated, allowing the interior office red-light to be on without interfering with telescope viewing and keeping the office warm if needed.

A roll-off roof observatory with full astro-photography set-up.

Roof overhang: run off rails are secured by stanchion (left) and sturdy roof overhang (lower right image).

- Roof operation for roll-off roof observatories can be tricky but common sense should prevail. Weight is a major consideration: it has to be sufficiently lightweight for one person to operate but sturdy enough to keep rigid and secure in high winds and snowfall.
- Domes are manufactured in aluminium alloy (strong, light, and corrosion-resistant), galvanised steel (strong, heavy, and fixings tend to rust) or fibreglass, the last being the lightest and cheapest option. However, they can, and do, blow away if not sufficiently strong or not secured properly to their base rings – a circular rail of welded steel. The aluminium is therefore the better option. In my run-off design, the lightweight single-pitch roof measured 12sq ft (3.7m^2) with a 4in fall away from the adjoining office roof and sufficient overhang for water run-off.

Roof interior: and one example of exterior lighting.

- Made of 0.75in plywood supported by 3in joists at 16in centres, with an internal brace, it was partially constructed on the ground, had five 3in fixed (not swivelling) industrial castors – rubber on steel – screwed into place on each long side, and was hauled into position onto tracks made from open-top plastic cable ducting, 2in wide by 1in deep and 24ft long. (These can be angle-iron rails, but they expand and contract with changing temperatures. Another alternative is channelled concrete.) It is vital the tracks are fixed, level, and parallel. Mine were spirit-levelled and screwed into the timber.

A fixed industrial castor: for smooth and straight rolling of roof.

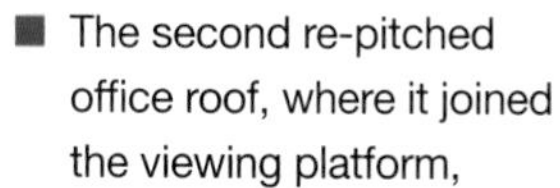

- The second re-pitched office roof, where it joined the viewing platform, provided a 6in 'corridor' beneath. This meant the sliding roof could fit snugly inside and slide smoothly on its similarly embedded track. It also gave added security, enabled the track to keep clear and dry, and helped prevent uplift in rough weather.

Roof overhang: the sliding roof fits snugly on the track, giving added security and weather protection.

- Two timber, track-embedded frames had to extend from each long side of the roof over the entrance where the old rotten conservatory had been, to enable the roof to open fully. Opening is done by simply pulling gently on a long, light pole that connects to one of four metal eyes on the interior joists.

Run-off roof interior: with metal 'eye', and pole, for easing open.

- Alternatively, more complex motorised winch systems can be used – a subject deserving a chapter in itself! In the case of domes, the opening roof slit needs to be at least twice as wide as the aperture of the telescope to offer the best views and avoid dome shadowing: little of the sky can be seen at any one time. When my run-off roof is closed, a cover of 12ft-long plastic inverted guttering is gently closed over the exposed track, giving protection from the elements and falling leaves, etc.
- All the joints were then covered with flashing tape to prevent leaking (unlike some domes!) and the whole double-coated with lightweight liquid felt. Remember to keep the pitch of the roof as low as possible so that it doesn't interfere with viewing but still facilitates effective water run-off.

Roof track: north side with protector.

- Flooring is also a consideration. I have concrete in my observatory but others may find that well-ventilated wooden decking, isolated from the telescope, is a better option, since concrete slowly re-radiates daylight heat, delaying cooling-down time. Consider using an old carpet for increased insulation and comfort. It also acts as a great cushion when you drop valuable lenses … as I have learned to my cost.
- Remember aesthetic and environmental considerations too: don't annoy the neighbours! A white dome reflects sunlight and keeps the interior cool, but a green one is equally functional and easier on the eye. Aluminium covering (sheet or paint) or white gloss paint is also effective. Wooden sheds using weatherboarding or shiplap, and shielded by foliage, are unobtrusive. In my case, feather-edge boarding tarted up the entire exterior without drawing attention to the building.
- It's obvious, but make sure the exterior is weatherproofed. Dampness and dew are a big problem for domes, run-off sheds, and run-off roof observatories just as they are for telescope operation itself. Timber constructions last a lifetime but they need help to do so, otherwise dry rot, wet rot, fungus, and mould move in. Most timbers are already treated with a basic preservative, but apply more to be on the safe side. For timber not exposed to water, a biannual coating of any creosote-type solution should suffice. Keep the colour light to promote interior coolness. Alternatively, water-based exterior treatments, although more expensive, work well. Avoid exterior or yacht varnishes: these are less effective and, like all varnishes, crack after a couple of years. Some varnishes even dissolve in water, while others react badly to ultraviolet light. Wooden decking is different and requires effective sealants. In my case, the entire exterior and interior viewing area had a double coating of top-quality antique pine and rich mahogany wood preserver.
- Temperature regulation must also be considered. Observatories, of whatever kind, must not get too hot, since it delays cooling-down time for the night's viewing and is detrimental to the equipment. Domes, especially metal ones, tend to retain heat and then funnel it out through the roof slit, like a chimney flue, when opened: this affects seeing. They can have air-conditioning installed. However, an easier option is not to have windows. If you do, shutter them or paint them out: this eradicates the greenhouse effect. Use reflective roof paint to speed up the cooling-down time. If not possible, because of environmental restrictions, make sure you ventilate for a long period before observing. Remember the flooring: wood is best since concrete loses heat slowly, which interferes with incoming light. In winter, the temperature variation is negligible. To control damp after a night's viewing, consider using wall-mounted fans, small dehumidifiers, or a thermostatically controlled electric convection heater on a low setting (as I do).

Interior lighting: fit two types – white light (above) but also optional red or red LED strips, with alternate sockets, for night-time use.

- With today's motor drives, 'GoTo' systems, and laptop planetariums etc., power is essential. Telescope mounts can operate on batteries, or be happily powered all night long from a 12V DC car cigarette lighter with the appropriate power cord. However, with a more permanent set-up comes a need for permanent mains 230V AC (UK), 110V AC (USA), power. With the plethora of equipment requiring juice, battery packs can be used, but they are heavy and tend to run down at inappropriate moments. You must hire a qualified electrician (if not one yourself) to do the installation. In the UK, it is illegal for a new mains circuit to be installed without certification from the Institution of Engineering and Technology. Regulations vary from country to country and state to state, so always check.
- Various accessories may require different power. Desktop computers, extension blocks and dew-defying hairdryers must be protected from damp, especially in run-off roof designs. Laptops have well-sealed power supplies and use low voltage so they are safer, but since their voltage requirements vary always check. Power installation in general will vary for run-off sheds and domes. Play safe: always consult a professional qualified electrician. In my case, a 230V AC mains supply, with full cut-off facility, already existed in an armoured cable underground. Plentiful wall-mounted weatherproof sockets were installed in the viewing area and outside the building for initial ease of access. Red low-level lighting and optional, but useful, IP55-rated (EU recognised 'Ingress Protection') white lighting was also installed in the viewing platform and office.
- Security is a matter of common sense. Always keep the dome, roll-off shed, or run-off roof observatory locked when not in use. Don't advertise its location in any medium and, if possible, consider installing a security system.

A large commercial dome: up and running.

Tents, domes, run-off sheds and run-off roof permanent observatories are *not* essential when starting out. There are numerous variants for each, way beyond the scope of this book. There are other halfway options too: outside fixed telescope piers and mountings on raised, ventilated wooden decking where the scope is simply covered when not in use … and there are many more possibilities. Check astronomy periodicals for advertisements of the latest products, shelters, dome suppliers, and purpose-built observatory builders. Read specialised books and magazines too: some include CDs offering instruction and detailed plans. Check websites. Astronomy forums and newsgroups are also very useful, or consult with astronomers in your local astronomy club or society. Refer to Appendix 3 as a first step.

AND FINALLY...

Above all, and to return to the beginning of this chapter, remember … astronomy can be enjoyed on any cloudless night, from any location, with or without additional equipment or paraphernalia, domed or 'undomed', in or out of a shed or observatory, or on a lounger in the garden. Don't be obsessed with gadgetry at the expense of the sky. Resist being embroiled in techno one-upmanship or gizmo-geekiness! The essence of astronomy is context and perspective, grandeur and wonder. Take time to grasp the fact that astronomy is too vast to grasp. Familiarise yourself by taking the grandest journey of all with the naked eye. Be humbled. Look up, not down, outward, not inward … and hours of fun and excitement await.

NGC 346: a star cluster containing some 2,500 newly born stars in a massive stellar nursery in the Small Magellanic Cloud. Starlight scatters off dust particles giving rise to the ghostly pink haze. It is 210,000ly away and 200ly in size.

Professional viewing

As we celebrate the 400th Anniversary of Galileo turning his telescope towards the night sky, it still seems inconceivable that it took mankind until possibly the early 17th century to combine magnifying lenses into such an instrument. It seems even more incredible that it took until the 1930s to realise invisible 'energies' as well as 'objects' were 'out there', and that we had to wait for the advent of 1950s technology to enable us to escape Earth's atmosphere and read and understand those energies. Yet the knowledge has always been within us and around us, awaiting discovery. All we had to do was catch up – and now, astronomers are…

'The dark is light enough'

Christopher Fry, 1907–2005

Invisible light

How can light be invisible? It is a contradiction in terms. When we look up, we see innumerable stars, ghostly nebulae and the odd giant galaxy, all else is black and seemingly void. Right? Wrong! It just appears so ... in visible light, the radiation our eyes can see, but what about other radiation, the energy we cannot see? It litters the Universe, how do we find it? We build specialised Earth and space-based telescopes and detectors. By doing so, professional astronomers have viewed beyond the realms of the visible range. Out there are the 'invisible' radio waves revealing great clouds of interstellar hydrogen gas; the infrared radiation opening doors of stellar nurseries; the exploding X-rays and gamma rays punching from supermassive black holes and the legacy background radiation of Universal birth… and a great deal more besides… out there in the endless 'empty' dark…

THE ELECTROMAGNETIC SPECTRUM

All astronomical objects emit radiation (energy) in the form of waves across the entire electromagnetic spectrum. The distance between each wave crest is known as the 'wavelength'. These lengths vary and the number of times each crest passes a fixed point is known as the 'frequency': the shorter the wavelength, the higher the frequency, or energy.

At the low-frequency end of the spectrum (see diagram below) are the 'cooler' astronomical objects transmitting at low energies with long wavelengths – radio waves (anything from one millimetre). As an object's energy level, and thus temperature, increases, the wavelength shortens to less than a millimetre. Eventually longer radio waves become shorter, higher-frequency microwave wavelengths, the legacy 'afterglow' energy of the Big Bang. With increasing energy, wavelengths shorten further and we enter the infrared (meaning 'below red', since red is the longest wavelength visible to the human eye) or heat region of the spectrum. From here, radiation wavelengths shorten further until they become sufficiently short to be seen with the human eye – visible (optical) waves. Beyond this narrow range of visible waves

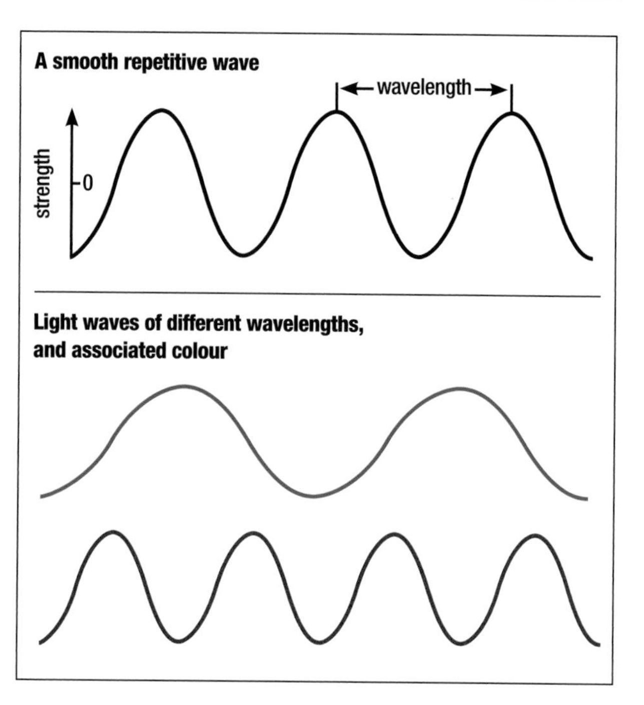

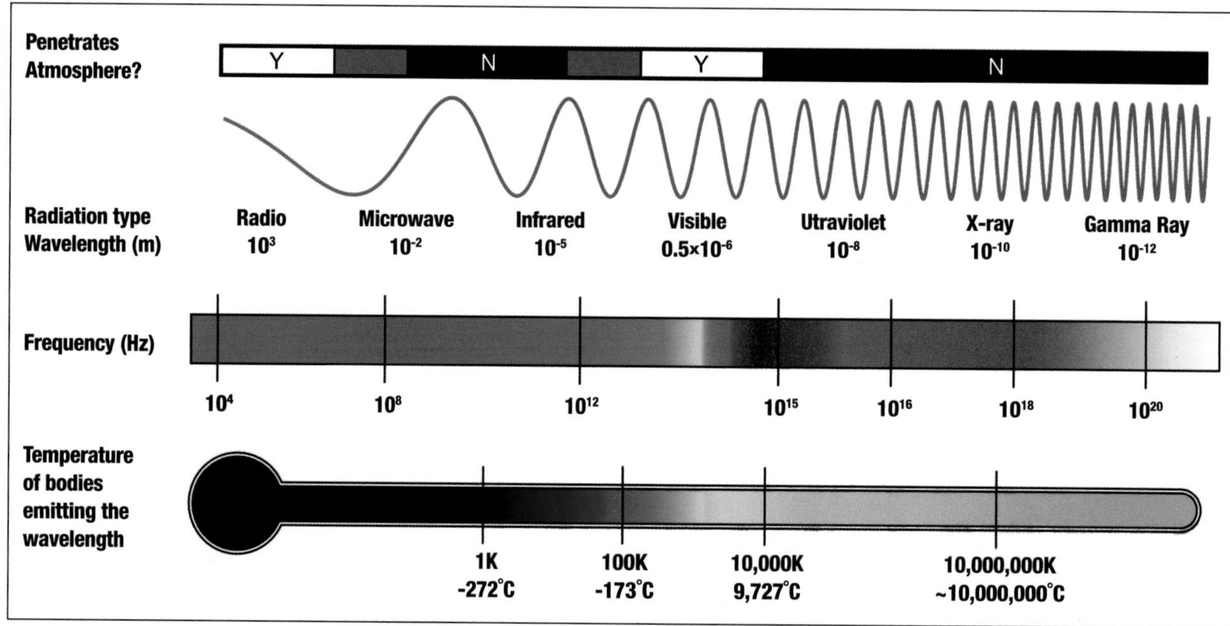

are: the other 'invisible' ones; those emitted by hot high-energy objects; the ultraviolet (meaning 'beyond violet', since violet is the shortest wavelength visible to the human eye); X-rays; and ultimately exceedingly high-energy short-wavelength gamma rays. Until the late 1950s, infrared, ultraviolet, X-rays, and gamma rays were undetectable: they were curtained by Earth's atmosphere. Today, a range of highly sophisticated telescopes and detectors enables us to open this curtain and 'see'.

To understand why objects are hidden in space, imagine a forest with a giant tramping through it. The trees are so small the giant can stride through unimpeded. To an ant, the space between tree trunks is so vast he scrabbles through also unhindered. For both, it's as if the trees are not there. But if an astronomer runs through, sooner or later he/she will smack into a trunk or branch. This analogy applies to differing wavelengths of radiation in dusty interstellar clouds.

INVISIBLE LONGER-WAVELENGTH RADIO WAVES

Longer-wavelength radio radiation passes through clouds like the giant in the forest – the dust, like the trees, is too small to notice. With the right ground-based telescope, however, this radiation is captured to reveal the splendours of gaseous nebulae, supernova remnants, pulsars (rapidly spinning neutron stars), and the cores of distant galaxies. Incredibly, having journeyed billions of light years and been stretched by the expanding fabric of space itself, these energies are barely detectable. Indeed, as Dr Carl Sagan once famously said, all the radio energy ever received by all the radio telescopes on this planet amounts to no more than the energy of a single snowflake settling on the ground!

INVISIBLE SHORTER-WAVELENGTH RADIO WAVES

Moving through the shorter, sub-millimetre radio wavelengths of the spectrum to the higher-energy, higher-frequency infrared radiation waves (felt as heat at temperatures ranging from -430 to 1,800°F (-260° and 1,000°C), these are still larger than dust grains, so, like the giant in the forest, they also pass unhindered through the dust clouds. Near and mid-infrared wavelengths can be detected from the ground: balloons, rockets, or satellites are used to locate others from above Earth's atmosphere.

VISIBLE (OPTICAL) WAVELENGTHS

Both radio waves and infrared waves are invisible to our eyes because we can only see objects whose wavelengths are emitted within the visible (optical) mid-range of the spectrum – from red (wavelength 700 nanometres/0.0007mm) to blue (400 nanometres). As visible light has a wavelength about the same size as a speck of dust, any object emitting energy in this range is stopped, like the astronomer hitting a tree in the forest. It is hidden from us, but sophisticated instrumentation can 'see' it at other wavelengths.

INVISIBLE ULTRAVIOLET, X-RAY, AND GAMMA RAY WAVELENGTHS

Beyond the visible range, at the high-frequency, high-energy end of the spectrum, are the invisible ultraviolet, X-rays and gamma rays: their radiation is emitted at wavelengths smaller than dust particles, therefore they pass easily through dust clouds like the ant between the trees. Their received energy, having travelled billions of light years and been expanded by space, is minuscule: a detected gamma ray has a millionth the kinetic (moving) energy of a flying mosquito! Sophisticated satellite-based detectors seek them out and reveal hot gas around black holes and neutron stars, cosmic rays, and the composition of stellar atmospheres.

Technology has come a long way since Thomas Harriot (1560–1621) and Galileo Galilei (1564–1642) magnified the sky with their rudimentary instruments. Let's take a closer look at our 21st-century renaissance machines.

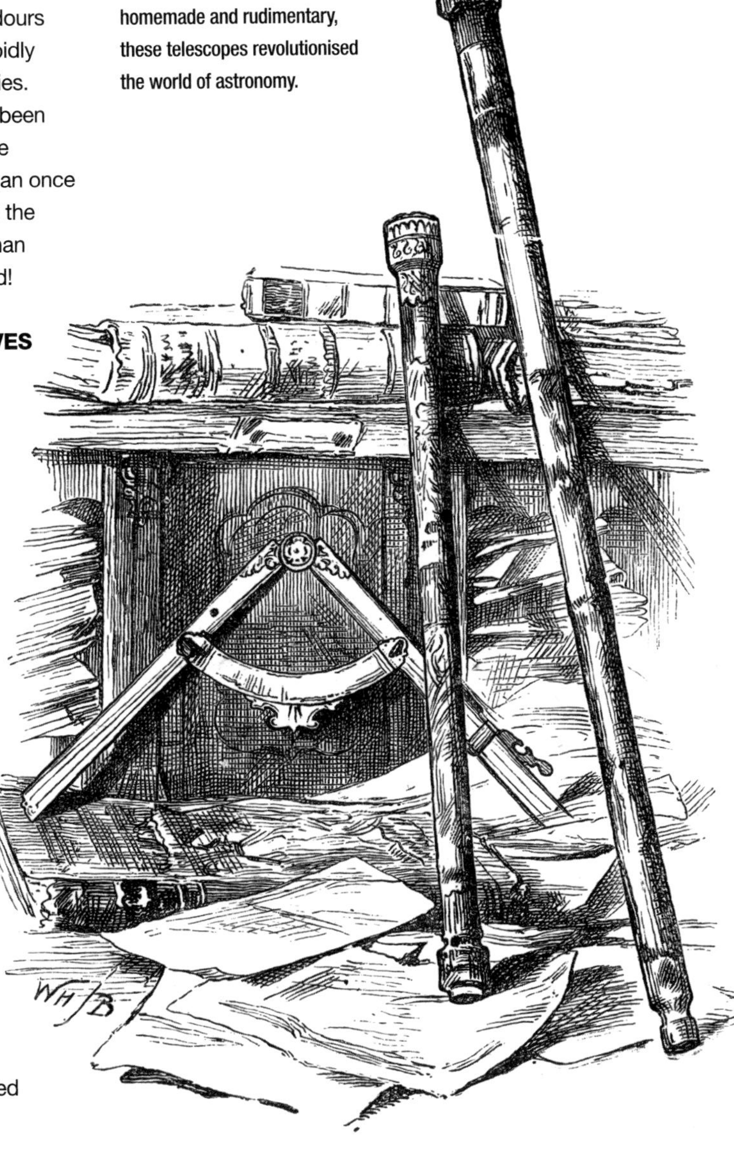

Galileo's instruments: homemade and rudimentary, these telescopes revolutionised the world of astronomy.

Ground telescopes

Supporting the awe-inspiring space-based observatories are the ground telescopes; optical instruments, many with mirrors over eight metres in diameter and assisted by ingenious 'light correctors'; giant radio antennae and roaming radio telescope arrays, their angled receivers attuned to signals beyond our 'ears'; the exquisitely innovative ultra high-energy particle and photon detectors sifting space itself in the deep beyond... all are a legacy of those simple magnifying lenses Galileo once turned to the skies.

THE KARL G JANSKY VERY LARGE ARRAY

Radio waves were discovered accidentally in the 1930s. If their weak signals were converted to sound, we would hear crackling and hissing. To enable astronomers to 'tune in', radio telescopes are needed. These vary widely, but all have two basic components: a large radio antenna (the dish) and a sensitive receiver. The sensitivity of the telescope – its ability to measure weak sources of radio emission – depends on the area and efficiency of the antenna and the sensitivity of the receiver that detects and amplifies the signals. Since radio sources are very weak, radio telescopes are usually very large.

An example is the Karl G. Jansky Very Large Array located just outside Socorro, New Mexico. This is one of the most powerful radio telescopes in the world today. Fully operational since 1980, it comprises 27 antennae, with their 25m (82ft) diameter parabolic dishes each weighing around 230 tons (100 tons of which is in the moveable reflector). Their aluminium-panelled surfaces are accurate to 0.5mm (20 thousandths of an inch). This enables the antenna to focus radio waves shorter than one centimetre and to feed them to the rear receiver. Cryogenically cooled amplifiers with very low internal noise are used to enhance sensitivity. To produce an image equalling those of optical telescopes, a single radio antenna would have to be 17 miles (27km) in diameter! However, by placing many small antennae in a 'Y' pattern (see above image), with each arm of the 'Y' 13 miles (21km) long, this level of image is achievable. In fact, by combining data from all the antennae, and utilising Earth's rotation, a radio picture can be produced that would be equivalent to using a single 22-mile (36km) wide antenna with the sensitivity of a dish 130m (422ft) in diameter!

The ability to discern detail (resolution) depends on the wavelength of observations divided by the size of the instrument. Since radio wavelengths are much longer than optical wavelengths, the telescopes must be much larger to obtain the same resolution. The VLA therefore changes the separation of the antennae, normally every four months: a high-resolution image will reveal a galactic core whereas lower resolution reveals overall structure. Moving the antennae in and out on each arm has the same effect as moving the zoom lens on a camera.

The array is normally found in one of four standard

VLA: The Karl G Jansky Very Large Array at Socorro, New Mexico, United States, configured in its smallest 'Y' pattern.

VLA: one of the Karl G Jansky Very Large Array's radio antennae, or dish, measuring 25m (82ft) in diameter.

configurations ranging from the smallest, where the antennae are crowded to within 2,000ft (0.6km) of the array centre, to the largest, where the antennae stretch out to 13 miles (21km) from the centre, the latter giving the most detailed image. The signals are amplified several million times and sent to the Control Building along a buried waveguide – a very accurately formed hollow pipe. This is a two-way system: command and reference signals are sent out to the antennae while the cosmic radio waves and equipment monitoring information are sent back to the Control Building. Here, the radio signals are extracted by a sophisticated digital computer, amplified again, and then converted into numbers that represent signal strength. A special-purpose computer – the correlator – combines the signal from one antenna with all the other antennae at a rate of a million times per second! The images are then stored as numbers, analysed, and finally displayed in colours or shades of grey. This system of synthesising a very large effective aperture from a number of smaller elements is known as interferometry. Radio astronomers combine telescopes spread out across continents too – known as very long baseline interferometry (VLBI) – whereby recorded signals are synchronised to within a few millionths of a second. The result has provided some of the most stunning observations in the history of astronomy.

VLA: Tracks for moving the dish antennas are visible in the foreground. The dishes can only be moved in very calm wind conditions.

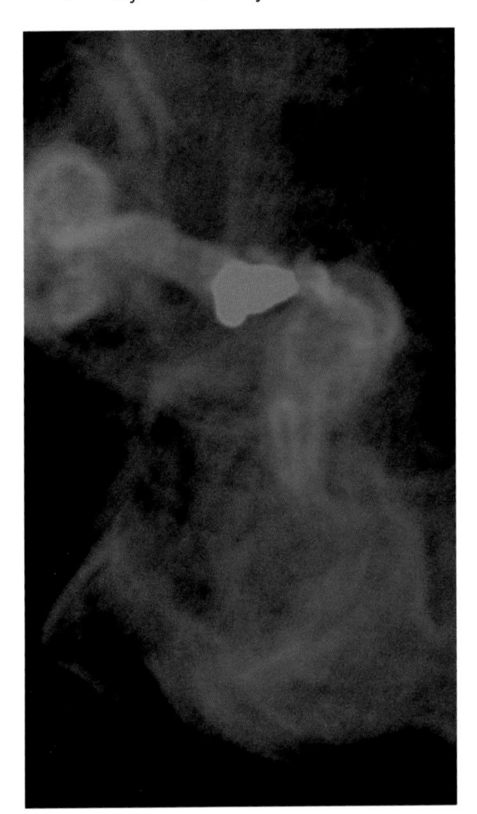

VLA: Very Large Array radio image of giant elliptical galaxy M87: evidence that repetitive outbursts from the central supermassive black hole have affected the entire galaxy for a hundred million years or more.

NOTABLE RADIO TELESCOPES

SINGLE DISH BUT MAY OPERATE IN AN ARRAY

- **National Astronomy and Ionospheric Centre, Arecibo, Puerto Rico:** 305m fixed dish (radio and radar). Listens and watches for extraterrestrial signals

- **National Radio Astronomy Observatory, Green Bank, West Virginia:** 7,000-tonne, 100m steerable dish (the heaviest steerable object on Earth). Listens for extraterrestrial signals.

- **Jodrell Bank Radio Observatory, Cheshire, England:** 76m and 66m steerable dish – used to track *Sputnik 1*'s carrier rocket in 1957, to confirm the Russians' *Luna 2* probe had reached the Moon, and to contact NASA's first deep-space probe, *Pioneer 5*.

INTERFEROMETERS

- **Parkes Radiothermal Telescope, Parkes, New South Wales, Australia:** 64m steerable dish. Used to relay communication and telemetry signals to NASA for the *Apollo 11* landing when the moon was on the Australian side of Earth (with assistance from 64m Goldstone antenna, California, and 26m antenna at Honeysuckle Creek, Canberra, Australia). Also used to track *Mariner 2* and *Mariner 4*, *Voyager*, *Giotto*, *Galileo*, and the *Cassini-Huygens* probes. Major world centre for pulsar research.

- **Atacama Large Millimetre/Sub-Millimetre Array (ALMA), Chajnantor Plateau, Atacama Desert, Chile:** 66 x 12m and 7m diameter radio telescopes observing at millimetre and sub-millimetre wavelenths and moveable over distances from 150m to 16km.

- **Australia Telescope Compact Array, Narrabri, New South Wales, Australia:** 6x22m dishes sited over 6km area. Premier radio interferometer in southern hemisphere.

- **Westerbork Synthesis Radio Telescope, north-east Netherlands:** 14x25m antennae in straight line 3km long: ten fixed permanently, four moveable. Used in tandem with other global radio telescopes in VLBA mode.

THE VERY LARGE TELESCOPE (VLT)

Operational since 1999, the European Southern Observatory's Very Large Telescope, sited 2,600m high at Cerro Paranal in the Atacama Desert, Chile, consists of four optical telescopes in an L-shaped formation, each with a flexible mirror 8.2m (26.9ft) wide and 170mm (6.7in) thick, cast from 45 tonnes of glass-like material known as Zerodur. Referred to as Unit Telescopes 1 to 4, they are respectively called Antu (Sun), Kueyen (Moon), Melipal (Southern Cross), and Yepun (Sirius) in the indigenous Mapuche language.

During the casting process, to maintain the mirror's shape as it solidified, each was spun until the temperature dropped to 800°C (1,472°F), and then allowed to cool for three months. The surfaces were then ground to an accuracy of a millionth of a metre and finally coated with aluminium. During the day, the air in each dome enclosure is cooled to the temperature predicted for sunset, allowing rapid 'equalisation' of the mirrors (see Chapter 3). Mirror maintenance is ongoing, and to ensure there is no degradation each is annually removed and resurfaced. Weighing some 400 tonnes, the telescopes rapidly slew around the entire sky, their instrumentation observing at near-ultraviolet to mid-infrared wavelengths and utilising many techniques, including high-resolution spectroscopy and imaging.

Very Large Telescope: aerial view showing the tunnels housing the delay lines.

ACTIVE OPTICS

However, no matter how slim a primary mirror, it still sags under its own weight (when pointing to different regions of the sky), which distorts an image. The deformity has to be corrected and this is done by a system of 'active' optics. At the VLT the mirrors are actively corrected with continuous computer control. By measuring a guide star (see opposite) in the field of view, corrections are sent to electromechanical actuators on the back of each primary mirror. These actuators push or pull on a section of the primary to change its shape at a rate of once per second: the result is much sharper images.

Kueyen's active optics: a view of Kueyen's mirror cells from behind. The bank of the mirror houses the active optics system.

Active optics systems are used on many large telescopes, including the giant twin 10m mirrors on the Keck I and II scopes at the Keck Observatory – the largest optical and infrared telescopes on Earth. Their primary mirrors are made from 36 interlocking hexagonal segments with each segment stabilised by rigid support structures and adjustable warping harnesses. All have been 'stressed mirror' polished so that if they were the width of Earth any imperfections would be only three feet high, and together they form one piece of reflective glass. However, like the VLT, when tracking objects throughout the night the primary mirrors slightly deform. Even a deviation of just a few tens of nanometres (a thousandth the thickness of a human hair) significantly distorts an image. To compensate a computer corrects each segment twice a second so that each is repositioned relative to its neighbour as the instrument moves on its mount.

Very Large Telescope: four 'eyes on the sky' with, at right, the sweeping arm of our Milky Way galaxy.

■ ADAPTIVE OPTICS

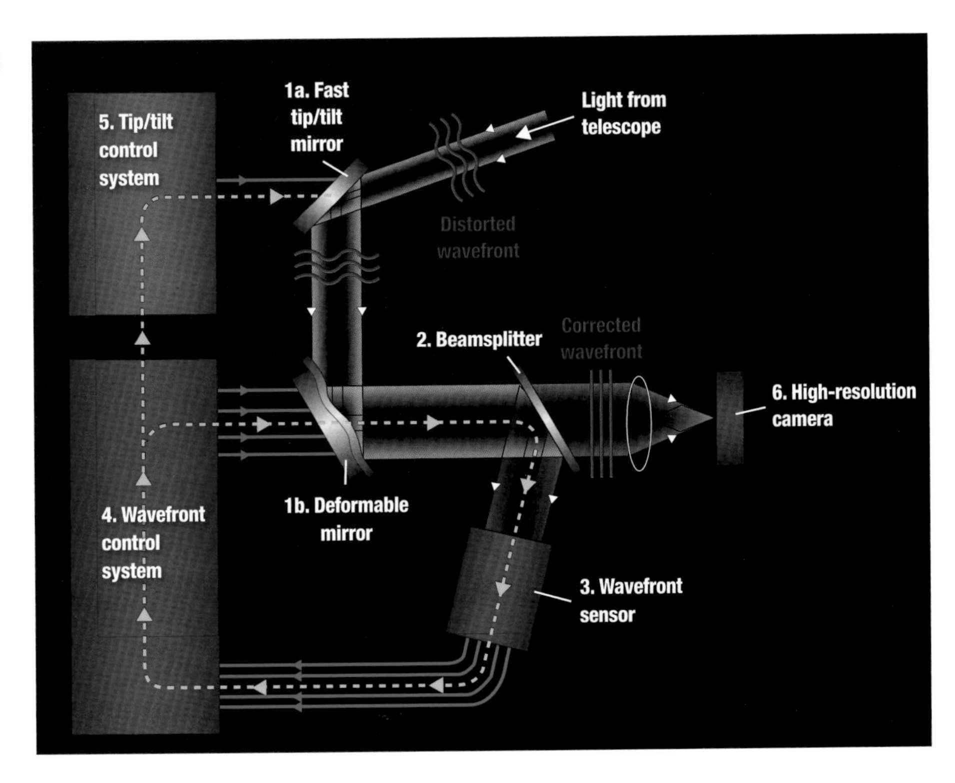

However, to gain even sharper images there is an even more ingenious system, adaptive optics (AO), first developed and used in 1999 on the Keck II telescope. In order to produce a sharp image, starlight should arrive at a telescope with a flat 'wavefront', but tiny atmospheric cells of varying temperatures act like miniature lenses to disfigure it. The 'wavefront' ends up resembling crinkly paper, and the larger the light-gathering area of the telescope, the greater the problem.

Operating at frequencies around 1,000Hz – a thousand cycles per second – (too fast for altering some large primary mirrors), AO systems employ smaller, rapidly flexible auxiliary mirrors and additional optical elements to correct this distorted incoming light.

Firstly, astronomers require a bright star to act as an ongoing 'guide', or reference, for making corrections. Initially AO relied on a star that was sufficiently close and sufficiently bright, but the number of stars up to this task amounted to a meagre 1% of the sky. With Keck's Laser Guide Star Adaptive Optics system astronomers now create an artificial star as and when required, enabling 70 to 80% of objects in the northern hemisphere sky to be targeted. To create the 'guide' a laser beam is used: this excites sodium atoms in an atmospheric layer of meteoritic dust some 55 miles (90km) up, and the result is a glowing artificial 'star'. Light from this star then enters the telescope. The diagram (above) shows what happens next, using the Keck AO system as a typical example.

The distorted wavefront enters the telescope and is sent to the AO system, where it is reflected off two mirrors, a fast tip/tilt mirror (1a), and a deformable mirror (1b), before reaching a beamsplitter (2). The beam splitter allows the longer wavelength near-infrared light to pass on to the high-resolution science camera (6) and diverts the shorter wavelength visible light to the wavefront sensor (3). The wavefront sensor measures the distortions in the wavefront over a thousand times per second and sends the data to the wavefront control system (4). The wavefront high-speed computer control system computes the corrections necessary to rapidly move the fast tip/tilt mirror (1a) to stabilise the image on the high-resolution camera and also to position the deformable mirror (1b) to correct for the wavefront distortions. The deformable mirror is made of a very thin sheet of glass that can be computer-controlled to bend and adapt into a mirror-image shape of the wavefront distortions: the light is then 'still' and the blurring effects cancelled. The result is a corrected wavefront that passes on to the high-resolution science camera and allows images to be recorded at the full resolving power of the telescope.

Adaptive optics is still in its infancy and confined to infrared wavelengths, but the ultra-sharp images obtained with the Keck II telescope have an angular resolution of 1/20 of an arcsecond (1/40,000 the diameter of the full Moon): if the human eye had equal acuity it would read a book a mile distant! Indeed, its accuracy equates to hitting the bullseye of a dartboard sited 8,000 miles away. The result is amazing discoveries: after combining data from NASA's Hubble and Spitzer Space Telescopes (of which more later) and the Keck II DEIMO instrument (Deep Imaging Multi-Object Spectrograph), astronomers discovered a remote galaxy, 12.3 billion light years away, punching out stars at a rate of 4,000 per year – compared to our own Galaxy's average of just ten per year. With advancing technology comes increasing resolution: wavelengths as tiny as 0.7 microns (one micron is one-millionth of a metre) are correctable. The Keck Observatory's envisioned Next Generation Adaptive Optics (NGAO) will feature multiple laser beacons and hugely increased sensitivity over a wider field of view. Who knows what comes next?

An artificial laser guide star: in action.

■ VERY LARGE TELESCOPE INTERFEROMETER (VLTI)

Just as with radio telescopes, optical telescopes can work in unison in the region of infrared/optical wavelengths, and the VLT operates the largest optical interferometric mode array on Earth – the Very Large Telescope Interferometer (VLTI). At least two scopes 100m (330ft) apart can have the resolving power of a single telescope with a 100m (330ft) mirror – a size impossible to construct at present. Theoretically, they could distinguish the lunar rover's wheels on the Moon! Due to the distance between the telescopes, light from a target object enters one telescope marginally ahead of another. Although tiny, this has to be compensated for.

Enter 'interferometry', a technique that exploits the wave nature of light. When astronomers combine two beams of light that are in 'phase' – their wavelength crests and troughs line up – the beams interfere with each other. When two beams arrive out of phase, they cancel each other out. The arrival of the light from each scope must be timed precisely to arrive no more than a few dozen nanometres apart.

To do this astronomers use 'delay lines' and these have been built into a long tunnel inside an underground facility. Delay lines comprise four small carriages on tracks that reflect beams to a common focal point. Changing the positions of these carriages alters the time it takes for each telescope's light to reach a focal point – it is synchronised – before reaching a detector, known as a beam combiner. The resultant interference pattern is known as a 'fringe'. Furthermore, since Earth's rotation ensures one scope tracks a star's movement before another, the carriages must also be positioned to equalise the light path lengths – sometimes to within 50 nanometres (50 billionths of a metre) – to keep the 'fringe' on the detector. The VLT, along with a set of four 1.8 metre diameter telescopes dedicated to interferometric observations, the VLTI, offer an instrumentation programme that is the most ambitious programme ever conceived for a single observatory. It includes large-field imagers, adaptive optics corrected cameras and spectrographs, as well as high-resolution and multi-object spectrographs and covers a broad spectral region, from deep ultraviolet (300 nm) to mid-infrared (24 µm) wavelengths. Such exquisite technology has enabled astronomers to image and measure the size and shape of stars in previously unprecedented detail.

As we have seen, the larger the light-collecting area of a telescope the better the performance, but gravity limits the size of mirrors: the 10m (30ft) diameter of the Keck I and II telescopes holds the record. So what about combining mirrors on a truly epic scale? Enter the Large Binocular Telescope (LBT).

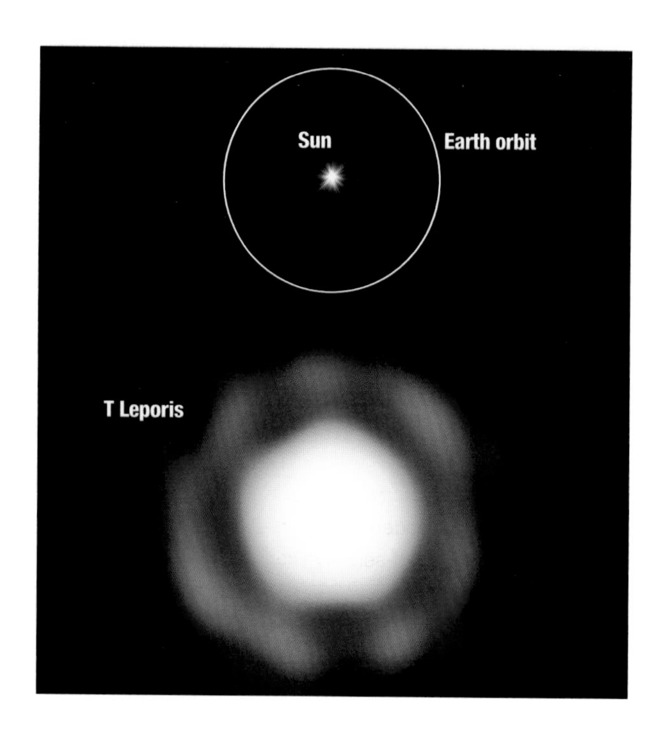

VLTI image: comparison between the Mira-like star T Leporis and the size of the orbit of the Earth around the Sun. The VLTI observations reveal the presence of a spherical molecular shell surrounding the star, which is about 100 times as large as the Sun.

(Above) Very Large Telescope Interferometer: the rails for the first VLTI Delay Line, as seen through a 'window' in the Retroreflector Carriage.

(Left) Very Large Telescope Interferometer: one of the delay lines in the Interferometric Tunnel that serves to keep the lengths of light paths constant from the telescopes.

■ THE LARGE BINOCULAR TELESCOPE (LBT)

With its two honeycomb-design 8.4m (27.5ft) primary mirrors (parabolic in shape and made of borosilicate glass) co-aligned on a single mount, this instrument, sited atop 3,270m Mount Graham in south-eastern Arizona, is one of the world's most powerful skygazing weapons and probes deep space at infrared wavelengths in a bid to answer our most fundamental question about the origin of the Universe and other planetary systems.

The all-steel moveable mount, despite weighing 600 tonnes, is easily steered by a one-horse-power electric motor, and the structure floats like a rocking chair on an oil pressure pad! To go from a vertical to a horizontal position takes anywhere from one to twelve minutes, depending on speed selection. One scope is fitted with a red-sensitive camera, the other with a blue, each with a field of view 23 arcminutes square – that's around ten times the width and 100 times the area of the Hubble Space Telescope's camera. The combined mirrors provide a light-gathering area equivalent to a record-breaking 11.8m (38.7ft) mirror.

A highly sophisticated adaptive optics system compensates for atmospheric distortions. A pair of f/15 curved adaptive secondary mirrors, each with 672 actuators, combine with two tertiary flat mirrors to provide smaller optics that swing on arms in and out of the light path. The tertiary mirrors rotate to direct the light to several central instrument locations (see below image). The red T-shaped instrument in the centre of the telescope contains the optics for combining and phasing the beams from the two telescopes: this add-on LBT Interferometer (LBTI) is an advanced system utilising a technique similar to that in the Very Large Telescope. The result is a resolution equivalent to a single 22.8m (74.8ft) telescope.

The LBTI provides unprecedented imaging in its 'nulling' mode, where light waves from particular stars are cancelled out (akin to lowering a vehicle sun-visor when driving on a sunny day): the overwhelming glare from stars is reduced to permit detection of their orbiting planets or dust discs. Additionally, by having a larger FOV the need for a 'guide' star is removed – light from several faint stars can be combined instead. It is hoped this technology will enable astronomers to find monster black holes or Earth-sized exoplanets in the nearby Milky Way neighbourhood.

LARGE BINOCULAR TELESCOPE

Roof height:	53m
Rotating enclosure dimensions:	25m x 28m x 29m: there are 16 storeys in total but only the top ten rotate
Rotating enclosure mass:	1,600 tonnes
Mirrors:	Duel 8.4m primary mirrors: polished to an accuracy 3,000 times thinner than a human hair; secondary mirrors and adaptive optics
Resolution:	0.005 arcseconds in visible light

Large Binocular Telescope (LBT): with its two honeycomb-design 8.4m primary mirrors.

Space telescopes

Earth's atmosphere is a cataract to telescope vision, its air turbulence veiling the great beyond. But much more is crystallised when the scales are removed by orbiting space observatories operating outside our planetary smoke screen. Functional 24/7 and with the Universe all around, these are the technological marvels returning tack-sharp images and capturing elusive radiation. Welcome to a small selection of some of the more than 100 launched since the 1960s... the explorers of our time...

■ HUBBLE SPACE TELESCOPE (HST)

More than 100 space observatories have been launched since the 1960s but the telescope that has captured the public imagination, and which is, arguably, the greatest invention of the 20th century, is NASA/ESA's Hubble Space Telescope. Its 2.4m (100in) mirror seems quite small: amazingly, it is almost the same size as the Hooker Telescope on Mount Wilson, California, considered the largest telescope in the world until 1948 and used by the eponymous Edwin Hubble to discover the expansion of the Universe.

The 100-inch mirror Hooker Telescope, Mount Wilson, California: used by Edwin Hubble to demonstrate that 'spiral nebulae' were distant spiral galaxies similar to our own Milky Way.

Launched: April 1990 from Space Shuttle *Discovery*.

Orbit: Low Earth orbit – 370 miles altitude.

Mission: To be the first large optical telescope to obtain images of the Universe with a resolution superior to those viewing through Earth's obscuring atmosphere.

Instruments: Despite an initial problem with spherical aberration of the primary mirror, a subsequent repair mission in December 1993 was a complete triumph. Subsequent servicing missions and a final servicing/upgrade mission in early 2009 has ensured that the HST's various cameras (including the repaired Advanced Camera for Surveys – ACS, the cryogenically cooled Near Infrared Camera and Multi-Object Spectrometer – NICMOS, the new Wide Field Camera 3), spectrographs (including the new Cosmic Origins Spectrograph and Imaging Spectrograph),

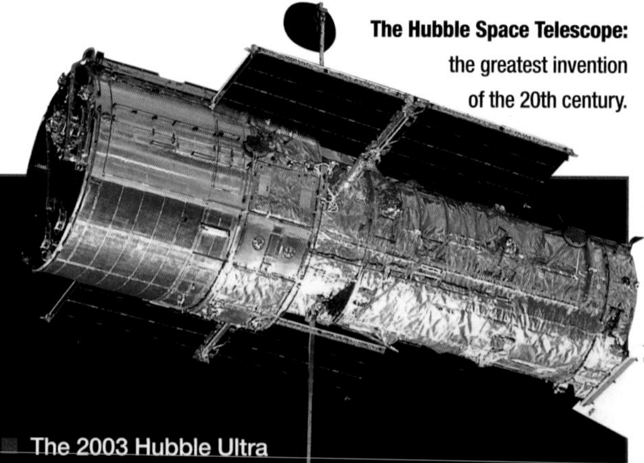

The Hubble Space Telescope: the greatest invention of the 20th century.

HUBBLE HIGHLIGHTS

- The spectacular impact of Comet Shoemaker-Levy 9 in Jupiter's atmosphere in July 1994 and the asteroid, or cometary, impact of July 2009.
- The life cycle of stars, from birth in dust-laden gas clouds to their deaths as beautiful planetary nebulae or titanic supernovae explosions.
- The sites of star formation, such as the famous gaseous 'Pillars of Creation' in the Eagle Nebula, where starbirth exposes evaporating gaseous globules (EGGs) containing embryonic stars.
- The sources of new solar systems – protoplanetary discs (proplyds).
- The December 1995 Hubble Deep Field '100-hour' exposure of an area a tenth the size of the Moon near the Handle of the Big Dipper. It yielded a portrait of the Universe as it existed when half its present age.
- The 2003 Hubble Ultra Deep Field exposure of 10,000 remote galaxies: the deepest visible-light image of the cosmos ever created, crossing billions of light years of space and revealing galaxies when the Universe was just 800 million years old.
- Detecting telltale planet-inducing 'wobbles' and dimming in parent stars in the collective search for exoplanets.
- The 2012 eXtreme Deep Field image exposing 5,500 galaxies.

Southern Pinwheel galaxy, M83, undergoing rapid star formation: captured in unprecedented detail by the HST's Wide Field Camera 3.

and guidance systems work collectively at visible, ultraviolet, and near-infrared wavelengths to continue to provide the most stunning images of the distant Universe ever seen. Its successor, the James Webb Space Telescope (JWST), will take over the baton in 2018.

The HST was one of four instruments in NASA's 'Great Observatories' programme. The others included the Compton Gamma-Ray Observatory, the Chandra X-ray Observatory and the Spitzer Space Telescope. Compton was retired in 2000 and was succeeded in 2008 by the Fermi Gamma-ray Space Telescope (formerly GLAST – Gamma-ray Large Area Space Telescope), currently detecting gamma ray signatures of supermassive black holes, quasars, gamma-ray bursts, pulsars, cosmic ray origins, and other high-energy sources. Space observatories are 'seeing' the Universe from gamma rays to infrared in a spectral range analogous to a piano keyboard some 25 octaves wide.

The Hubble eXtreme Deep Field (HDF): the HST pointed at a patch of southern sky for a total of 50 days and captured 5,500 galaxies spanning back 13.2 billion years.

■ CHANDRA X-RAY OBSERVATORY

Alongside ESA's more sensitive XMM-Newton space telescope that concentrates on X-ray spectroscopy, Chandra also produces spectacular X-ray images of the Universe by 'seeing' into clouds of gas heated to millions of degrees.

Launched: July 1999 from Space Shuttle *Columbia*.

Orbit: Eccentric 64-hour Earth orbit, perigee 18,000 miles (29,000km), apogee 120,000km.

Instruments: High Resolution Camera (HRC); Advanced CCD Imaging Spectrometer (ACIS); two transmission gratings for spectroscopy.

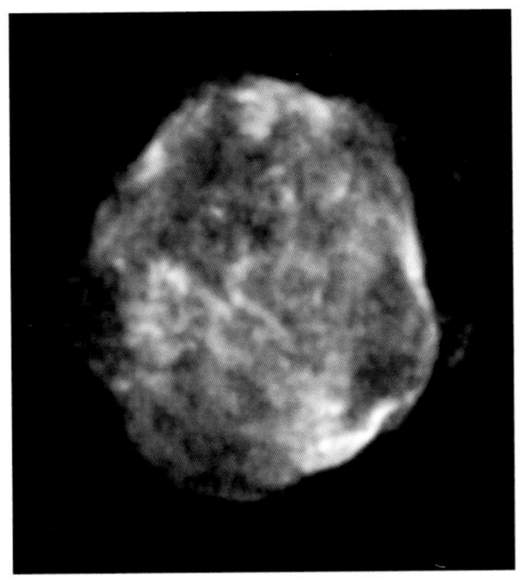

A Chandra X-ray image of N49B: the remains of an exploded star with a cloud of multimillion -degree gas that has been expanding for around 10,000 years.

CHANDRA HIGHLIGHTS

- Resolved diffuse X-ray emission into supermassive black holes with attendant accretion discs in distant galaxies.
- Revealed high-energy particles and magnetic fields around neutron stars.
- Mapped heavy elements around supernovae.
- Discovered new black holes.
- Mapped the temperature and distribution of hot gas in galaxy clusters, enabling measurements of the mysterious dark matter.
- Contributed to more in-depth studies of stars, star formation, and objects in the Solar System.

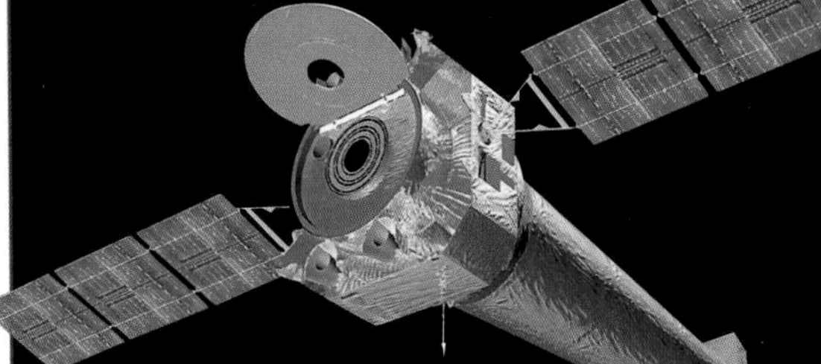

SPITZER INFRARED SPACE TELESCOPE

Launched: August 2003.

Orbit: Earth-trailing (62 million miles) heliocentric orbit, currently receding from Earth at rate of 0.1AU per year. This is necessary to ensure Spitzer operates at just a few degrees above absolute zero – its own heat does not interfere with its infrared detectors.

Instruments: Infrared Array Camera (IRAC); Multiband Imaging Photometer for Spitzer (MIPS).

Mission: Revised in July 2009, mission objectives are to refine Hubble's Constant (the expansion rate of the Universe), search for primordial galaxies, analyse asteroids in Earth's vicinity, and study gas-giant planetary atmospheres.

SPITZER HIGHLIGHTS

- The study of proto-planetary systems around nearby young hot stars.
- Being the first observatory (with Keck II) to see light from an exoplanet and split it into various infrared wavelengths to reveal its atmosphere.
- Creating the first temperature map of an exoplanet and witnessing a colossal exoplanet storm.
- Studying young galaxies in the early Universe.
- Mapping in greater detail nearby galaxies, revealing such features as spiral structure, filaments, and bubbles.
- 'Seeing' the Milky Way's dust-obscured mid-plane and revealing the longer central bar.
- Discovering a myriad of dust-enshrouded black holes billions of light years distant.
- Mapping starlight and glowing interstellar dust.
- Recently discovering the nearly invisible largest ring around Saturn: starting from 3.7 million miles (6 million km) and extending some 7.4 million miles (12 million km), the ring's diameter equivalent to 300 Saturns lined up side by side.

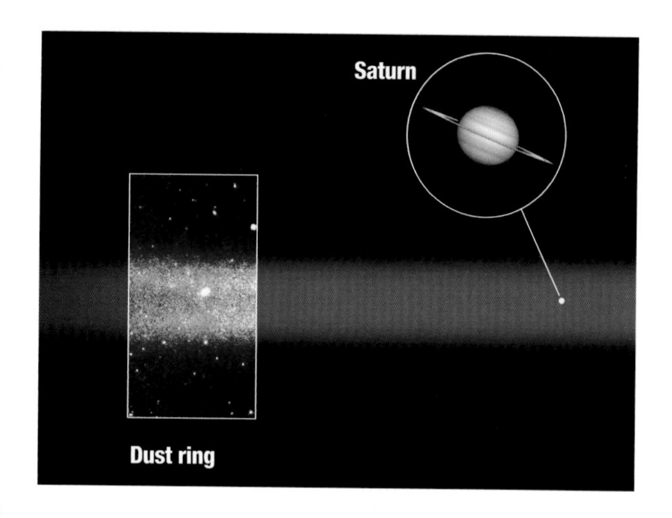

Nearly invisible ring spotted around Saturn: by NASA's Spitzer Space Telescope. The ring is so diffuse it reflects little visible light, but dusty particles shine with infrared light, or heat radiation.

GAIA

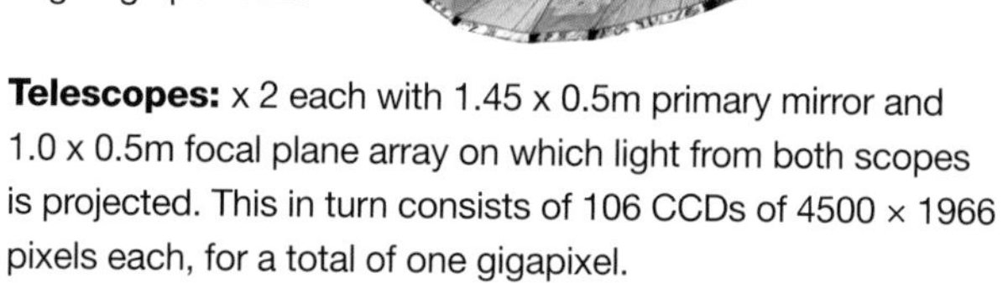

Launched: December 2013 and operated by the European Space Agency.

Orbit: Sun-Earth Lagrange point L2.

Telescopes: x 2 each with 1.45 x 0.5m primary mirror and 1.0 x 0.5m focal plane array on which light from both scopes is projected. This in turn consists of 106 CCDs of 4500 × 1966 pixels each, for a total of one gigapixel.

Instruments: ASTRO astrometry instrument to precisely determine positions of stars of mag 5.7 to 20 by measuring their angular position; BP/RP blue and red photometric instruments to take luminosity measurements and determine stellar temperatures, mass, age and elemental composition: RVS Radial-Velocity Spectrometer to determine velocity of celestial objects along line of site.

GAIA'S GOALS

- Determine the position, distance and annual proper motion of 1 billion stars with an accuracy of about 20 microseconds.
- Measure 20 million stars with a distant precision of 1% and about 200 million to precision of better than 10% – to as far away as the Galactic centre.
- Accurately measure the orbits and inclinations of 1000 extrasolar planets.
- Detect the bending of starlight by the Sun's gravitational field.
- Detect up to 500,000 quasars

■ SWIFT GAMMA RAY BURST EXPLORER (SWIFT)

Gamma ray bursts (GRBs) were discovered in the 1960s but remained mysterious. Now, astronomers know this radiation is emitted as 'torch beams' throughout the Universe and is the result of massive explosions releasing more energy than the Sun could produce at its current rate for 80 billion years! They illuminate the entire Universe and then disappear in seconds without a trace. There are two types: long-duration bursts lasting from more than two seconds to days or weeks, and short bursts lasting from a millisecond to two seconds. The former were known to indicate enormous hypernovae (collapsars) explosions in short-lived stars where the cores have imploded into a black hole. The latter, discovered with the aid of SWIFT, are theorised to be the dying gasps of energy emitted when two neutron stars (or a neutron star and black hole) merge after orbiting each other for millions or billions of years.

Launched: November 2004. Duration: anticipated 2011.

Orbit: Circular, 373-mile (600km) altitude.

Mission: Dedicated to the study of gamma ray bursts (GRBs) and their afterglows observable in the gamma ray, X-ray, ultraviolet, and optical wavebands. SWIFT is designed to automatically slew to pinpoint a source within 20 seconds of a GRB being identified by its on-board system.

Telescopes: Burst Alert Telescope (BAT): detects initial bursts using shadows, not images. Incoming rays pass through a metal shadow mask and cast a unique pattern onto the detector, allowing SWIFT to locate its source. X-Ray Telescope (XRT): studies X-ray afterglows. X-rays cannot be focused with lenses or mirrors: they pass straight through without interaction. But they will reflect off mirrors at narrow angles so this telescope employs a series of mirrors in concentric circles, all of which are almost parallel to the inbound ray. 300mm f/12.7 modified Ritchey-Chretien Ultraviolet/Optical Telescope (UVOT): obtains images in visible and ultraviolet light that are the result of gamma rays impacting clouds of gas and dust. This radiation eventually becomes less energetic infrared and finally radio energy. The UVOT detects the source of an optical glow more accurately than the other instruments and is essential for immediate global ground-based follow-up observations.

SWIFT HIGHLIGHTS

- Determining the origin of gamma ray bursts (GRBs).
- Classifying and seeking new types of GRBs.
- Determining how GRB blast waves evolve and interact with their surroundings.
- Using GRBs to understand the early Universe.
- Undertaking a gamma ray survey of the sky.

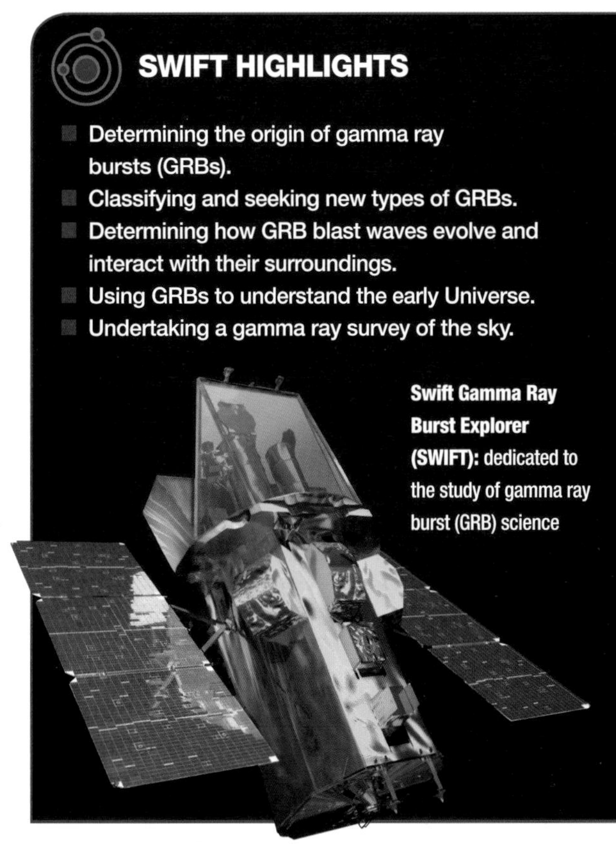

Swift Gamma Ray Burst Explorer (SWIFT): dedicated to the study of gamma ray burst (GRB) science

SWIFT'S 500 GAMMA-RAY BURST

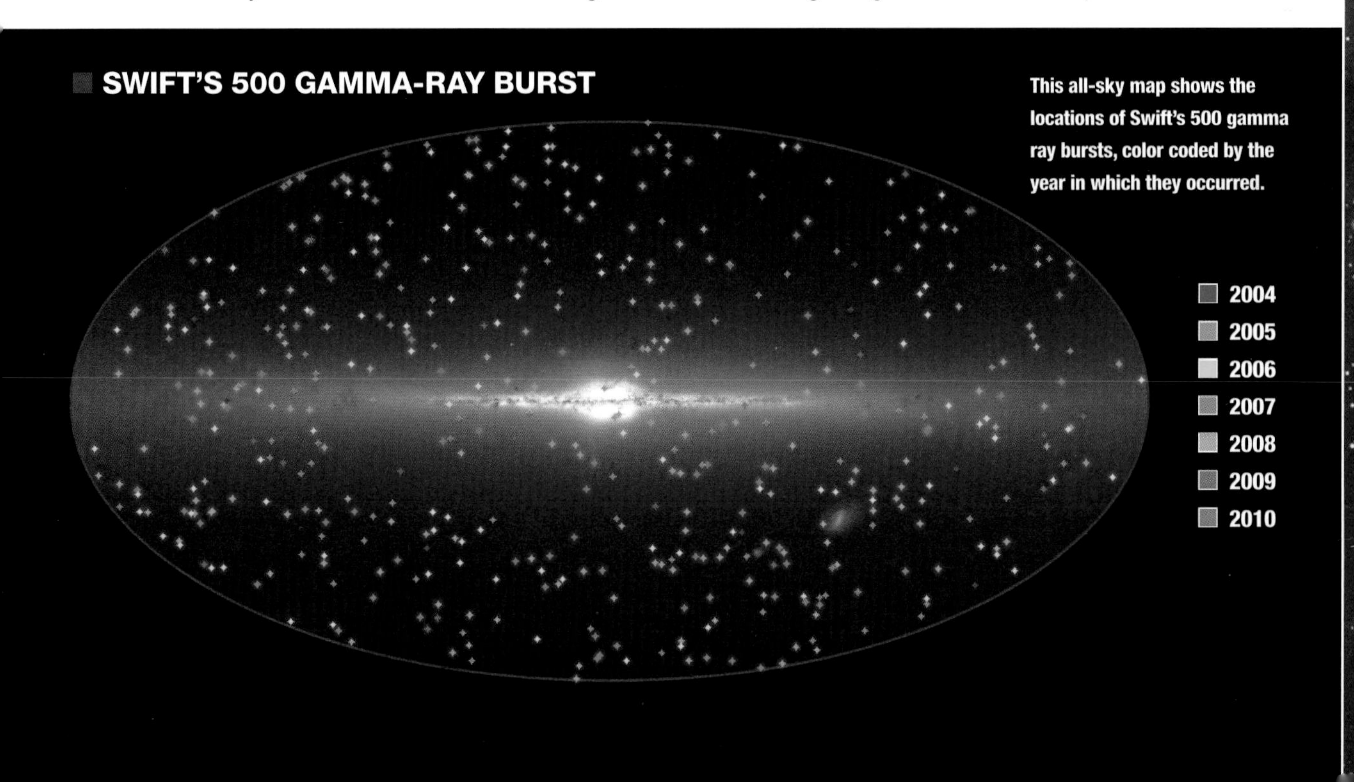

This all-sky map shows the locations of Swift's 500 gamma ray bursts, color coded by the year in which they occurred.

■ SOLAR AND HELIOSPHERIC OBSERVATORY (SOHO)

Following the successful solar studies made by Skylab (launched in 1973), SOHO is a collaborative project between NASA and ESA to understand the structure, dynamics and Earth–Sun interaction of our Star. Why does the corona exist? How is it heated to such phenomenally high temperatures? Where is the solar wind produced and how is it accelerated? SOHO has answered many of these questions.

Launched: On an Atlas II rocket from Cape Canaveral, Florida, in December 1995.

Orbit: Slowly orbits around the First Lagrangian Point (L1) in the direction of the Sun, averaging 1.5 million km from Earth (four times the distance to the Moon), and receiving uninterrupted daylight views of our Star.

Mission: To provide unprecedented breadth and depth information about the Sun.

Design: A spacecraft with two main elements: the payload module and the service module, the latter providing essentials such as thrusters, power, and communications. The payload module houses a scientific payload of 12 instrument packages, five of which are:

- Coronal Diagnostic Spectrometer (CDS): detects emission lines from ions/atoms in the corona and transition region to give information on solar atmosphere, especially plasma (a gas of highly charged particles) in the 10,000 to 1,000,000°C range.
- Extreme ultraviolet Imaging Telescope (EIT): provides full disc images at four selected colours in extreme UV.
- Energetic and Relativistic Nuclei and Electron experiment (ERNE): measures high-energy particles originating from the Sun and Milky Way.
- Global Oscillations at Low Frequencies (GOLF): studies internal structure by measuring velocity oscillations over the entire solar disc.
- Large Angle and Spectrometric Coronograph (LASCO): observes the corona from near the solar limb to a distance of 21,000,000km (one-seventh the distance between the Sun and Earth). It blocks direct light from the Sun's surface with an occulter to create an artificial eclipse, 24/7, and is the key instrument for finding comets.

■ SOLAR DYNAMICS OBSERVATORY

In 2010 NASA launched its Solar Dynamics Observatory (SDO), the first mission as part of NASA's 'Living With A Star' (LWS) program. The program is designed to understand the causes of solar variability and its impact on Earth. SDO itself is designed to help astronomers understand the Sun's influence on Earth and Near-Earth space by studying the solar atmosphere on small scales of space and time and in many simultaneous wavelengths.

Launched: Cape Canaveral Air Force Station Space Launch Complex 41, utilizing an Atlas V-401 rocket with a RD-180 powered Common Core Booster, on February 2010.

Instruments: SDO is a 3-axis stabilized spacecraft, with two solar arrays, and two high-gain antennas. The spacecraft includes three instruments: the Helioseismic and Magnetic

SOHO HIGHLIGHTS

- Studying seismic waves in the Sun's outer shell (which appear as ripples on its surface): the first images of its turbulence, and the sunspots below its 'surface', have been seen.
- Providing the most detailed measurements of the temperature structure, interior rotation, and interior gas flows.
- Measuring the acceleration of the fast and slow solar wind.
- Identifying the source areas and acceleration mechanism of the solar wind in the magnetically 'open' regions at the Sun's poles.
- Discovering coronal waves and solar tornadoes.
- Revolutionising space weather forecasting: three days' notice of Earth-directed disturbances.
- Monitoring the total solar irradiance (total energy from the Sun across different wavelengths) and studying UV flux – both vital in the understanding of the Sun's impact on Earth's climate.
- The observatory is becoming the most prolific comet discoverer in astronomical history.

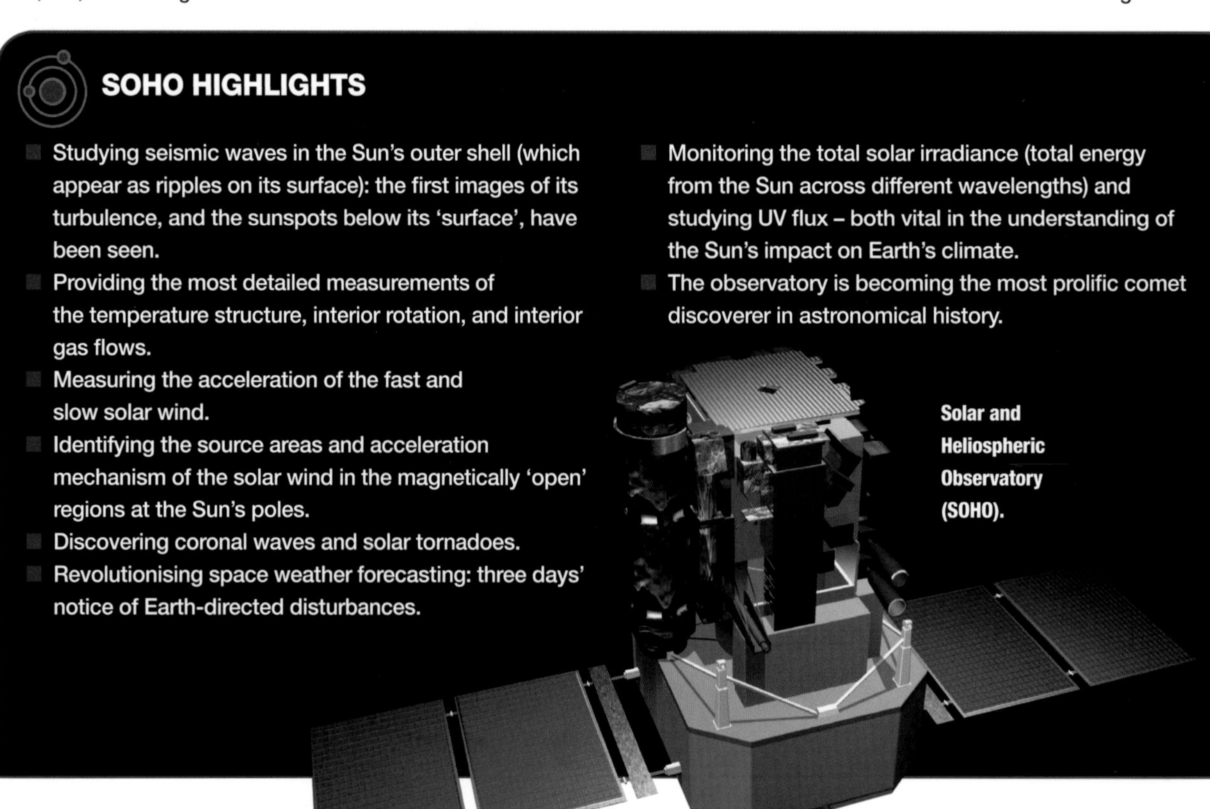

Solar and Heliospheric Observatory (SOHO).

Imager (HMI), the Atmomspheric Imaging Assembly (AIA), and the Extreme Ultraviolet Variability Experiment (EVE).

Helioseismic and Magnetic Imager (HMI): studies solar variability and characterizes our Star's interior and its various components of magnetic activity. It also produces data to determine the interior origins and mechanisms of solar variability and how such physical processes relate to the surface magnetic field and activity. Other data enables estimates of the Sun's coronal magnetic field with regard to variability in the extended solar atmosphere.

Atmospheric Imaging Assembly (AIA): this provides continuous full-disk observations of the solar chromosphere and corona in seven extreme ultraviolet (EUV) channels, spanning a temperature range from approximately 20,000 Kelvin to in excess of 20 million Kelvin.

Extreme Ultraviolet Variability Experiment (EVE): measures the Sun's extreme ultraviolet irradiance with improved spectral resolution, 'temporal cadence', accuracy, and precision.

SDO'S Goals: The goal is to understand, with a view to predictive capability, the solar variations that influence life on our planet and humanity's technological systems, and this is being done by determining how the Sun's magnetic field is generated and structured and by understanding how this stored magnetic energy is converted and released into the heliosphere and geospace in the form of solar wind, energetic particles and variations in the solar irradiance.

SDO Plasma Push and Pull: Dark strands of plasma hovering above the Sun's surface began to interact with each other in a form of tug of war over two and a half days (June 28-30, 2015). At times, strands of plasma extended a tenuous connection between one area and the other. Twice the small tower of plasma to the lower left shot a burst of energy over to the quivering filament higher up. We are seeing the push and pull of magnetic forces revealed in a wavelength of extreme ultraviolet light.

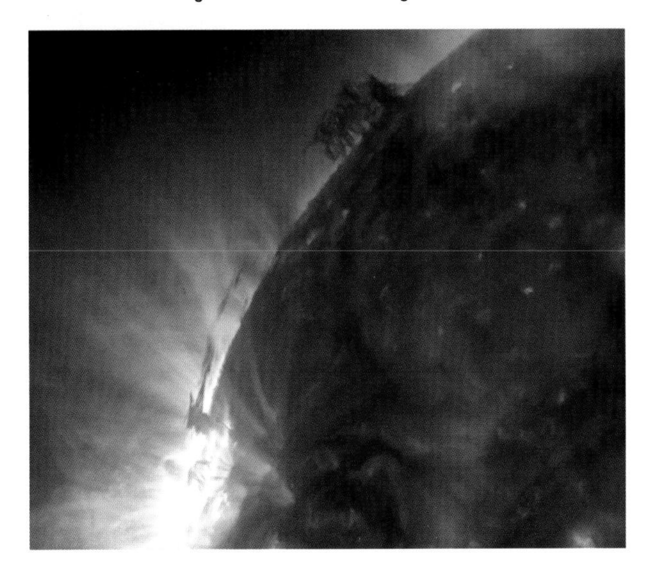

HMI (Helioseismic and Magnetic Imager): The Helioseismic and Magnetic Imager extends the capabilities of the SOHO/MDI instrument with continual full-disk coverage at higher spatial resolution.

AIA (Atmospheric Imaging Assembly) (image top left): The Atmospheric Imaging Assembly images the solar atmosphere in multiple wavelengths to link changes in the surface to interior changes. Data includes images of the Sun in 10 wavelengths every 10 seconds. PI: Alan Title, PI Institution: Lockheed Martin Solar Astrophysics Laboratory.

EVE (Extreme Ultraviolet Variability Experiment) (above image): The Extreme Ultraviolet Variability Experiment measures the solar extreme-ultraviolet (EUV) irradiance with unprecedented spectral resolution, temporal cadence, and precision. EVE measures the solar extreme ultraviolet (EUV) spectral irradiance to understand variations on the timescales which influence Earth's climate and near-Earth space. PI: Tom Woods, PI Institution: University of Colorado.

NASA's Solar Dynamics Observatory spacecraft.

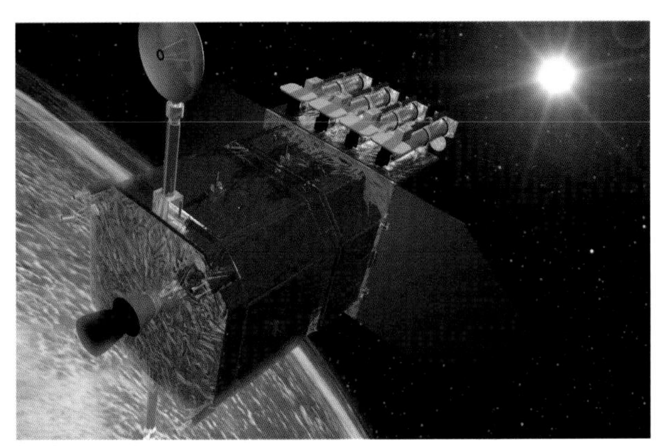

■ NUSTAR (NUCLEAR SPECTROSCOPIC TELESCOPE ARRAY)

NuSTAR is a space-based X-ray telescope that focuses high-energy X-rays from astrophysical sources, especially for nuclear spectroscopy. It is the eleventh mission of the NASA Small Explorer satellite program and the first space-based direct-imaging X-ray telescope at energies beyond those of the Chandra X-ray Observatory (see page 155) and XMM-Newton.

Launched: On 13 June 2012 on a Pegasus XL rocket dropped from the L-1011 'Stargazer' aircraft, about 117 nautical miles south of Kwajalein Atoll, part of the Republic of the Marshall Islands.

Instruments: Unlike visible light telescopes – which employ mirrors or lenses – NuSTAR has to employ grazing incidence optics to be able to focus X-rays. For this two conical approximation Wolter telescope design optics with 10.15 metres (33.3 ft) focal length are held at the end of a long deployable mast. A laser metrology system is used to determine the exact relative positions of the optics and the focal plane at all times, so that each detected photon can be mapped back to the correct point on the sky even if the optics and the focal plane move relative to one another during an exposure.

NuSTAR's GOALS:

NuSTAR's primary goals are to conduct a deep survey for black holes a billion times more massive than the Sun, to investigate how particles are accelerated to super high energy in active galaxies and to understand, by imaging their remains, how radioactive elements are created in supernova explosions.

In February 2013, NASA revealed that the Observatory, along with the XMM-Newton Observatory, had measured the spin rate of the supermassive black hole at the centre of the galaxy NGC 1365.

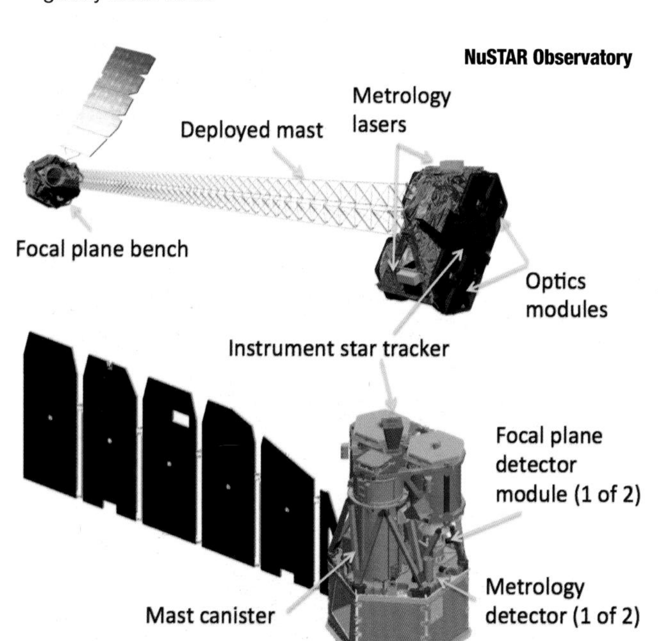

NuSTAR Observatory

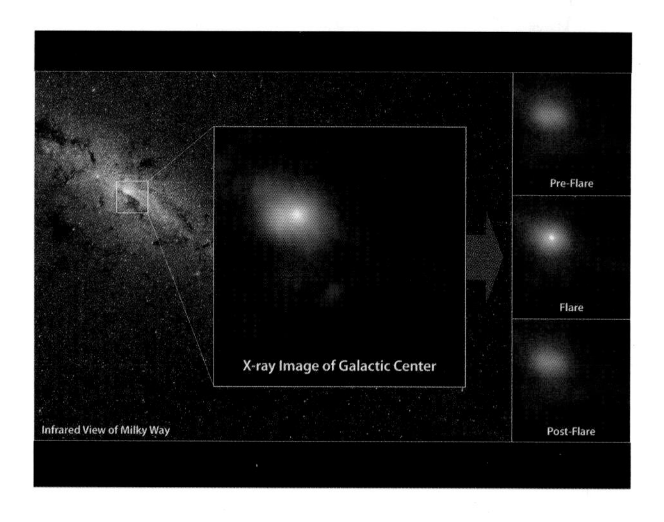

NuSTAR View of Black Hole: NuSTAR captured these first, focused views of the supermassive black hole at the heart of our Milky Way Galaxy in high-energy X-ray light.

■ KEPLER SPACE TELESCOPE

A number of ground- and space-based telescopes are employed as 'exoplanet hunters' – they seek out planets in orbit around other stars in the Solar System using the ingenious methods outlined in Chapter 2. The Canadian Space Agency's Microvariability and Oscillations of Stars (MOST) space telescope and the French Space Agency/ESA's, Convection Rotation and Planetary Transits (CORoT) satellite have paved the way, but NASA's Kepler Space Telescope and ESA's GAIA spacecraft, along with other planned missions, are mankind's next leap on the road to discovering the Holy Grail in astronomy – a planet most resembling Earth.

Kepler space telescope: an artist's impression.

Launched: March 2009. Failures in two of four reaction wheels crippled its extended mission in 2013. Without three functioning wheels, the telescope could not be pointed accurately. NASA announced the completion of Kepler's primary mission, and the beginning of its extended K2 mission, which, at time of going to press, is still ongoing. Duration: 3½ years primary mission.

Named: After the German mathematician and astronomer, Johannes Kepler (1571–1630), who constructed the famous three laws of planetary movement.

Orbit: The K2 mission observes a specific portion of the distant sky for approximately 80 days, until it is necessary to rotate the spacecraft to prevent sunlight from entering the telescope. The spacecraft orbits the sun every 372 days as it trails Earth, allowing for four full campaigns per orbit or year

Mission:
KEPLER's Goals:
The scientific objective of KEPLER has been to explore the structure and diversity of planetary systems. It has observed a large sample of stars to achieve several key goals:
Determining how many Earth-size and larger planets there are in or near the habitable zone (often called 'Goldilocks planets') of a wide variety of spectral types of stars.
Determining the range of size and shape of the orbits of these planets.
Estimating how many planets there are in multiple-star systems.
Determining the range of orbit size, brightness, size, mass and density of short-period giant planets.
Identifying additional members of each discovered planetary system using other techniques.
Determining the properties of those stars harbouring planetary systems.

Achievements:
As a result of failures with two of the four reaction wheels, in November 2013 a new mission plan, named K2 'Second Light', was presented for consideration. This new mission would involve using KEPLER's remaining capability, photometric precision of about 300 parts per million, compared with its earlier 20 parts per million, to collect data for the study of supernova explosions, star formation and solar-system bodies such as asteroids and comets, and for finding and studying more exoplanets, searching a much larger area in the plane of Earth's orbit around the Sun.
In May 2015, KEPLER observed a newly discovered supernova, KSN 2011b (Type 1a), before, during and after explosion. Details of the pre-nova moments could help scientists better understand dark energy.
In July 2015, NASA announced the discovery of Kepler-452b, a confirmed exoplanet that is near-Earth in size and found orbiting the habitable zone of a Sun-like star.
In September 2015, the number of exoplanet candidates numbered 4,696 and the confirmed number of planets 1,030, of which 12 were less than twice Earth's size and in the 'habitable zones' of their parent star. Operating as the new two-wheeled K2 Mission, a further 24 exoplanets have also been confirmed.

Kepler Telescope: a view of the backside of the solar array on the left. The spacecraft and photometer without the sunshade are shown on the right.

Telescope: 0.95m telescope with a photometer consisting of 42 CCDs, each with 2,200x1,024 pixels.

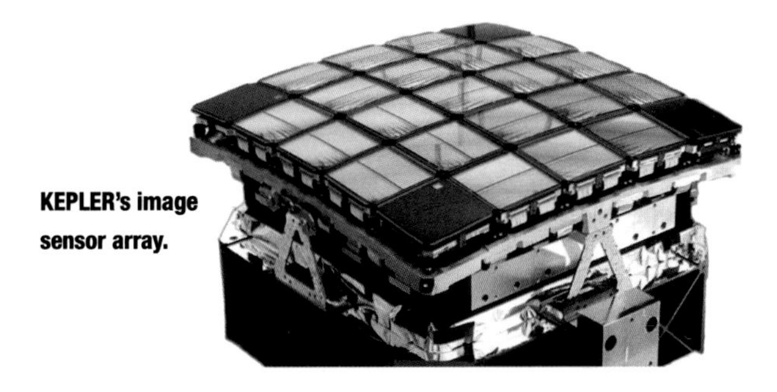

KEPLER's image sensor array.

Kepler K2: This artistic impression shows NASA's planet-hunting KEPLER spacecraft operating in a new mission profile called K2. Rather than giving up on the stalwart spacecraft, a team of scientists and engineers crafted a resourceful strategy to use pressure from sunlight as a virtual reaction wheel to help control the spacecraft while observing the sky in the ecliptic plane, the orbital path of Earth around the Sun, depicted by the grey-blue line marked by opaque cross-like shapes. Each shape represents the field-of-view of an observing campaign.

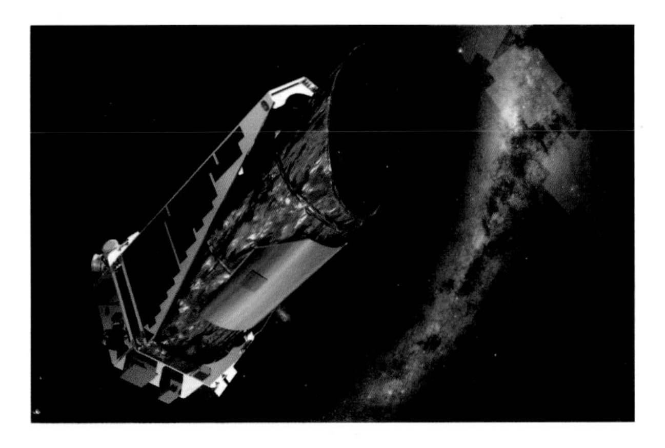

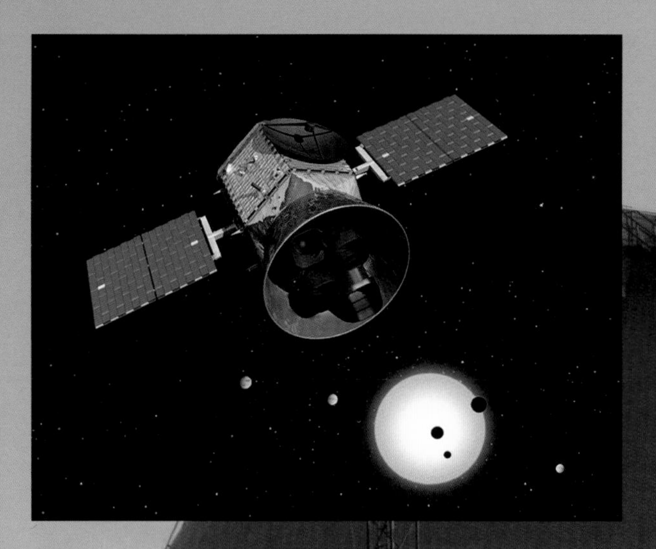

This is a conceptual image of the TESS mission. Designed as the first all-sky survey, the Transiting Exoplanet Survey Satellite, or TESS, would spend two years of an overall three-year funded science mission searching both hemispheres of the sky for nearby exoplanets.

The Gemini Planet Imager is the next generation adaptive optics instrument being built for the Gemini South Telescope in Chile (background image). The goal is to image extrasolar planets orbiting nearby stars.

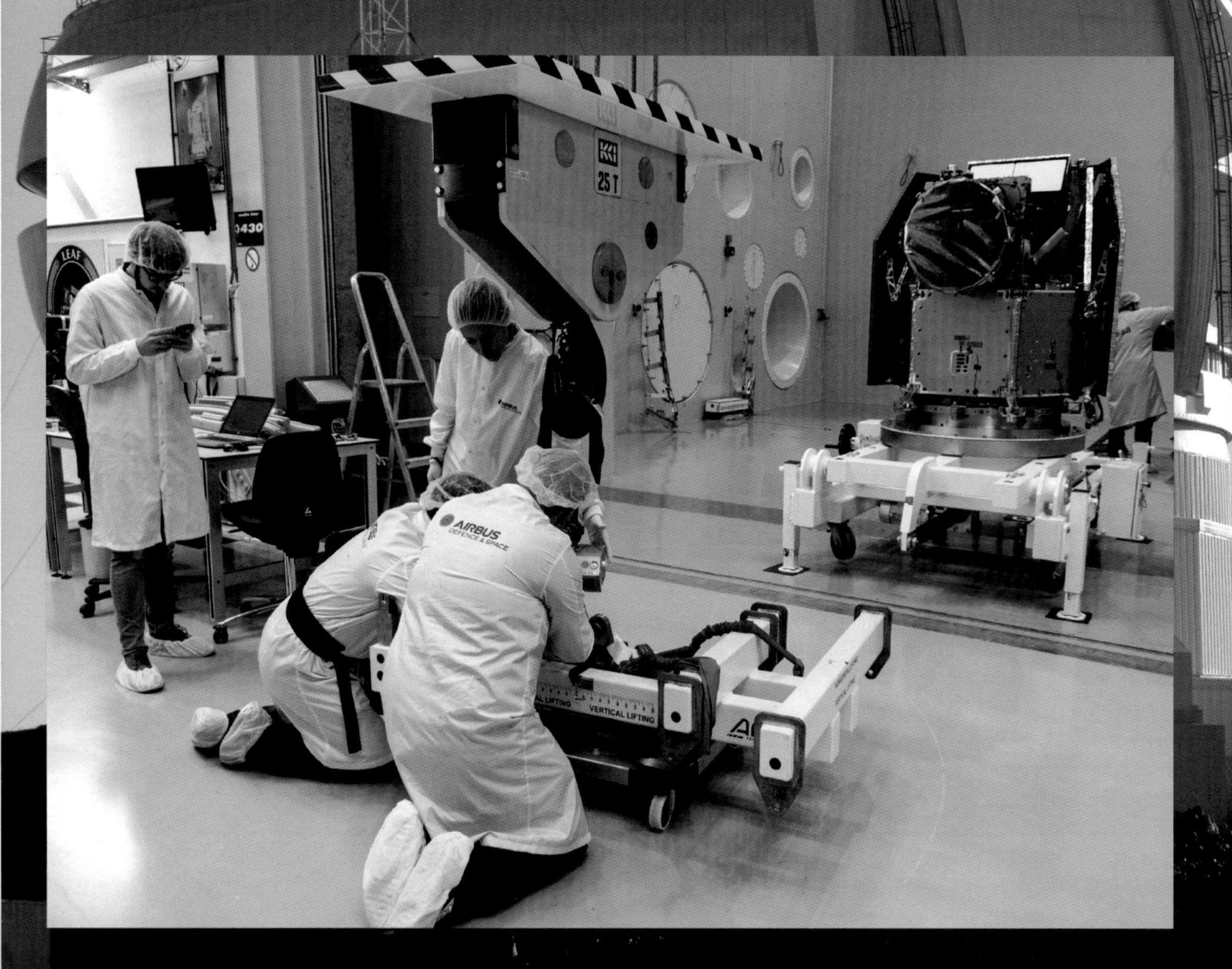

A test model of ESA's exoplanet-watching CHEOPS satellite being placed in an acoustic chamber in Europe's largest spacecraft testing centre. The Characterising ExOPlanet Satellite is ESA's first small science mission. Selected in October 2012, it will track the crossings of known planets across the face of their parent stars, to make detailed deductions of their size and composition. The telescope will detect tiny shifts in stellar brightness with ultra-high precision

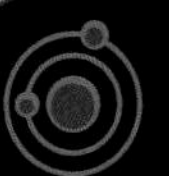

BEYOND KEPLER: OTHER MISSIONS TO SEARCH FOR ALIEN PLANETS

Future:
As well as NASA's GAIA mission (see page 156) a handful of other spacecraft are poised to join the search for exoplanets, as follows:

Characterizing Exoplanets Satellite, or CHEOPS: operated by ESA with a planned launch for 2017. This satellite will stare at nearby stars known to host planets, watching for these worlds to transt their host stars. High precision measurements made by the satellite should help astronomers detail planet sizes. Data gathered from the ground should then provide masses for newly discovered worlds, allowing astronomers to calculate their density.

Transiting Exoplanet Survey Satellite, or TESS: operated by NASA with a planned launch for 2017. This satellite will also use the transit method to search for exoplanets orbiting nearby stars, with a focus on Earth-size planets that may be capable of supporting life.

James Webb Space Telescope, or JWST: slated for launch in 2018, NASA's powerful telescope will also be aimed at newfound worlds, scanning their atmospheres for water vapour and gases that may have been produced by living organisms, such as oxygen, nitrous oxide and methane. (See page 164)

Wide-Field Infrared Survey Telescope, or WFIRST: many researchers are hopeful that NASA will build and launch this observatory. Its telescope would not only hunt for exoplanets but also probe the mysteries of dark energy and galaxy evolution.

Current
Wide Angle Search for Planets (WASP and SuperWASP): uses two continuously operating robotic telescopes, one in La Palma, Canary Islands, and the other near Sutherland, South Africa. Both observatories use eight wide-angle cameras that simultaneously monitor the sky for exoplanetary transits. By monitoring millions of stars simultaneously, they are able to capture rare transit events.

Next-Generation Transit Survey (NGTS): sited at the European Southern Observatory's Paranal Observatory in Chile, this is a wide-field photometric survey designed to discover transiting exoplanets of Neptune-size and super-Earth size around stars brighter than those surveyed in NASA's KEPLER mission, providing prime targets for characterization by the VLT (see page 152), E-ELT (see page 166) and the JWST (see page 164). It employs an array of fully-robotic small telescopes operating in the 600-900nm band.

Evryscope: Greek for 'wide-seeing', this is an array of telescopes, at the Cerro Tololo Inter-American Observatory (CTIO) in Chile, pointed at every part of the accessible sky simultaneously and continuously, together forming a gigapixel-scale telescope monitoring an overlapping 8,000 square degree field every 2 minutes. It will be capable of searching for transiting giant planets around the brightest and most nearby stars, where the planets are much easier to characterize; it will also search for small planets around nearby M-dwarfs, for planetary occultations of white dwarfs, and will perform comprehensive nearby microlensing and eclipse-timing searches for exoplanets inaccessible to other planet-finding methods.

The Evryscope contains 27 61-mm telescopes fitted onto a single German Equatorial mount.

Gemini Planet Imager (GPI): this is a high contrast imaging instrument built for the Gemini South Telescope in Chile, which allows for direct imaging and integral field spectroscopy of exoplanets around nearby stars, specifically young gas giants, via their thermal emission.
A number of research groups around the world have employed other Earth-based instruments – the **High Accuracy Radial Velocity Planet Searcher (HARPS)** spectrograph, attached to a 3.6m telescope at ESO's La Silla site in Chile, and the **High Resolution Echelle Spectrometer (HIRES)**, on Hawaii's Keck Telescope, are two examples — to spot exoplanets. These scientists often use the radial velocity method, which detects the tiny gravitational wobbles orbiting worlds induce in their parent stars.

The future

How far we have come since Lippershey, Harriott, Galileo, and others peered through tiny eyepieces at objects now considered to be on the cosmic doorstep! The desire to image and study Earth-like exoplanets, detect the earliest supernovae, dissect dark energy, and isolate stars in distant galaxies to determine how they formed, drives the development of larger and more sensitive instruments. Below is just a selection of the exciting new space- and ground-based leviathans currently in development or being planned.

JAMES WEBB SPACE TELESCOPE (JWST)

Set for launch in 2018, and named after a former NASA administrator, this US/European/Canadian collaboration will be seven times more sensitive than its predecessor, the HST, and will image the Universe in infrared light to a distance and accuracy not seen before. It should shed light on the first objects formed in the Universe, witness primordial galaxy formation, and search for new planetary systems.

It will be in stable orbit at the Lagrange 2 point some 1.5km away from Earth where the gravity of our Star and planet cancels out. Consisting of a 6.5m lightweight beryllium mirror constructed from 18 separate hexagonal segments, it will be the sixth largest science project ever undertaken. Its open truss design requires a large sunshield (the size of a tennis court) to protect it from interference from the Sun, Earth, and Moon, stray infrared light, and its own thermal infrared emissions.

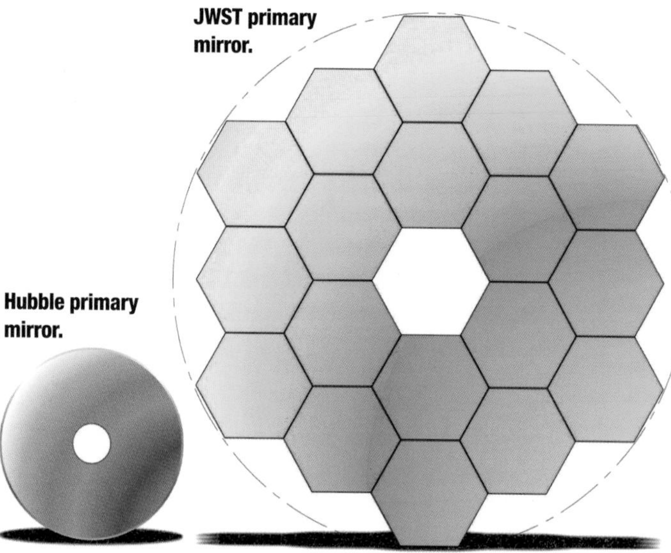

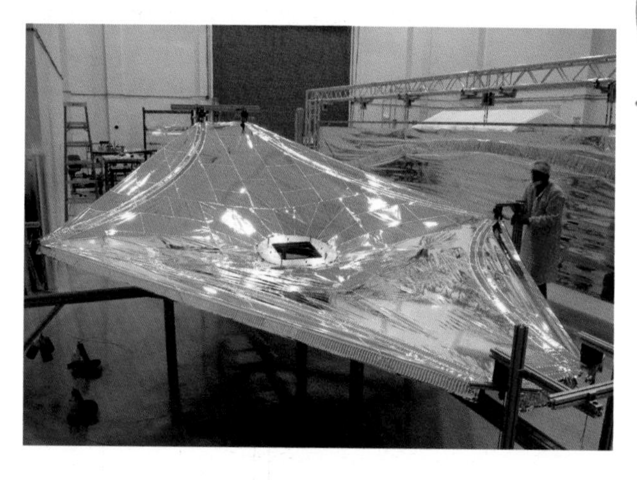

Other revolutionary space telescopes are also being proposed. The Advanced Technology Large-Aperture Space Telescope (ATLAST) would be an 8 to 16.8 metre Ultra Violet -Optical-Near Infrared space telescope and would be a replacement for the Hubble Space Telescope, with better

James Webb Space Telescope: the JWST's one-third scale sunshield under tension testing at the Nexolve facility in Huntsville, Alabama.

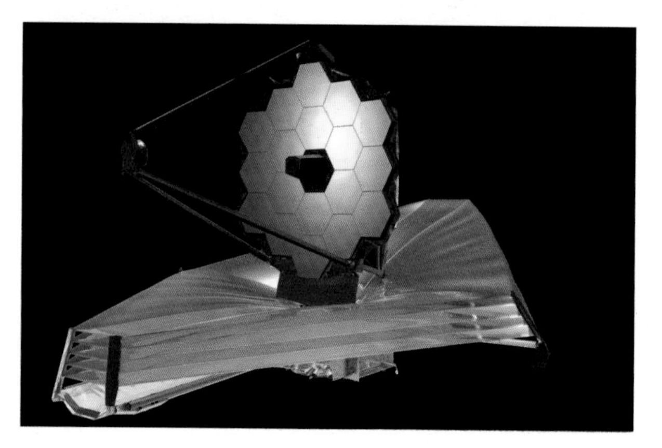

ATLAST: One proposed ATLAST concept, a design based on an 8 meter monolithic mirror.

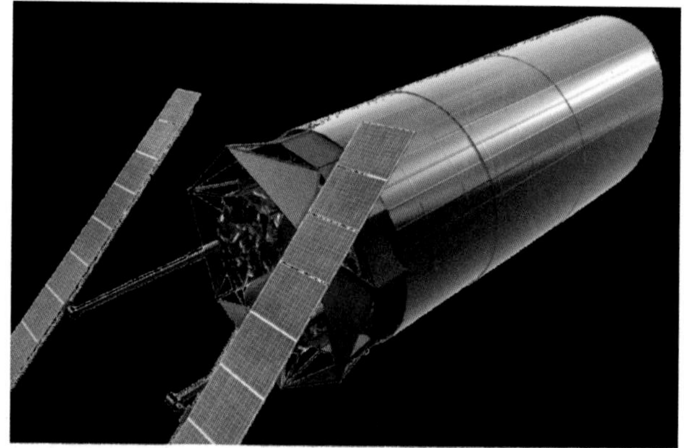

Northern leg (x-arm) of LIGO interferometer on Hanford Reservation.

resolution than either the HST or the JWST. Like both these scopes it will be launched to the Sun-Earth L2 Lagrange Point. Envisioned as a flagship mission, it will search for biosignatures (such as molecular oxygen, ozone, methane and water) in the spectra of terrestrial exoplanets. Yet another is the High-Definition Space Telescope (HDST), a space observatory also to be located at the Sun-Earth Lagrange Point. It would be a megascope, comprising 54 mirror segments with an aperture of 12 metres, offering images 24 times sharper than the HST as a result of being five times as big and 100 times as sensitive – large enough to detect the possible dozens of Earth-like planets lurking in our nearby neighbourhood.

Already fully operational is the ground-based Advanced Laser Interferometer Gravitational-wave Observatory (LIGO), its goal to directly observe gravitational waves created at the birth of the Universe. Measurable emissions of gravitational waves are also anticipated from binary systems (collisions of neutron stars or black holes), supernova explosions of massive stars (which form neutron stars and black holes), accreting neutron stars and the rotations of neutron stars with deformed crusts. The observatory could also observe exotic hypothetical phenomena, such as gravitational waves caused by oscillating cosmic strings.

LIGO currently operates two observatories in unison, the LIGO Livingston Observatory in Livingston, Louisana, and the LIGO Hanford Observatory near Richland, Washington. Both sites are separated by 1,865 miles. Gravitational waves travel at the speed of light so this distance equates to different wave arrival times of up to ten milliseconds! Through triangulation, this difference can determine the source of the wave in the sky.

Based on current models of astronomical events, and the predictions of Einstein's general theory of relativity, gravitational waves that originate tens of millions of light years from Earth are expected to distort the 4 kilometer mirror spacing by about 10–18 m, less than one-thousandth the charge diameter of a proton.

Each observatory has an L-shaped ultra high vacuum system, measuring 2.5 miles on each side and up to five interferometers can exist in each vacuum system. The primary interferometer at each site consists of mirrors suspended at each of the corners of the 'L'. A pre-stabilized laser emits a beam of up to 200 Watts. This passes through an optical mode cleaner before reaching a beam splitter at the vertex of the 'L'. Here, the beam splits into two paths, one for each arm of the L. When a gravitational wave passes through the interferometer, the space-time in the local area is altered. Depending on the source of the wave and its polarization, this results in an effective change in length of one or both of the cavities. The effective length change between the beams will cause the light currently in the cavity to become very slightly out of phase with the incoming light. The cavity will therefore periodically become very slightly out of resonance. The beams, which are tuned to destructively interfere at the detector, will then exquisitely lightly periodically vary the detuning. This results in a measurable signal.

The LIGO Laboratory is currently building Advanced LIGO wherby the detectors at both observatory sites are being replaced to further improve the sensitivity of the original LIGO detectors by at least a factor of 10.

■ GIANT MAGELLAN TELESCOPE (GMT)

Being constructedat the Cerro Las Campanas Observatory in Chile – home to the twin 6.5m Magellan Telescopes – this Goliath will have seven 8.4m (28ft) honeycomb-backed mirrors (the same size as the LBT) arranged like flower petals and should see 'first light' in 2025. The telescope is expected to have over five to ten times the light-gathering ability of existing instruments. Three mirrors have been cast. Even though seven primary mirrors are planned, it can begin operation with four. Utilising the most advanced adaptive optics to correct atmospheric distortion, image sharpness and quality should exceed that of the Hubble Space Telescope.

■ THIRTY METER TELESCOPE (TMT)

The futuristic Californian Thirty Meter Telescope (TMT), due for completion around the middle of the next decade, will collect nine times more light than a single 10m Keck scope, capture objects ten times fainter, and have a spatial resolution three times greater. Its enormous f/1 primary mirror will comprise 492 segments and, as its name suggests, be 30m wide. Complementing this will be a concave 'active' deformable secondary one, providing minuscule corrections as necessary to compensate for mirror flexure. The tertiary mirror will then direct the corrected light to a suite of instruments sited on solid platforms along the altitude axis. This mammoth's primary goal will be to peer to the most distant horizons at near-infrared wavelengths.

Thirty Meter Telescope (TMT): a leviathan in the making.

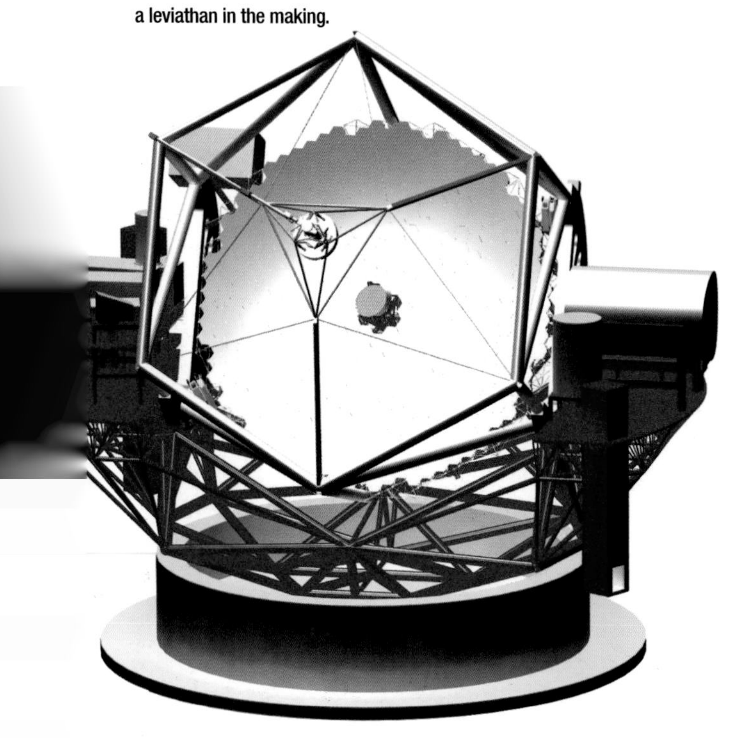

■ EUROPEAN EXTREMELY LARGE TELESCOPE (E-ELT)

The European Extremely Large Telescope (E-ELT) is currently being built by the European Southern Observatory (ESO) atop Cerro Armazones in the Atacama Desert of northern Chile and should see 'first light' in 2024. With its gargantuan mirror, 42m (138ft) in diameter and consisting of 906 hexagonal segments, it is a staggering operation. Of necessity, the complementing mirrors in its light path will be similarly supersized, with the secondary measuring 6m across – as large as the primary in many of today's telescopes! A 4.2m tertiary mirror will relay the light beam to an adaptive optics system that will comprise a 2.5m active mirror, with a stunning 5,000 actuators tweaking its shape 1,000 times a second, and a 2.7m mirror for the final image corrections. Such a phenomenal telescope will ramp up the search for exoplanets by probing the earliest stages of planetary formation, as well as detecting water and organic molecules in the protoplanetary discs around infant stars.

European Extremely Large Telescope (E-ELT): a gargantuan undertaking.

Have we reached a size threshold? Certainly, if telescopes are to be larger then the very manufacturing tools will need to be redesigned. Just as JFK's mission to put man on the Moon took over 400,000 personnel and the creation of tools to make the tools to make the instruments to achieve that goal, so bigger and better scopes will require new power mechanisms, tolerances, metal strengths, and a whole host of other issues yet to be considered. As we have seen, the bigger the mirror, the greater the atmospheric challenge: catching starlight over a wider area and varying altitudes increases wavefront distortion. So, optical scientists are devising a new technique: multi-conjugate adaptive optics (MCAO), where up to five artificial 'guide' stars can be created from sodium lasers on the ground.

And what about moveable mirrors? Canadian scientists have already constructed liquid-mirror telescopes in which starlight is reflected by the naturally curved surface of a rotating reservoir of mercury. They are inexpensive and relatively straightforward to build, but their disadvantage is that they only 'see' overhead. These could be used in the study of remote galaxies anywhere in the sky. There are even thoughts on a Large Aperture Mirror Array (LAMA): a collection of eighteen 10m mercury telescopes working in unison with a light-gathering power equal to the European Extremely Large Telescope.

Atacama Large Millimetre Array (ALMA): (pictured above) 66 12-metre and 7-metre dishes will target the shortest wavelength radio waves to revolutionise our understanding of star formation.

■ THE ATACAMA LARGE MILLIMETER/ SUB-MILLIMETRE ARRAY (ALMA)

Size is less of a concern, however, for radio telescopes: simply achieve the power of a giant receiver by 'connecting' numerous smaller ones. For 21st-century astronomers-artists, the radio wavelength palette stretches far and wide.

Now fully operational, the Atacama Large/Sub-millimetre Array (ALMA) is a collection of 66 12-metre and 7-metre diameter radio telescopes observing at millimetre and sub-millimetre wavelengths and has been constructed, at an altitude of 5,000 metres, on the Chajnantor Plateau near the Llano do Chajnantor Observatory in the Atacama Desert. A giant transporter moves the dishes to reconfigure the array, enabling zoom-lens imaging into places of star birth in the early Universe, as well as providing an insight into nearby star and planetary formation.

The European Extremely Large Telescope (E-ELT)

The LOFAR core near Exloo, Netherlands.

■ LOW-FREQUENCY ARRAY (LOFAR), LONG WAVELENGTH ARRAY (LWA), AND MURCHISON WIDE-FIELD ARRAY (MWA)

Operating at longer wavelengths (10mm to 5m) is the Low-Frequency Array (LOFAR), in the Netherlands, the Long Wavelength Array (LWA), under construction in New Mexico, and the Murchison Wide-field Array (MWA), also under construction in Western Australia.

Traditional radio telescopes use dish antennae but these are only effective if they are at least five times wider than the incoming radio signal wavelength: wavelengths measured in metres would necessitate much larger dishes. To compensate, LOFAR, LWA,under construction and MWA, also under construction comprise thousands of dipole receivers resembling the old 'rabbit ear' TV antennae. Instead of being aimed in a fixed direction, these 'ears' seek signals from all over the sky. High-speed supercomputers, known as correlators, combine the signals to give simultaneous coverage of many celestial objects. In the case of LOFAR, the largest connected radio telescope ever built, the project utilises an interferometric array of radio telescopes; about 25,000 small antennas concentrated in at least 48 larger stations. Forty of these stations are distributed across the Netherlands, five stations in Germany, and one each in Great Britain, France and Sweden. Fibre optics network them to a central supercomputer, enabling observations in eight different directions at any one time! And many astronomers seek a far more distant vantage point ... the Moon. An array on the far side of our satellite would provide the perfect zero-Earth-interference base.

■ AUSTRALIAN SQUARE KILOMETRE ARRAY PATHFINDER (ASKAP) AND SQUARE KILOMETRE ARRAY (SKA)

The Australian Square Kilometre Array Pathfinder, or ASKAP, is a new radio telescope currently being commissioned at the Murchison Radio-astronomy Observatory (MRO) in the Mid West region of Western

Antennas of CSIRO's ASKAP telescope at the Murchison Radio-astronomy Observatory in Western Australia.

Australia, and will pave the way for the Square Kilometre Array (SKA), the largest and most sensitive telescope, consisting of thousands of antennas linked together by high bandwidth optical fibres, due to see first light around 2020. The telescope will be implemented across two main sites: the Murchison region in Western Australia and southern Africa and will be 50 times as sensitive as the best existing telescopes, with a survey speed 10,000 times faster than any of its current day rivals. It aims to address fundamental questions about the evolution of the Universe including the formation of black holes, the origins of the first stars and the generation of magnetic fields in space.

ASKAP comprises 36 identical antennas, each 12 metres in diameter, all working in unison as a single instrument to achieve a total collecting area of approximately 4,000 square metres. These exquisitely sensitive antennas, located in a low-populated, and thus extremely 'radio quiet' region of Western Australia, will be used to study galaxy formation and the evolution of gas in the nearby Universe, as well as investigating the evolution, population and formation of galaxies across cosmic time.

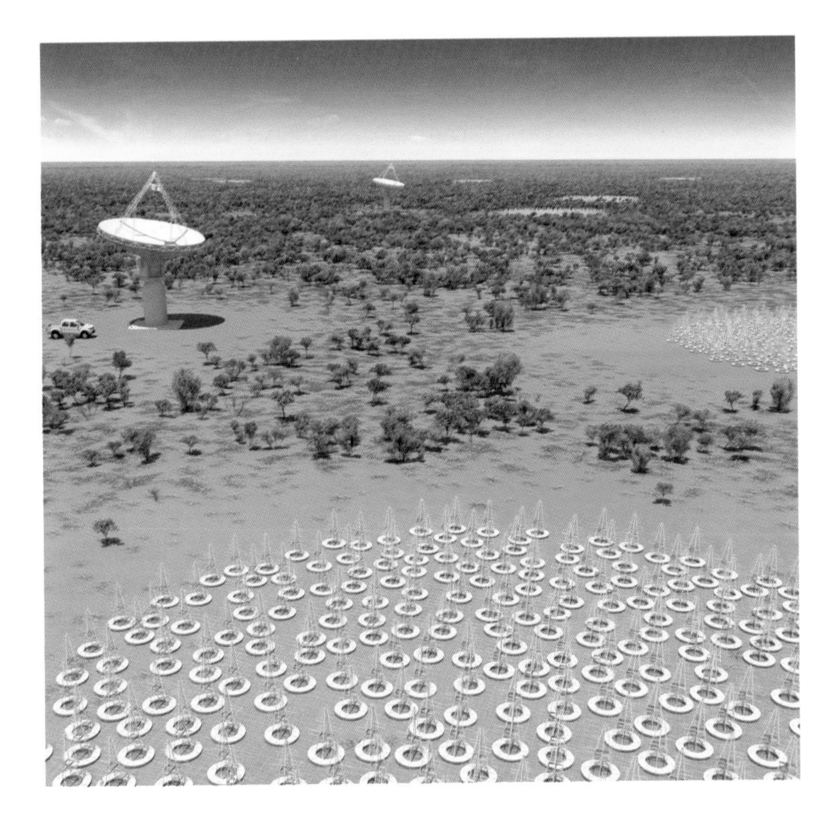

The SKA will comprise a number of different receiver systems, which will enable observations over a large frequency range. Credit: CSIRO.

Artist's impression of the 15m x 12m Offset Gregorian Antennas within the central core of the Square Kilometre Array.

Global survey telescopes

Surveying the entire sky is not a new concept. Astronomers as far back as Hipparchus (*c*130 BC) have catalogued the stars. William Herschel may also have made a full-scale sweep, detecting as much as his 7ft focal length reflector could allow. Around a century ago the Carte du Ciel project tried to record the entire sky on photographic plates, a feat successfully achieved in the 1950s by the Palomar Observatory Sky Survey.

SLOAN DIGITAL SKY SURVEY (SDSS)

The Sloan Digital Sky Survey (SDSS) is one of the most ambitious and significant global surveys in the history of astronomy. Over eight years (SDSS-I, 2000-2005; SDSS-II, 2005-2008), it obtained deep, multi-colour images covering more than a quarter of the sky and created 3-dimensional maps containing more than 930,000 galaxies and more than 120,000 quasars. To date, it has created the most detailed three-dimensional maps of the Universe ever made.
Building on the SDSS and SDSS-II legacy, the SDSS-III Collaboration works to map the Milky Way, search for extrasolar planets, and solve the mystery of dark energy.
SDSS-III's recent release was Data Release 12 (DR12) and this contained the final data from SDSS-III, including the first spectra of the Multi-object APO Radial Velocity Exoplanet Large-area Survey (MARVELS), additional data from the APO Galactic Evolution Experiment (APOGEE), and additional sky coverage and improved galaxy parameter estimates from BOSS (Baryon Oscillation Spectroscopic Survey).

MARVELS: Characterizing Extrasolar Planets

The Multi-object APO Radial Velocity Exoplanet Large-area Survey (MARVELS) monitored the radial velocities of 11,000 bright stars, with the precision needed to detect gas giant planets that have orbital periods ranging from several hours to two years.

APOGEE: Probing the evolution of the Milky Way

The APO Galactic Evolution Experiment (APOGEE) is using high-resolution, high signal-to-noise infrared spectroscopy to penetrate the dust that obscures large fractions of our Galaxy's disk and central bulge. APOGEE will survey over 100,000 red giant stars across the full range of the Galactic bulge, bar, disk, and halo. Precise radial velocities and detailed chemical abundance 'fingerprinting' will provide unprecedented insights into our Galaxy's structure and chemical history.

The Apache Point Observatory Galactic Evolution Experiment (APOGEE) focuses on the structure and evolution of our own Milky Way galaxy using high-resolution infrared spectroscopy.

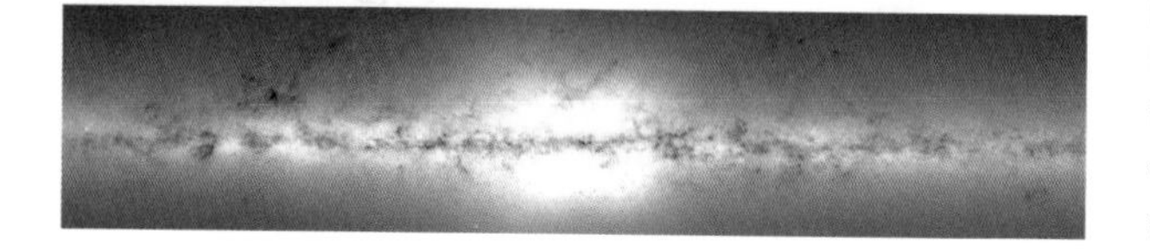

Both SEGUE-2 and APOGEE data will play a central role in near-field cosmology tests of galaxy formation and the small-scale distribution of dark matter.

BOSS: Dark energy and the geometry of space

BOSS will map the distribution of luminous red galaxies (LRGs) and quasars to detect the characteristic scale imprinted by baryon acoustic oscillations in the early universe. Sound waves that propagate in the early Universe, like spreading ripples in a pond, imprint a characteristic scale on cosmic microwave background fluctuations. These fluctuations have evolved into today's walls and voids of galaxies, meaning this baryon acoustic oscillation (BAO) scale (about 150 Mpc) is visible among galaxies today.
However, from the Northern Hemisphere, our planet blocks our view of a quarter of the Milky Way, and obscures our view of the Galactic centre so, to complete the picture, the Irénée du Pont Telescope at Las Campanas Observatory in Chile, will also be used to study stars in the nearby Magellanic Clouds, to give astronomers a better understanding of our Galaxy's immediate neighborhood. Dubbed SDSS-IV, this new program will comprise three surveys, enabling astronomers to: explore the composition of motions of stars across the entire Milky Way in unprecedented detail (APOGEE-2); measure the expansion of the Universe during a five-billion-year epoch that is still poorly understood (eBOSS); produce detailed maps of the internal structure of 10,000 neighbouring galaxies using Sloan spectrographs to understand how they have evolved over billions of years (MaNGA).

THE PANORAMIC SURVEY TELESCOPE AND RAPID RESPONSE SYSTEM (PAN-STARRS)

Pan-STARRS 1 (PS1) and Pan-STARRS 2 (PS2) are two 1.8m survey telescopes sited on the summit of Haleakala in Hawaii that form part of the Panoramic Survey Telescope and Rapid Response System that will ultimately comprise four identical telescopes. By utilising astronomical cameras, telescopes and a computing facility, Pan-STARRS's mission is to continually scan the sky for moving celestial objects, as well as to survey previously detected objects using photometry and astrometry. By detecting differences from previous observations covering the same area of sky, it is able to detect asteroids, comets, variable stars and other celestial interlopers and capitalising on the largest digital camera ever built, is the most powerful sky

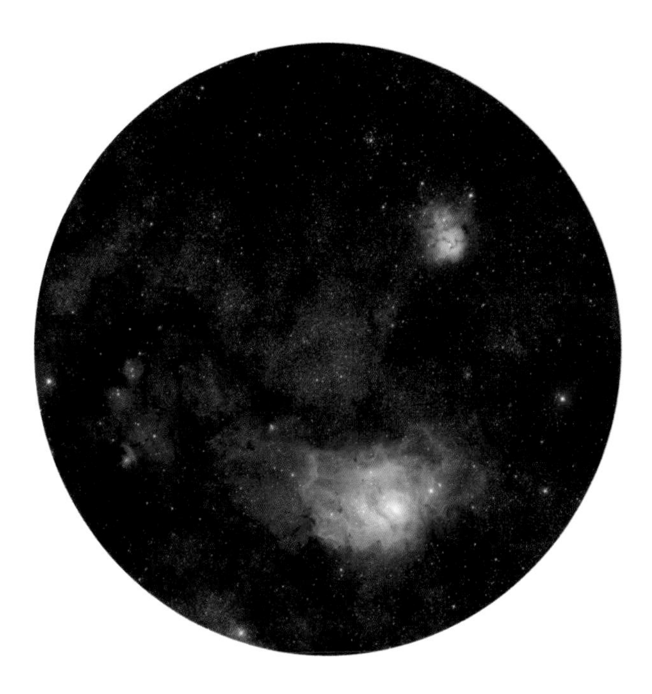

M8, The Lagoon Nebula (top right) and M20, The Triffid Nebula (bottom). This image, taken by PS1, captures a star forming nursery comprising a cluster of young stars, an emission nebula, seen here glowing green due to hot hydrogen gas, and a blue reflection nebula, where starlight from the cluster is reflected off dust grains.

survey to date. With its main mission being to detect Near Earth Objects on a possible collision course with our planet, its 'first light' has not been a day too soon.

The initial telescope, PS1, saw full operation in May 2010. It has a 3° field of view, large for a telescope of this size, and is equipped with the largest digital camera ever built, recording almost 1.4 billion pixels per image. The focal plane has 60 separately mounted close packed CCDs arranged in an 8 × 8 array and each CCD device, called an Orthogonal Transfer Array (OTA), has 4800 × 4800 pixels, separated into 64 cells, each of 600 × 600 pixels.

The very large field of view of the telescope and the short exposure times enable approximately 6000 square degrees of sky to be imaged every night, so the entire sky can be imaged in a period of 40 hours (or about 10 hours per night over four days).

■ LARGE SYNOPTIC SURVEY TELESCOPE (LSST)

With the advances in wide-field optical design, detectors, and supercomputer technology, the success of survey telescopes will increase. Engineering 'first light' is anticipated in 2019, science 'first light' in 2021 and a fully-operational 10-year survey should commence in January 2022. At the core of the LSST will be a single glass disc 8.4m across, offering two optical zones: the outer ring comprises the f/1.18 primary mirror, and the inner zone hosts the f/0.83 tertiary. The convex secondary will be 3.4m in diameter – the largest ever made. Having been reflected three times, light passes through three big lenses, the largest of which is 55% bigger than the objective lens of the 40in Yerkes telescope! How times change: such designs were inconceivable a century ago. Stars will be pinpoint sharp across a 3.5° FOV. To go one step better than Pan-STARRS, the LSST will be home to a refrigerator-sized digital camera offering a staggering 3.2 billion pixels. In just three clear nights, the LSST will image the entire heavens to a magnitude of 24.5.

Operated by the European Southern Observatory, the VISTA (Visible and Infrared Survey Telescope for Astronomy) is the largest telescope in the world currently dedicated to surveying the sky at near-infrared wavelengths. It is a wide-field reflecting telescope with a 4.1 metre mirror and is located at the Paranal Observatory in Chile. With its 3-tonne InfraRed camera, totalling 67 million pixels, astronomers are able to study celestial objects that are cool, obscured by dark dust clouds or whose light has been stretched to dimmer red wavelengths by the expansion of space itself. And there are others: SkyMapper is a fully automated 1.35 metre optical telescope operated at Siding Spring Observatory in New South Wales. Its 268-million pixel imaging camera will image the entire southern hemisphere sky several times over a period of five years. Housed next to the Unit Telescopes at the VLT at Paranal is the VLT Survey Telescope, its most recent addition. This is a wide-field survey telescope with a FOV twice as wide as the full Moon – the largest scope in the world designed to survey the sky in visible light. And the Discovery Channel Telescope (DCT) at the Lowell Observatory in Happy Jack, Arizona, is anticipated to perform deep imaging surveys even during the bright phases of the Moon.

What would Hipparchus, Lippershey, Harriott, Galileo, Herschel … and so many others … say? How far have we come? How far will we go?

A 2010 rendering of the LSST, a proposed 8.4-meter ground-based telescope that will survey the entire visible sky deeply in multiple colours every week from a mountaintop in Chile.
Courtesy of LSST Corp./NOAO

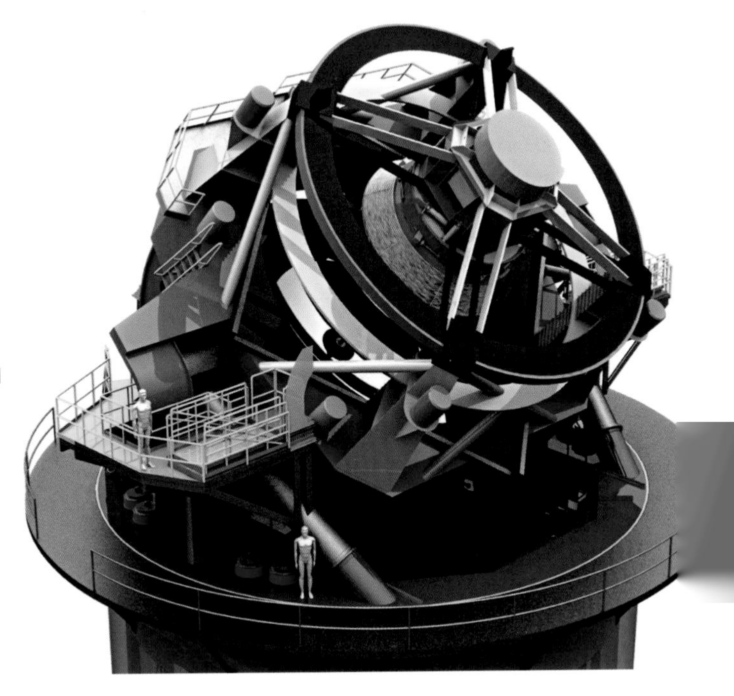

The rotating sky above ESO's Very Large Telescope at Paranal: in this 45-minute exposure the stars leave trails in their dance around the Celestial South Pole.

5 Astrophotography

With the relentless interference of light pollution, initial views through an amateur telescope may seem disappointing: only the very brightest deep-sky targets exhibit any colour and most other objects appear as faint ghost-like wisps. So why bother taking pictures? Well, why not? By attaching a modern camera, or webcam, to a telescope, images can be taken rivalling those captured by professionals a few decades ago. Mountain-top observatory panoramic vistas are within the grasp of modest astrophotography equipment beneath dark rural skies. It is a hobby within a hobby, a symbiosis of art and science, catering for casual dabblers through to advanced practitioners. Once bitten by this astro-bug, you may never look back.

'A time will come when men will stretch out their eyes. They should see planets like our Earth'

Sir Christopher Wren, 1632–1723

Digital SLR cameras

Astrophotography allows for faint light emitted by deep-sky objects to be focused through a lens on to a surface – either film (although rarely these days) or computer chip – from where it is then collected and stored. A shutter opens, light enters and hits the film or chip, and the shutter closes. The greater the lens aperture and the longer the shutter remains open, the deeper the final image. Typically, astro-images require the shutter to be open for a few seconds to a few minutes ... to quite a few minutes!

Based on the specifications of the EOS 60D and succeeding the EOS 20Da, the EOS 60Da is designed to capture rich red colours produced by emission nebulae, with a modified low pass filter that makes it more sensitive to hydrogen-alpha (H) wavelengths of light. It offers high performance and creative flexibility. The 18 Megapixel APS-C CMOS sensor is ideal for capturing rich detail in star clusters, and advanced performance at high ISO speeds up to ISO 6,400 (expandable to ISO 12,800) ensures images are low in noise, reducing the need for longer exposures and star tracker mounts.

Digital SLR cameras have surpassed film SLR models. Film, when used for astrophotography, requires long exposures to maximise light entering the camera: the shutter has to be kept open for hours with a locking cable-release. Digital cameras, however, capture images in a fraction of the time. A digital image also reveals greater detail, with sharper stars and a less 'grainy' appearance. Since digital is typically the preferred option for astrophotography, we will concentrate on these cameras for starters.

WHAT IS A DIGITAL SLR CAMERA?

A DIGITAL SINGLE LENS REFLEX (DSLR) CAMERA OPERATES AS FOLLOWS:

Light from the single lens is directed to a reflex mirror. This mirror reflects the light to a focusing screen – hence 'single lens reflex'. When you look through the optical viewfinder, this is the image you are seeing. The mirror and shutter shield a light-sensitive sensor so it is protected from damage and dust when the lens is removed. When an image is taken the sensor receives light: the mirror flips up, the shutter opens in front of the sensor, and light enters.

A light-sensitive chip, rather than film, is used – hence 'digital'. Leading camera manufacturers, including the arguably superior Canon models, offer full-frame sensor chips the size of 35mm film frames (36mm x 24mm). Many affordable DSLRs offer smaller APS-sized (Advanced Photo System) sensor chips (15mm x 22mm) that are more forgiving with telescope optics – some larger-chipped DSLRs will magnify every lens or optic imperfection. Sensor chips in a DSLR typically measure 15mm x 22mm but the actual pixels (photosensitive elements) in each chip measure 5 to 8 microns (1 micron = 1/1,000mm) and can number up to 22 million in more expensive models. Larger pixels operate like larger-aperture telescopes: the bigger the aperture, the more photons of light can be received. These are then recorded and turned into a signal of electrons. The greater the light, the better the signal. Smaller pixels receive less light: their signal is smothered with 'unused' electrons, known as 'noise', and this affects the quality of the final image. DSLR cameras produce images with a greater signal-to-noise ratio, so their images are sharper.

Like the old film SLR cameras, DSLRs allow you to swap the main camera lens with another lens – macro, wide-angle, fish-eye for panoramic views, ultra-wide, or non-zoom – offering a much wider FOV. By removing the lens, attaching a telescope, and fully exposing the chip, the light-gathering aperture increases, making DSLRs ideal for astronomy. An adaptor is needed to fit the camera to the 1.25in or 2in eyepiece but these are readily available.

Many DSLRs now have LCD monitor screens that fold out from the camera and can be tilted or turned – very useful when affixing them to telescopes.

With its 20.2 MP APS-C CMOS sensor, the EOS 70D camera delivers smooth and accurate autofocus (AF) when shooting Full HD movies and fast AF acquisition when shooting in Live View mode.

COMPACT DIGITAL CAMERAS

Compact 'point-and-shoot' digital cameras differ from DSLRs in the following ways:

- Unlike DSLRs, they have an integral lens that is not interchangeable. Some may have lenses with large (x12 or more) optical stabilised zooms but their image quality is inferior to DSLRs.
- They operate as a contained unit: light from the lens directly hits the sensor chip. This chip feeds the display at the rear of the camera, offering a 'live view' – you see what the camera sees.
- Sensor chips in compact digital cameras measure typically 4mm x 5mm across, but they still contain eight to ten million pixels in each chip! To accommodate these pixels, a typical one measures just two to three microns across. Their tiny apertures mean less light is collected – fewer photons are received – and any signal is drowned by 'unusable' electrons, ie their signal-to-noise ratio is much lower than a DSLR.

A compact digital camera can also be used for afocal astrophotography by adding an extension tube and adaptor ring.

However, compacts can be used for limited astrophotography using the afocal method described below.

AFOCAL ASTROPHOTOGRAPHY FOR COMPACT DIGITAL CAMERAS

At night, when light photons are minimal, long-exposure photography with compact 'point-and-shoot' cameras reveals 'noise': images end up covered in specks. However, it is possible to take images. Afocal astrophotography is a simple method for taking short-exposure photographs with any camera and is ideal for imaging the Moon and planets, especially since the Moon requires exposures similar to daylight (1/250th to 1/30th). Simply:

- Set the camera to automatic.
- Focus the telescope's eyepiece.
- Hold the camera up to the eyepiece.
- Take the picture.

The crescent Moon and Venus: imaged with a simple point and shoot compact digital camera.

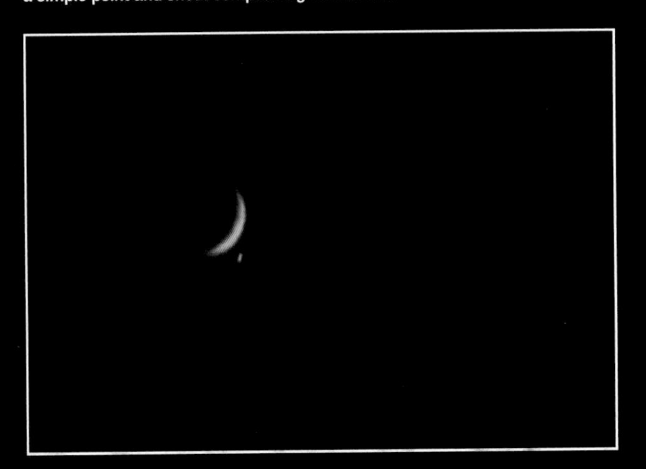

Assess the result and make minor adjustments: refocus the telescope's eyepiece and the camera's live-view LCD screen. Whether using the automatic or manual option, a picture will result. It is a little 'home-made', but it works.

Avoid using zoom with digitals as it can reduce resolution and image quality. If you are using high magnification, or taking images of planets, you will need a tripod to keep the camera steady at the eyepiece. Better still, fix the camera to the eyepiece. This can be done with compacts by screwing an extension tube on to the threads around the camera lens, attaching an adaptor ring to the other end, and inserting this into a Barlow lens (optional) or straight into the telescope's eyepiece for a secure eyepiece-to-camera set-up. Alternatively, use a bracket clamped around the eyepiece. Both set-ups enable you to utilise the telescope's auto-guiding (tracking) features, if you have them, permitting longer exposures at higher magnification.

However, the DSLR is the way to go for astrophotography.

A compact digital camera: with extension tube and adaptor ring, easily inserts into a telescope focuser for quick afocal astro-imaging.

■ WHAT CAN I PHOTOGRAPH?

Whatever you wish, even if you cannot see it! A vigilant observer should be able to see ice haloes, sundogs, and coronas throughout the year. These form from the interplay of light and ice crystals in Earth's atmosphere. They are relatively easy to photograph with, in most cases, the camera being left on its automatic setting. Do not, of course, point the camera at the Sun (see *Solar viewing*, page 128): if you spot a halo, it's easy to position a building, tree or telegraph pole to block it out. In most cases, the camera's own exposure setting will reduce obstructions to dramatic silhouettes.

Clouds also offer dramatic photographic effects, such as crepuscular (Latin for 'twilight') rays and dark dusty bands, especially at dawn or sunset. In June or July, capture the rare high-altitude (60 miles) noctilucent (meaning 'night shining') clouds. These are typically seen as silvery or white with a feathery texture. Fix a DSLR to a static tripod and it will do the job. As with most forms of twilight and night photography, it is a good idea to 'bracket' exposures: this is done by using a range of shutter speeds to vary exposure times.

At twilight and during the night, DSLRs are great for astrophotography. Many targets, such as the constellations, aurorae, lunar and planetary close-ups, dusk and dawn conjunctions, lunar-illuminated landscapes, and the brighter deep-sky targets, can all be captured with short exposures.

If you wish to image the phases of Venus, however, or the rings of Saturn or Jupiter's belts, then opt for a cheaper webcam (see later): although some DSLRs take images at a rate of five to thirty per second, they cannot keep pace with a webcam's 600 a minute, or faster. Moreover, most planetary images feature a tiny planet and copious dark sky – this takes longer to process.

Crepuscular rays: where below horizon cloud partially blocks the Sun.

The Pleiades and Hyades above the UK's William Herschel Telescope: Canon 20D DSLR and 28mm f/2.8 lens on AstroTrac TT320 motorised tracking drive.

The Canon TC-80N3 Timer: a remote shutter release that can be set from one second to 99 hours.

WHAT ELSE MIGHT I NEED?

- A remote electronic shutter is an essential. These can be simple devices: you just press a button, the shutter opens and it can then be locked for longer exposures. A more sophisticated version, for higher-priced DSLRs, allows a sequence of exposures to be preset for any length or interval – these are great for time-lapse shooting or unattended shots.
- Several spare batteries, or use an AC power supply if available, especially in cold conditions – long exposures eat up the juice!
- Additional lenses: most lenses work well, but those in the 18mm to 75mm range are ideal for photographing constellations.

■ WHICH CAMERA DO I PURCHASE?

Your choice is, of course, determined by available funds and your intended targets. It is generally accepted within the astrophotography fraternity that low-noise-level Canon DSLRs, closely followed by Nikon, are currently the best tools for the task. Models change and are regularly enhanced, so do your research. Stock cameras are those optimised for domestic tasks and will have integral filters to block poorly focused 'unwanted' light, such as infrared. However, additional to 'stock' cameras is the 'modified' camera. Here, the DSLR camera will have been filter-modified for astrophotography – the standard infrared cut-off filter sited in front of the sensor is replaced with one that maximises on the deeper, hydrogen-emitting light, ie light from faint nebulae. These are not great cameras for daytime shooting but they do a wonderful job for astrophotography!

■ FOCUSING THE DSLR CAMERA

Whether using a DSLR during the day or at night, with or without a tripod, on or off a telescope or platform, the camera will need to be focused. For the more advanced close-up shots, it will need to be precisely focused.

DAYLIGHT SHOOTING

To focus a DSLR for a daytime shot switch on the auto-focus and then press the shutter button. The camera does the rest.

NIGHT SHOOTING

With the exception of twilight shots, it is difficult to focus a DSLR at night since there is little upon which to focus; but no focus, no photo, and focusing must be precise otherwise targets appear distorted. Originally, with older film cameras, the lens had to be focused at infinity to ensure stars did not appear 'blobby'. Nowadays, DSLR cameras focus past infinity, which is not necessarily the best point of focus. To achieve precise focus, consider the following:

- Right-angle magnifier finders: these attach to a camera's viewfinder and act like a Barlow lens by offering 2x power. First, make sure you turn the dioptre setting on the camera to suit your eye. Then turn the dioptre setting on the magnifier to achieve precise focus. Finally, focus the camera's lens (or telescope), using a bright star. It is initially trial and error, but by moving the focus back and forth and taking shots each time it will become clear, from the camera's playback option, when best focus is achieved – the target will be sharpest. Additionally, like an eyepiece diagonal, their right-angled position makes for more comfortable viewing.

Canon right-angle magnifier finder: for precise focusing.

- Some astrophotographers focus using computers and specialised software (for example, MaxDSLR and ImagesPlus). The computer operates the shutter and the downloaded images appear on the screen along with graphic or numeric displays of the star's intensity. This is a lengthy process involving constant picture snapping and subsequent adjustments. The data is also affected by atmospheric 'seeing' disturbing the figures.
- The best focusing option for DSLRs is 'live view', where the mirror is flipped out of the way and the chip exposed. Here, the camera's LCD screen displays a magnified view of what the camera is seeing. You can zoom in on a bright star, focusing as if with a telescope's eyepiece until the target is tiny and sharp. A higher ISO (see the Camera basics boxout) and wider aperture helps the process. Make sure you refocus throughout the night – changing temperatures lead to defocusing. Allow equipment to cool and refocus every couple of hours.

DSLR 'live view': what you see is what you get.

CAMERA BASICS

APERTURE

As with telescopes, the aperture is like the iris in a human eye – it opens and closes to allow varying amounts of light to enter. It is usually stated as a focal ratio, or f-ratio, which represents the focal length of the camera lens divided by the diameter of the iris opening. These f-ratios come as standard numbers (f/1, f/1.4, f/2, f/2.8, f/4, f/5.6, f/8, et al.). The smaller the number, the wider the opening and the more light enters. Each step from a larger f-ratio to a smaller one, eg f/5.6 to f/4, doubles the quantity of light entering and is equivalent to doubling the duration of the shutter speed.

ISO (INTERNATIONAL STANDARDS ORGANISATION) SPEED

This term dates back to the days of emulsion film photography, where there was a trade-off between film sensitivity and the grain structure of the emulsion. High-quality images were obtained using 'slow' film with fine grain, such as 64 ISO film. Low-light-level photography was executed using 'fast' (more sensitive) film with correspondingly larger grain. Films are rated by this ISO speed: an ISO 400 film is four times as fast (requires a quarter the exposure time) as an ISO 100 film. Typical values are: 100, 200, 400, 800, 1600, and 3200. Digital detectors in DSLR cameras boost their sensitivity electronically, allowing fainter targets to be photographed but with an increase in noise, not grain size, ie higher ISO settings require shorter exposures but at the cost of worsening noise. For astrophotography, it is best to start off with 'slower' settings, such as 100 to 400 ISO, otherwise light pollution swamps the image. If working from dark sites, change the ISO rating to higher values.

SHUTTER CONTROL

The shutter control determines how long light will hit a chip. Doubling exposure time doubles the light recorded.

■ TAKING PICTURES USING A TRIPOD

With a straightforward tripod set-up, preferably one adjustable for height and equipped with a manoeuvrable pan-and-tilt head, many stunning images can be achieved with a DSLR.

AURORAS

When the Sun is most active at solar maximum, auroras may appear. Again, a static tripod and standard lens will capture the 'curtains' of colour. Use foreground objects to set a sense of scale, use an f-ratio of 5.6 from light-polluted locations and keep exposures short.

Aurora: simple but spectacular.

Stability: a digital single lens reflex (DSLR) camera and tripod.

COMETS

Many comets come and go unseen by the naked eye, but, somewhat rarely, one appears for weeks on enc like Hale-Bopp (see Chapter 2). Thes can be captured with just standard, telephoto and zoom lenses.

Comet Holmes: imaged with a Starlight Xpress SXV-H9 CCD camera and Pentax 75mm SDHF apochromatic refractor.

CONSTELLATIONS

In summer, the zenith constellation of Cygnus, the Swan, with its rich star fields and some of the brightest regions of the Milky Way, makes an ideal target. (For southern hemisphere observers, use Sagittarius.) Likewise, the winter constellation of Orion, the Hunter, with its prominent bright stars and famous deep-sky targets, is also an alluring target. Try exposures of 30 to 40 seconds with an ISO of 400. The ISO can increase to 1600 – but don't forget, increasing the ISO picks up fainter targets but noise will correspondingly increase. Keeping exposures to less than 20 seconds will prevent Earth's rotation drawing the stars out into 'trails'.

The Scorpius constellation: note the red supergiant star, Antares, upper right, and the dark dust clouds towards the centre of the Galaxy.

INTERNATIONAL SPACE STATION OR IRIDIUM FLARES

(See Chapter 3.) First check for their visibility. The heavens-above.com website is excellent: its predictions of azimuth (an angle on the horizon that increases from north 0°, to east 90°, to south 180°, to west 270°) and altitude (an angle from horizon 0° to zenith 90°) are extremely accurate. Ensure your time is accurate too: precise time signals are available from many sources. Then use these azimuth and altitude indications to aim your camera in the flare's anticipated direction. As with a film camera, a digital exposure of at least 15 to 20 seconds, with ISO 100 and at least a 50mm lens set to f/2.8 or faster, will capture a streak across the sky from a flare of -2 magnitude, or greater. Remember to add foreground – trees, buildings. Use sufficient exposure to silhouette them (a one to two-minute exposure with a first or last quarter Moon can assist, but be careful not to 'fill in' the silhouette) and brighten the sky, but try not to overexpose and wash out stars and other objects of interest. For the International Space Station, exposures of 30 to 40 seconds at ISO 400 will capture its passing.

METEORS

Sporadic meteors (those not associated with showers) can also be captured, although rather serendipitously! To have greater success, shoot during an active meteor shower, such as the popular Perseids in August or Geminids in December (see *Meteor showers*, page 49), when meteors appear from a specific area of sky – the radiant. Set your camera to its maximum setting, fully open the lens, focus at infinity, and aim it in the radiant's direction. Take short 30- to 60-second exposures with minimal delay between shots – one-second intervals are ideal. Exposure time should be sufficient to capture the shooting stars but short enough to avoid light pollution washing out the image. Remember, sodium or mercury filters can counteract this problem. Have a plentiful supply of charged batteries and at least two memory cards – you will be trigger-happy once you start – and consider a remote, or programmable, shutter release for automation.

MOON

This is another great target for tripod photography in the evening or morning sky, especially with a telephoto or zoom lens and using the manual settings described above. The best times are the few days centred around the first- and last-quarter phases: craters and mountains close to the terminator – the divide between the dark and light areas – cast long shadows and general detail is thrown into sharper relief. At crescent phase, it is easy to capture 'Earthshine' – the faint glow on the Moon's non-illuminated region as a result of Earth's clouds reflecting sunlight back on to its surface. When the moon is full, select an alluring night-time landscape, use ISO 400, f/2.8 and around a 30-second exposure, and you will capture a bright sky paradoxically littered with stars.

The Moon: an easy and stunning target.

PLANETARY CONJUNCTIONS

This is a great way of capturing planetary conjunctions at dawn or dusk. When taking simple shots of planets, set the camera to manual and keep the exposures short. Set the aperture to the f/2 to f/5 range, and the ISO to 100 to 200. Of course, no detail will be seen with standard lenses but, with good seeing, you may capture the Galilean moons of Jupiter. Don't forget to set a recognisable foreground – trees, buildings, houses – to establish a sense of scale and enhance the target.

Star trails at dawn: caused by Earth's 24-hour rotation.

STAR TRAILS

Star trails, caused by Earth's 24-hour rotation, will appear after only 20- or 30-second exposures. Shorter focal-length lenses exhibit less trailing in a given period than greater focal-length or zoom lenses. These can be captured: simply reduce the ISO to 100, use an f-ratio of f/4 to f/8, and extend the exposure to up to 60 minutes. Alternatively, use the preset remote control shutter to take multiple 30- to 60-second shots in rapid succession. These can be 'stacked' (of which more later) to create longer trails. The best method to reduce noise is to take a dark frame at the beginning and end of the sequence. For single long-exposure shots, the use of the Long Exposure Noise Reduction option works well too.

■ TAKING PICTURES MOVIE STYLE

Pioneered by NASA's Dr Robert Layton, time-lapse photography – the process of shooting numerous frames and 'stitching' them together to make movies – shocked the world when a moon of Jupiter was captured transiting the planet's disc. Today, similar fun can be had with a DSLR camera and a preset remote control shutter. These can be set to snap hundreds of automatic exposures – batteries and memory card permitting – resulting in 'movies' capturing the movement, due to Earth's rotation, of the night sky. To do so, use ISO 1600 with around a 30-second exposure at f/2 to f/2.8. Keep the intervals short (around one second) to enable the frames to knit seamlessly in the movie. A folder of sequentially numbered images can be downloaded to your computer, and then – with appropriate software – all you need to do is isolate the first image, establish how many frames you would like per second of footage, and make the necessary adjustments for size.

■ TAKING PICTURES USING A TELESCOPE (PRIME FOCUS ASTROPHOTOGRAPHY)

Why not utilise the telescope as an even larger telephoto lens by connecting it to your camera? This is known as prime focus photography. By removing the DSLR camera's lens, connecting a T-ring (a metal ring with a standard interior screw thread) to a 1.25in or 2in (depending on the size of the scope's focuser) threaded prime focus adaptor (available from telescope dealers) to the camera, and then joining these with the telescope's focuser, superb close-ups can be taken, especially of bright objects such as the Moon and Sun (remember to observe safely!). Schmidt-Cassegrains require a T-adaptor that screws to the rear cell of the OTA, followed by a T-mount for the camera. The camera is then coupled to the scope. Other adaptors are supplied with projection adaptors: these extensions enable the insertion of a telescope eyepiece to yield higher magnification. However, greater magnification for close-up shots requires precise tracking, and sometimes auxiliary tracking, with a motor drive in order to keep targets centred – known as auto-guiding.

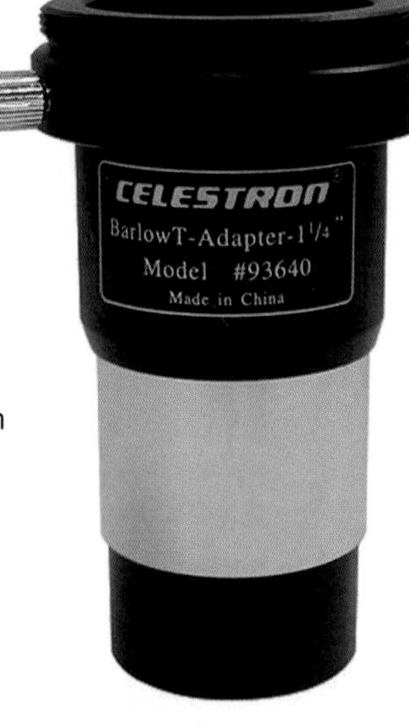

Prime focus adaptor: threaded for attaching to a T-ring and the telescope's focuser.

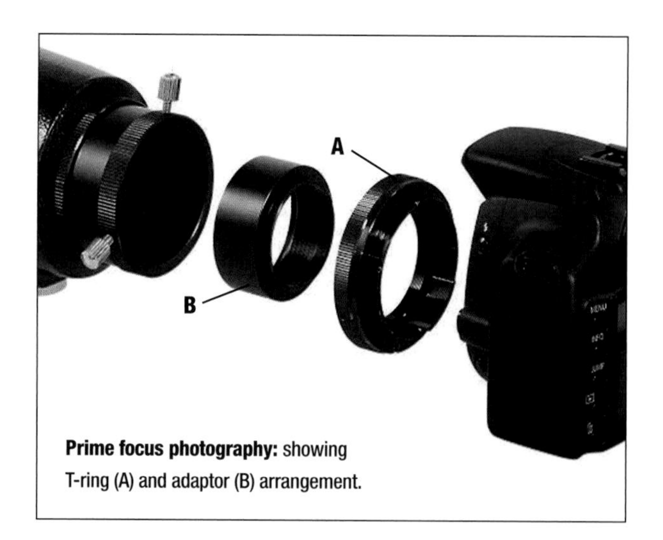

Prime focus photography: showing T-ring (A) and adaptor (B) arrangement.

PRIME-FOCUS TARGETS

Target		Exposure	ISO
Moon	(slim crescent)	1/4 second	100
	(quarter)	1/30 second	100
	(full)	1/250 second	100
Planets		1/4 to 2 seconds	200 to 800

TELESCOPES AND MOUNTS

WHICH TELESCOPE AND MOUNT DO I USE?

Telescopes and mounts have been covered in greater detail in Chapter 3. For prime-focus shooting, ideally use a lightweight, short focal-length (400mm to 600mm) refractor with an apochromatic lens design (60mm to 90mm in diameter), and a good-quality, portable German equatorial mount with an inbuilt polefinder scope that is parallel to the drive axis. Most operate from 12V battery supplies purchasable with the mount. Check when purchasing that the telescope is astro-imaging-friendly. The shorter focal length is more conducive to initial polar-alignment and any subsequent alignment or tracking problems. The faster optics (f/4 to f/6) also facilitates shorter exposures. Since a smaller focal-ratio wins out over a larger aperture, these scopes can also be used for distant shots of fainter, deep-sky targets, but close-up images require larger focal-lengths ... larger scopes ... larger wallets ... and a deeper obsession!

■ TAKING PICTURES PIGGYBACK STYLE

WITH A TELESCOPE

How do you capture stunning distant images of the Milky Way or fiery nebulae? Firstly, pick a night when the moon is absent and the sky is truly dark. Now we move into the advanced realms of long exposures and tracking systems. In this style of astrophotography, the DSLR camera and lens are generally mounted on the telescope 'piggyback' style: the camera shoots through its own lens and not through the telescope. In most cases, this means securing a sufficiently robust ball-and-socket tripod head to the telescope OTA's tube rings. Since the camera 'tracks' the sky, more light can accumulate on the sensor chip: this enables the camera to capture colour images of dim stars and nebulae too faint for the unaided eye.

To take these images, polar-alignment (aligning the scope's polar axis to within a few arcminutes of Earth's celestial north pole) is essential on your equatorial mount, especially if you are using a wide-angle lens. It is also essential to have a speed-control motor drive on the right ascension axis.

A universal piggyback mount: with bracket and fixings, for attaching all types of camera to Meade LX telescopes.

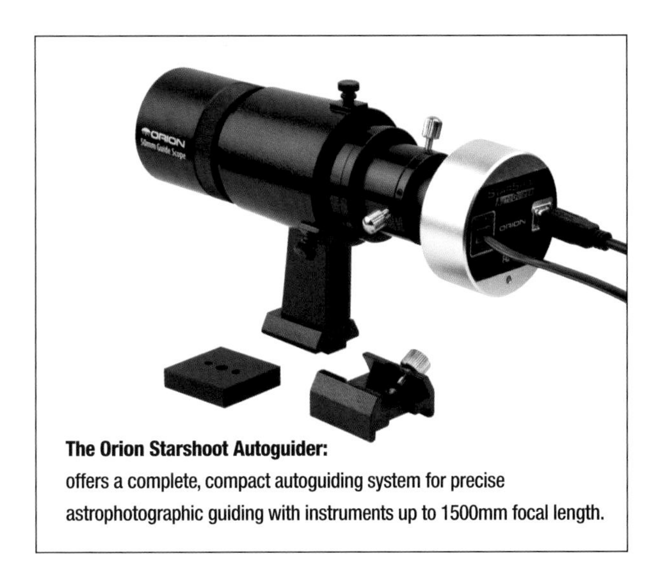

The Orion Starshoot Autoguider: offers a complete, compact autoguiding system for precise astrophotographic guiding with instruments up to 1500mm focal length.

A less efficient, but affordable, alternative to motorised tracking is manual guiding. Here a second telescope is used as a guidescope – insert an illuminated reticle into the eyepiece and use the scope's slow-motion controls to keep a bright 'guide star' centred in the cross hairs. All these topics have been covered in depth in Chapter 3.

WITHOUT A TELESCOPE

Here, the DSLR camera is set directly on a polar-aligned mount capable of precise tracking. It is ideal for taking shots with a wide-angle lens.

Another innovation is the use of a motorised base. Again, there is no telescope. Provided the tripod is sturdy this platform-style set-up can replicate an equatorial mount utilising the reticle and 'guide star' method outlined above.

■ PRIME-FOCUS, CLOSE-UP DEEP-SKY TARGETS

Taking images of deep-sky targets, such as the stunning Andromeda Galaxy or Great Orion Nebula, requires long exposures, and long exposures expose more than the image. It is not for the faint-hearted! No matter how accurately polar-aligned, or how precise the essential manual or motorised tracking, the very mechanical nature of mounts and telescopes leaves a legacy of 'squirmy' stars and star trails. What is the best way to proceed? To avoid aircraft and satellite trails, guiding errors when an auto-guider is not used, or electronic noise build-up, start by taking a large number of shorter exposures – typically three to five minutes at ISO 400, or 800 – sufficiently long that images are not underexposed but not overly long, to avoid star trails. Exposure times will vary, depending on how precise the polar-alignment and the amount of light pollution. Images can then be 'stacked' (see page 179) and 'averaged' into a single sharp shot with minimal noise.

If polar-alignment is not sufficiently precise for close-up deep-sky imaging, targets drift after just a few minutes, so be as precise as possible. Ideally, stars should not move more than two or three arcseconds! To refine tracking, there is the manual option of an illuminated reticle discussed earlier: this is placed in a separate guidescope or in an off-axis guider. An off-axis guider, together with a reticle eyepiece, enables 'through-the-telescope' monitoring to keep the telescope precisely positioned for long exposures. It couples the camera to the telescope: after a bright star has been selected, the off-axis guider diverts, at a 90° angle, a small amount of its incoming light where it is observed on the cross hairs of the reticle for tracking errors; any necessary adjustments can be made with the drive correctors.

Alternatively, for the serious photographer there is an auto-guider. This involves using a second telescope as a guidescope, a digital auto-guiding camera, and a laptop with auto-guiding software. The guidescope is mounted in a side-by-side scenario or on top of the main telescope. The camera can be a webcam or affordable CCD camera (see page 184): the latter is preferable since it also detects faint stars so offers a greater selection. Focus the camera on a bright 'guide star' near your target. If the star drifts, the software automatically responds by sending a pulse to the mount every few seconds, sufficiently nudging it to keep the 'guide star' locked in position. They are USB compatible for fast data transfer, have easy interface with most Windows-based laptops, and the software automatically sorts, aligns, and blends the best images for a resultant sharp, highly detailed shot.

An off-axis guider: this couples the camera body to the telescope and allows an observer to monitor an exposure as the photo is being taken (see above text).

Planetarium software: enabling a telescope to slew automatically to objects selected from a graphical display of the night sky and take stunning images of the Moon, planets and brighter deep-sky objects.

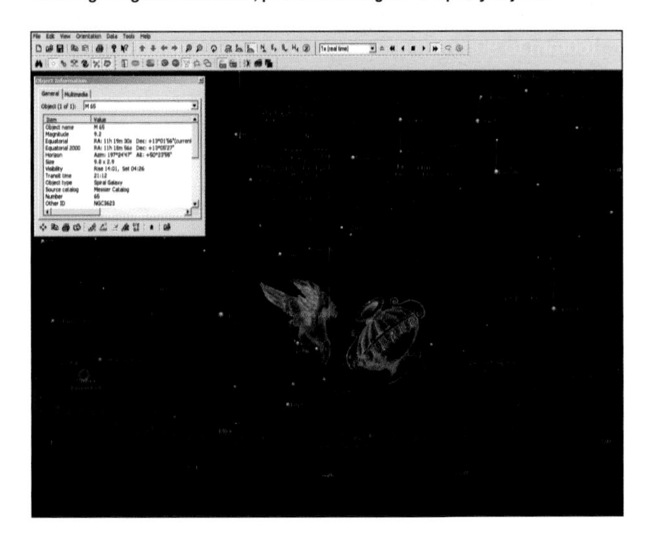

■ CUSTOMISING AND CAPITALISING ON THE DSLR

Before taking photographs, whether the camera is attached to a tripod, platform, telescope, or guidescope, be aware of the astro-image-enhancing functions on your DSLR. These can be used to gain the best from your camera and, therefore, the best from your images before later processing. Below is a list of astro-imaging jargon that may prove useful:

ASTRO-IMAGING JARGON

Colour balance Set this to a neutral value on your DSLR. If you have a filter-modified camera, shift the colour towards the blue for daylight imaging. This counterbalances the filter's red tint.

Colour temperature Typically leave the DSLR on 'Auto White Balance'. If 'movie-making', consider setting it to around 5,200K in the RAW file format to ensure all the frames are the same. If you need to make adjustments you can do so later when processing with image software.

Dark frame With a DSLR's Long Exposure Noise Reduction option turned on (see right), a secondary exposure, equal in length to the original exposure, is taken with the shutter closed. This is known as a dark frame: only noise is recorded but this is subtracted from the original long exposure to eliminate specks.

De-Bayering Pixels in a DSLR are covered by minuscule red, green, or blue filters arranged in a Bayer array, so named after its inventor, Bryce Bayer. Every pixel records a monochrome image. To create a coloured image, the camera decides what each pixel should be by taking individual data from one pixel and mixing it with data from surrounding pixels. This deciphering is known as de-Bayering.

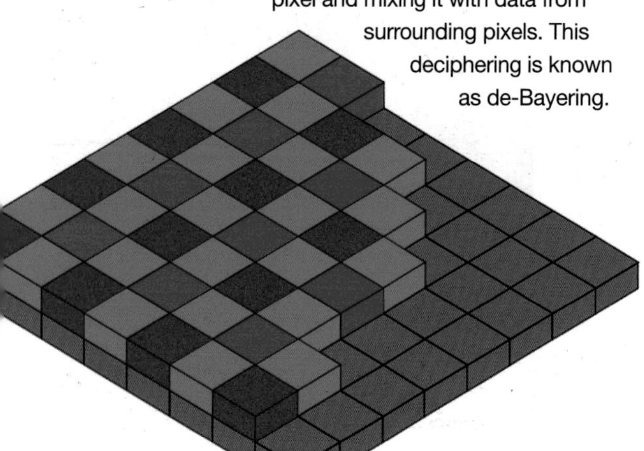

A dark frame: an example of a single 300-second dark frame taken with the telescope capped. Only noise is recorded.

Image parameters The contrast, sharpness, saturation, and colour tone of images can be enhanced, via pre-set programs, on the DSLR when using JPG file format.

LENR Long Exposure Noise Reduction. Typically located under 'Custom Function' on DSLRs, with this option 'on' the camera automatically snaps a dark frame (see above) after any exposure greater than one second, to reduce noise.

Noise The unwanted signal in CCD and DSLR images, for which there are two sources: noise in the image itself and thermal noise. Fans or a Peltier cooling system will reduce this.

RAW image files The large files created by DSLRs are saved in this file format and appear exactly as the camera's sensors 'saw' them: data is untouched, so recorded images are not compressed, re-coloured or de-Bayered, and keep the full range of monochrome information. As a result, captured images are generally more detailed. This RAW data can then be processed with imaging software.

■ IMAGE PROCESSING

You have taken multiple RAW images, so what's next? Processing – 'developing' – them. This requires a book in itself, believe me, so what follows are merely pointers which may be used as an initial guide.

After a night's imaging, you will be left with a collection of RAW files which must be converted – de-Bayered – into a format for subsequent software processing. See Appendix 3 for a list of software programs designed for this conversion. By opening up a set of images, these programs are activated and various image adjustments, such as contrast, colour, and noise reduction, can be made.

Once images have been converted, they will need to be 'aligned' and 'stacked' using specific software programs: see Appendix 3 for suggestions. Aligning using Adobe Photoshop, for example, involves first selecting all the images to register them. The next step is to 'stack' the registered frames – 'layers', in processing jargon. The prime reason for stacking is to average out image distorting 'noise'. This process involves selecting the best frame first, to act as a 'base layer' or 'background layer', and then selecting and copying all the other images with the aim of producing one multi-layered view. These multi-layers then have to be aligned with the 'background layer' to achieve the sharpest final image. With some recent software programs this is an automated process, completed in seconds and done at the press of a single 'auto align' option.

For those starting out, the free software 'Deep Sky Stacker' is excellent, as it automates the entire process. Photoshop can be used for later cosmetic enhancements. With older software versions, and for the more advanced imager, it is done manually by setting the 'background layer' to 100% opacity. Then, all the other layers can be hidden (by clicking the small 'eye' icon to the left of each layer) with the exception of the second stacked layer (Layer 1). Change the 'blending' mode from 'Normal' to 'Difference' and set Layer 1 to 50% opacity: it can then be slid and tweaked, using the tools and arrow keys, to highlight any overlapping areas and ultimately align the two layers, ie the layer is effectively cancelled out. The 'blending' mode can then be returned to 'Normal'. This process is then repeated for the third layer (Layer 2) to 33% opacity, the fourth layer (Layer 3) to 25% cent opacity, etc.

NGC 6559: a diffuse nebula in Sagittarius, its stunning colours enhanced by final software processing.

And it does not stop here. Various software programs offer palettes for further enhancing a stacked image. It is a veritable 'digital darkroom'. Always read the manual!

In the final stages of processing, remaining noise can be reduced further with noise-reduction filters, colour and contrast can be adjusted to bring out a wispy nebulae's true definition, the option of 'Curves' can be used to create further definition by utilising advanced algorithms to darken the sky, dark frames can be removed to counteract any lingering noise, the layers can be saved as a 'master', and finally this master can be 'flattened' to a single layer and sharpened with a 'High-Pass' filter option.

M13 before: a single 300-second exposure of the globular cluster M13 in Hercules taken with a Quantum Scientific Imaging 583wsg CCD camera operating at -22 degrees Centigrade. The image has been calibrated using dark frames, flat field and bias frames.

M13 after: this image is made up of twelve 300-second exposures of the globular cluster M13 in Hercules. The images have been added together to produce a much stronger signal-to-noise ratio. Using the popular FITS Liberator plug-in software for Adobe Photoshop the image was logarithmically scaled to show many of the faint stars not visible in the single 300-second exposure.

CCD Imaging

Conventional astronomical colour photography is more usually done by taking three exposures through red, green and blue filters and combining them into an image. Other methods of light detection rely on the photoelectric effect: light hits a metal surface, loses its energy, electrons are ejected and an electric current is formed, examples being television cameras and photomultipliers. CCDs are slices of wafer-thin silicon comprising multitudinous light-sensitive cells; light hits these cells, releases electrons and a charge, proportional to the light received, is emitted. And this is just the beginning...

■ WHAT ARE CCD CAMERAS?

For imaging the faintest deep-sky objects and distant galaxy clusters Charge Coupled Devices (CCDs) – once the domain of serious professional photographers and the precursor to consumer digital cameras and DSLRs – are not to be overlooked ... but they are expensive, and are for the more advanced astrophotographer. These are ideal for important research projects, such as searching for supernovae in other galaxies, asteroid hunting, variable star monitoring, and a host of other interesting collaborative projects with professional astronomers. However, for anyone keen to take stunning images, the CCD is a remarkable tool. When placed at the prime focus of even very modest small aperture telescopes, a good CCD camera can produce stunning colour images. This does not require an enormous amount of expertise or high end instrumentation. By adhering to a set of basic principles, anyone, after a brief period of familiarisation, can capture images rivalling those adorning pages of popular astronomy magazines and the internet.

WHAT IS A CCD?

A charge coupled device **(CCD)** is essentially an integrated circuit etched onto a silicon surface forming light sensitive elements called pixels. Photons incident on this surface generate a charge that can be read by electronics and turned into a digital copy of the light patterns falling on the device.

There are a number of factors to consider when choosing a CCD camera:

THE PHYSICAL SIZE OF THE CCD

This will determine the field of view provided with any given focal length telescope. The larger the CCD the wider the field of view.

The following calculation will determine the field of view of a given CCD camera.

Field of View in arc minutes = 3438 x CCD Size in mm/focal length in mm.

An example is as follows using a Starlight SXVR-H9 and 80mm F6 refractor.

3438 x 9/480 = 64.4 arc minutes or a little over 1°.

Starlight Xpress Trius SX-814 USB Hub Monochrome CCD Camera utilizes a Sony ICX814 Sensor with 9.2 million pixels.

THE RESOLUTION, OR NUMBER AND SIZE OF THE PIXELS ON THE CCD

As an example, the SXVR-H9 CCD camera has 1392 by 1040 6.45um pixels, which compares very favourably to early CCD cameras which typically used CCDs with 375 by 244 25um pixels. The result is far higher resolution images.

Many amateur astronomers who upgrade from a DSLR camera to a typical cooled CCD camera discover that its increased sensitivity, much lower noise, greater dynamic range and its ability to make full use of narrow band filters enable it to easily outperform a DSLR.

The resolution per pixel on the CCD is known as the sampling rate. The sampling rate in arcseconds can be determined by performing the following small calculation.

206265 x pixel size/focal length of telescope in mm

An example is as follows using an SXV-H9 and an 80mm F6 refractor:

206265 x .00645/480 = 2.77 arc seconds

Cameras with many small pixels produce high sampling rates and give much better performance when used with short focal length small aperture instruments.

It is far easier to obtain quality images with perfect tracking by using a high quality short focal length and small aperture instrument, providing you use a CCD camera that provides a decent sampling rate.

Imaging at much longer focal lengths is more demanding, particularly in terms of tracking accuracy required, and necessitates much longer exposures.

MONOCHROME IMAGES

CCD pixels receiving more light produce a greater signal than those that receive less, but overall most capture light with an efficiency of 60 to 80%. The final image is composed of discrete shades of grey – monochrome; white represents a strong signal, black a weaker one. These CCDs tend to be more sensitive than their colour counterparts and work well for narrowband imaging.

COLOUR IMAGES

To produce colour images with CCDs, a set of colour filters are used. This is popular for deep sky imaging. Colour images are produced using red, green and blue filters in turn and by taking a sequence of images of the same target. This works because the human eye sees colour as a combination of these three primary colours. The 'cones' in our eyes are of three types, sensitive to either red, green, or blue. Humans obtain primarily colour data from these cones but not spatial information (detail). The spatial information comes from the 'rods' in the eye, which are not overly sensitive to colour. By combining an unfiltered high-resolution black-and-white exposure which captures the greatest signal (known as a Luminance frame) with low-resolution colour images that have been combined into a single RGB file – with the use of graphics software (see page 190) – a high-resolution LRGB, or four-colour, image is achieved. Resolution data is not captured from the colour images – the images are monochrome but record significant differences between the three filters. The entire exposure process can take hours but the LRGB image result is far superior to a standard RGB file.

More commonly these days, astro-imagers use 'one-shot' colour CCD cameras which have an inbuilt Bayer matrix coloured microlenses placed above the CCD sensor: once 'de-Bayered' they produce a colour image with just one exposure and without the use of expensive filters and filter wheels.

The 12 megapixel sensor of the Atik 4120EX is the ideal solution for capturing stunning high resolution images. Available as one-shot colour (OSC), this is an ideal match for Hyperstar systems and short focal length telescopes.

TRIUS XS-814 rear view.

CCD NOISE

A benefit of the CCD over the DSLR is its 'cooling' option. DSLRs suffer from thermal 'noise' when exposure times extend to minutes. CCDs also suffer from electronic noise as a result of circuitry and sensors – known as 'dark current'. This varies from camera to camera and increases if exposures are lengthened, but can be overcome by cooling the CCD sensor to sub-zero temperatures. By cooling the CCD to approximately 30 degrees below, or even to -50 degrees C below the ambient temperature, these cameras are able to produce higher quality images. With every reduction of 7°C the dark current is halved. These separately powered cooling mechanisms take the form of fans that draw heat away from the back of the CCD. All cooled amateur-level CCD cameras rely on Peltier cooling that is augmented by cooling fins and small fans. At the more expensive end the Peltier cooling system is backed up by fan cooling, and at the even more expensive end by water-cooling.

These cooling mechanisms are very effective, especially in winter temperatures. Not all dark current will be eradicated, but remaining noise can be removed using the 'dark frame subtraction' procedure whereby an equal-duration, equal-temperature image is taken with the telescope capped – the dark frame. Once this dark frame is stored on computer it can be subtracted from the light frame using appropriate processing software, thus extracting the dark current signal from the image.

Most deep sky celestial targets are faint. Exquisite, far-reaching detail in well-known objects, like The Great Orion Nebula and Andromeda Galaxy, can be lost when using uncooled low-cost CCD imagers. High quality cooled CCD cameras are much lower noise devices and perform to a much higher level. Today's CCD cameras, such as Starlight Xpress or ATIK, using ultra-low noise Sony Exview sensors, have very low noise cooled CCDs; in typical conditions with these cameras it is not necessary do 'dark frame subtraction'.

CALIBRATING

A second, very important 'calibration' frame is also required when imaging with CCD cameras. This is known as a 'flat field' and is taken to record any optical aberrations in the telescope/camera set-up. There are many unwanted 'artefacts' recorded in a CCD image, not least of which is dust near a sensor or on a filter. Furthermore, if focal reducers are used they can produce 'vignetting', which darkens the field of view around the edge of the image. The flat field frame addresses this problem. To produce such an image, the scope and camera must be pointed at an evenly illuminated blank surface. This will record the artefacts and can then be used to successfully remove the problem from the light frame.

An internal view of the filter wheel with cover removed showing 5 x 48 mm filter disk with one filter in place. The USB connector is visible at lower left and the Serial/ Manual connector at lower right.

AUTOGUIDING

This is enormously important. CCD cameras are fiendishly sensitive and will expose even the tiniest of guiding errors. The best results are obtained by using an Autoguider which guides the scope's mounting. Autoguiders vary, from simple webcams to elaborate dedicated guiders, but selecting a CCD camera with an optimised Autoguider will make the difference between success or failure. Modern Autoguiders are available in various forms. Guiding with a Guidescope works well when imaging through telescopes with fixed and rigid optics. Highly sensitive guiders, such as the Starlight Xpress Lodestar, work well with Off Axis Guiders and using an OAG with a Lodestar is an excellent way of achieving a self-guiding camera without the need for a Guidescope.

MOUNTINGS

As with all astro-imaging, a quality mount is vital for CCD imaging. Place a modern CCD camera on a good mount and, even at the prime focus of an 80mm aperture refractor, the resultant, well-guided images will be amazing. Imaging with longer focal length telescopes is more demanding and will require greater guiding precision. Mounts such as the Skywatcher EQ Series offer precision and load-bearing capacity, and the choice is overwhelming. There are also the Avalon and iOptron mountings and, for the bigger wallet, Paramount have been the leaders in ultra-high quality mounts for many years, offering legendary performance with their MX+, MEII and MYT models.

FILTERS

Even in the worst light-polluted conditions, amateur astro-photographers can capture stunning images with small aperture instruments and a CCD camera equipped with narrow band filters.

These filters allow only narrow select wavelengths of light to pass but, when used with a CCD camera sensitive to these wavelengths, will produce stunning images regardless of light pollution or any interfering light from the Moon.

Any CCD camera should have good sensitivity at the important narrow band wavelengths of Hydrogen Alpha (656nm) and Oxygen III (499nm and 501nm). Many spectacular emission targets in the sky emit strongly in these wavelengths. If imaged using these filters, the results will far exceed those images captured when using conventional LRGB filters or colour cameras.

This top-of-the-range camera is the result of extensive design and development and incorporates a huge 37mm x 25mm sensor. Using the Kodak KAI 11002 CCD, and available in mono or one-shot colour versions, it is the camera for the uncompromising user who is looking to create stunning high-resolution images.

NARROWBAND FILTERS

Another option with monochrome cameras is the use of narrowband filters. These are used in a similar manner to RGB filters but record very specific wavelengths of light. Popular filters are Hydrogen Alpha and Oxygen III. Targets like the Great Orion Nebula emit very strongly in the light of Hydrogen Alpha: when these filters are used, it is possible to capture extraordinary images. One particularly remarkable advantage of narrowband filters is that, since they admit light of a specific wavelength, they block light pollution – ideal for urban observers! See also *Filters* on page 128.

CCD cooling: to cut thermal and electronic noise, or 'dark current'.

COOLING

All the same procedures apply when using a CCD camera: the telescope must precisely track the sky and be well focused. A series of, ideally, three- to five-minute exposures (light frames), depending on your equatorial mount, can be taken and stored on computer. These are then 'co-added' together to build a strong signal in the final image.

However, a benefit of the CCD over the DSLR is its 'cooling' option. As we have seen, DSLRs suffer from thermal noise when exposure times extend to minutes. CCDs also suffer from electronic noise as a result of circuitry and sensors – known as 'dark current'. It varies from camera to camera and increases if exposures are lengthened, but can be overcome by cooling the CCD sensor to sub-zero temperatures. With every reduction of 7°C the dark current is halved. These separately powered cooling mechanisms take the form of fans that draw heat away from the back of the CCD. All cooled amateur-level CCD cameras rely on Peltier cooling that is augmented by cooling fins and small fans. At the more expensive end the Peltier cooling system is backed up by fan cooling, and at the even more expensive end by water-cooling.

These mechanisms are very effective, especially in winter temperatures. Not all dark current will be eradicated, but remaining noise can be removed using the 'dark frame subtraction' procedure whereby an equal-duration, equal-temperature image is taken with the telescope capped – the dark frame. Once this dark frame is stored on computer it can be subtracted from the light frame using appropriate processing software and the procedures outlined earlier, thus extracting the dark current signal from the image.

A second, very important 'calibration' frame is also required. This is known as a 'flat field' and is taken to record any optical aberrations in the telescope/camera set-up. There are many unwanted 'artefacts' that will be recorded in a CCD image, not least of which is dust near a sensor or on a filter. Furthermore, if focal reducers are used they can bring about 'vignetting', which darkens the field of view around the edge of the image. The flat field frame will address this problem. To produce such an image, the telescope and camera should be pointed at an evenly illuminated blank surface and an image should be taken. This should record the artefacts and it can then be used to successfully remove the problem from the light frame.

The Rosette Nebula in the constellation of Monoceros: this image was taken with a Quantum Scientific Imaging 532s CCD camera operating at -22 degrees Centigrade. The image was acquired using Astronomik Hydrogen Alpha (red), Oxygen III (green) and Hydrogen Beta (Blue) narrowband filters and consists of 12x300-second exposures per filter.

Webcam and smartphone imaging

Webcams were originally designed for taking home movies or for attaching to your computer to make video calls, but when attached to an 8in or 10in telescope they can capture high-resolution astro-images of bright objects like the Moon, Venus, Mars, Jupiter, and Saturn. Instead of having to take individual images, you can record them at the rate of hundreds per minute. This is something DSLRs cannot do. Likewise, the explosion in smartphone technology has enabled mobile phones to enter the arena of astrophotography, easily producing some stunning results.

Webcam: when attached to an adaptor, a webcam can easily be inserted into a telescope eyepiece to capture a continuous video stream.

WEBCAMS

Although mostly superseded by other forms of astrophotography these days, webcams still have a place. Bad 'seeing' (atmospheric turbulence) is magnified when observing Solar System objects through a telescope; images are mostly blurred. However, there are moments when the 'seeing' is steady: when a webcam is attached to a scope, it captures and records myriad frames, among which will be these excellent moments. You can stop with one perfect frame, or, by combining several of the best with processing software, produce an even sharper image.

WHICH WEBCAM?

The Philips ToUcam used to be the webcam of choice but is no longer manufactured. However, cheaper alternatives are still available, such as the Logitech, QuickCam, the Microsoft LifeCam or the 1.25"USB Telescope Digital Camera Eyepiece Webcam TD130, among others. Whichever webcam you choose, you will invariably need to remove the front lens and replace it with an adaptor barrel, available from telescope suppliers. The webcam should then be inserted into the telescope eyepiece and then connected to a laptop to capture the images of the night sky.

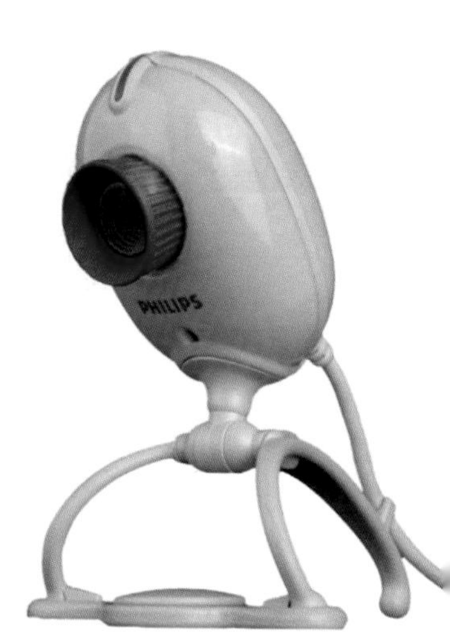

A ToUcam webcam : a Barlow lens can be added between the webcam and scope to increase image size.

WEBCAM IMAGING TIPS

- Use sharp optics and sufficient aperture to obtain the brightest images. Consider using a Barlow or Powermate lens to enlarge the image scale. Take shorter videos, around a minute long, to enable the post-processing computer program to create tighter images.
- Choose an observing site free of houses, trees, and other obstructions: these radiate heat, subsequent eddying air currents and distorted images. Ideally be near the sea or in the centre of a large field.
- Ensure the webcam is correctly focused: use the Moon's clearly defined terminator or, for planets, a nearby bright star.

An imaging camera and filter wheel: fixed to the rear port of a telescope's OTA.

■ CREATING A WEBCAM IMAGE

For the best webcam image, take short movie footage of, say, a minute's duration. Any footage will most likely be in the form of an AVI file. To process this footage on a laptop, you will need a good software program, like Registax, which is freely downloadable. This analyses your movie, ignoring those blurry frames caused by wind shake, your mount's tracking errors or atmospheric turbulence. The resultant 'best' frames are then stacked to enhance detail, quality and colour. The program then applies shifts, rotations, colour balances and other minor modifications to end up with a 'clean' superior image.

■ SMARTPHONES

Many smartphone users today employ planetarium apps to navigate around the night sky. However, these hi-tech phones can do so much more. They have made imaging easy and instantly enjoyable for anyone starting out in the bewildering realms of astrophotography, whether it's for solar system or deep sky objects. The resultant images are pretty sharp too.

WHAT YOU NEED

Smart phone adapter: this holds your smartphone in place above the scope's eyepiece. You can simply hold your phone at the eyepiece – afocal astrophotography – but this is rarely successful because it is difficult to centre the celestial object or achieve sufficient exposure. After first correctly attaching an adapter to a scope's eyepiece, and ensuring that the phone's camera lens is aligned with the eyepiece, a smartphone adapter will ensure an object is centred in the view screen, will keep the phone steady and offer correct focus and exposure. To aid accurate focus, live video apps, such as FiLMiC, which have a digital zoom function, enable a 'zoom in' facility which allows an observer to get 'up close and personal' to a much greater accuracy than the standard smartphone view.

Eyepiece Filter: many smartphones do not, as yet, have the necessary manual exposure control settings to capture solar system objects in their entirety, such as a full Moon, or to

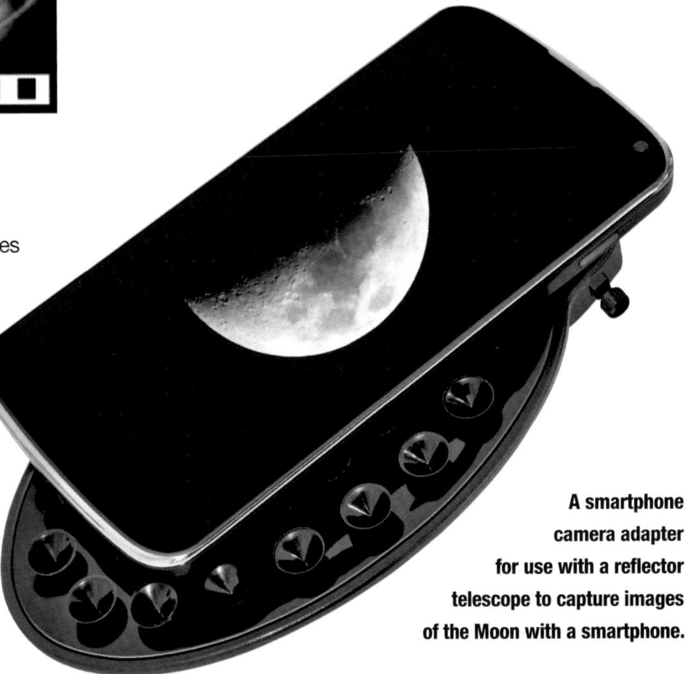

A smartphone camera adapter for use with a reflector telescope to capture images of the Moon with a smartphone.

A Duo Eyepiece Smartphone Adapter consisting of a case and adapter ring: This provides a solid attachment for imaging the Moon, solar system and deep sky objects.

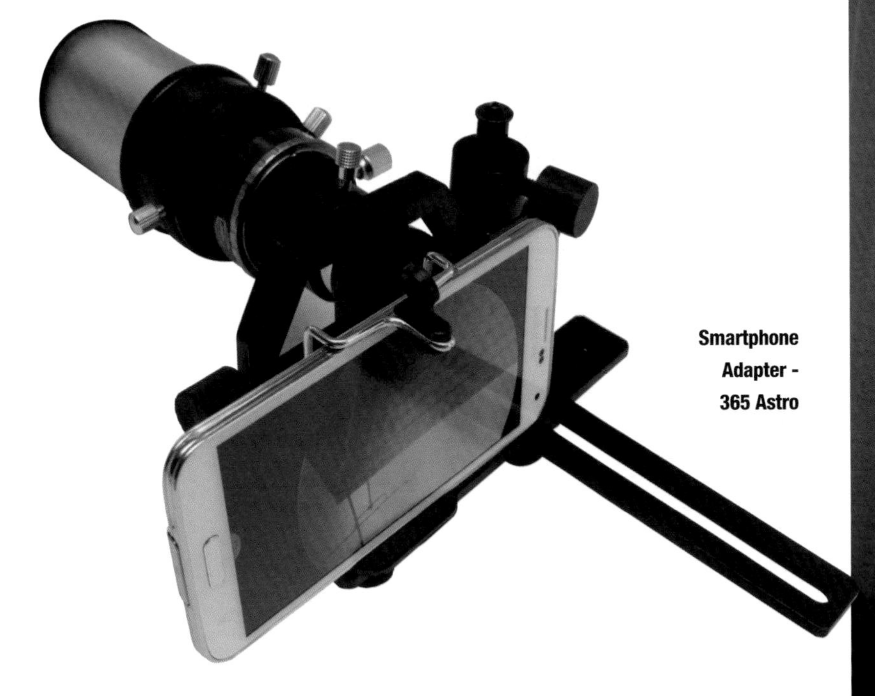

Smartphone Adapter - 365 Astro

image more subtle details. Without light reducing eyepiece filters, a planetary image can be over exposed and, therefore, featureless. Eyepiece filters can help overcome this (see page 129). A full Moon, for instance, through a low-powered eyepiece, requires a dark filter, whereas a light green filter (Wratten 56), used with high magnification, will enhance the features of Jupiter or Saturn. A transmission Moon filter will lessen Jupiter's brightness and draw out its cloud bands and Great Red Spot.

For deep sky objects, such as planetary nebulae, with the aid of phone Apps, such as NightCap, it is possible to simulate the long exposures customary with DSLR cameras to obtain images of extraordinary clarity and colour.

Stacking and Editing: to obtain even greater detail of a planet, a similar concept to using a webcam can be employed by using a smartphone but taking a short video clip using the phone's video camera. Ensure your smartphone battery is fully charged before a night's viewing, especially in cold weather. There are numerous websites on the internet which give introductory video tutorials about stacking and editing footage. *Registax, K3CCD Tools, AviStack, Deep Sky Stacker and AutoStakkert* are popular freeware software tools. For Apple iPhones, video footage can be imported directly into shareware programs to achieve the same results.

PROCESSING IMAGES

Webcam movie footage will invariably take the form of an AVI file and to process it requires software. An excellent, free-download program is Cor Berrevoets' RegiStax or Lucam Recorder. This automatically analyses the file, selects the sharpest of the myriad frames, and then aligns them on top of each other – stacking – to reduce noise and improve detail quality and colour. These images will still be subtly different, so the program then rotates, shifts, and colour-adjusts until a single, cleaned-up, brighter and sharper image is created. The program can also stitch together a mosaic of smaller frames to create stunning panoramic views.

Planetary imaging: with a webcam is easy. Processing with appropriate software will sharpen the final shot.

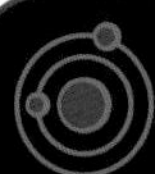

REMOTE ASTROPHOTOGRAPHY

If you wish to remotely image the night sky via the Internet, from the comfort of your own home, then it is possible. Having access to remotely controlled telescopes was once the preserve of professional astronomers. No longer! Today, there are several organisations offering amateurs, from beginners to advanced level, the opportunity to image using stunning scopes around the world – see Appendix 3.

There is normally a fee: this can be a one-off payment for an annual package, a monthly package, or payment for credits on individual scopes at certain locations. You must first register. All the systems are controlled via the Internet so a high-speed connection and up-to-date browser are vital. Most telescopes can be controlled via your usual web browser (Skylive will require specific software). Once you have logged on, the scope will be ready for you to use at will, ie you can select your astro-image target, select a camera, and choose the exposures and filters. Some organisations have more advanced controls, permitting customised imaging runs with multiple filters and exposures. Some images may be returned immediately, others may take a few days. Once downloaded, it is a simple matter of processing them.

All in all, astrophotography is an exhaustive subject. Everything I have mentioned is simply an overview, and more reading, or thorough research, is required beyond the scope of this book. For further resource information see Appendix 3.

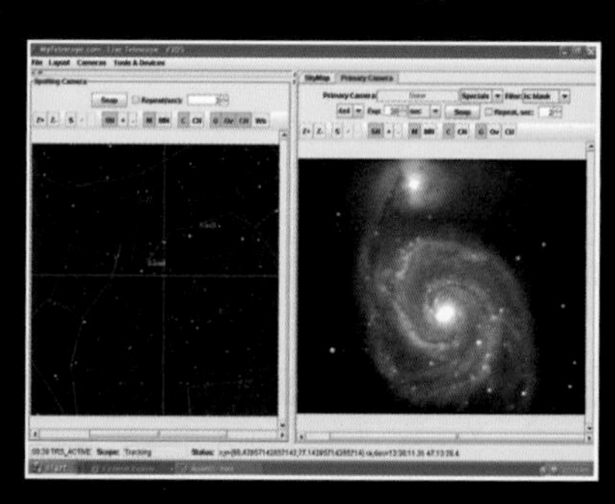

Remote access: puts global telescopes within the reach of amateurs.

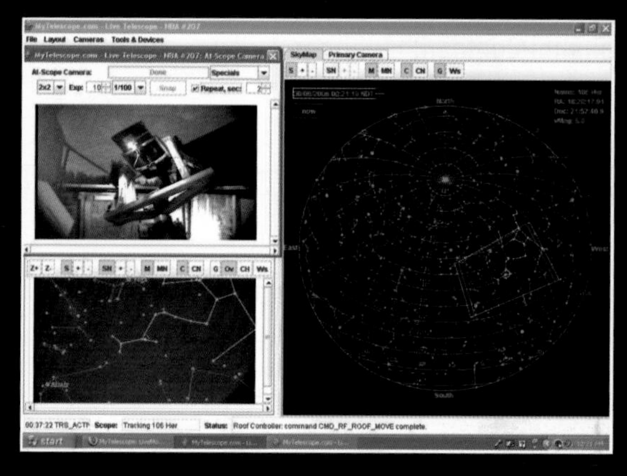

ASTROPHOTOGRAPHY GALLERY

Here are some beautiful examples of what can be achieved by enthusiastic and dedicated amateurs. All images captured by astrophotographer, Richie Jarvis.

M31 The Andromeda Galaxy
Skywatcher EQ6 Pro Mount
Scope: APM TMB 114mm @ F/5.6 using Televue TRF-2008 0.8x Focal Reducer
Main Camera: Starlight Xpress SXVR-H18
Filters: Astronomik Luminance/Red/Green/Blue
Guidescope: Williams Optics Zenithstar 66mm @ F/5.9
Guidecamera: Starlight Xpress SX Guidecamera
4 Pane Mosaic - Each pane took:
1 minute 40 second Luminance Integration Total Time: 20 minutes
5 minute Luminance Integration Total Time: 1 hour
RGB Integration Total Time: 1.5 hours

M81 (also known as NGC 3031 or Bode's Galaxy) - at left **and M82 (also known as NGC 3034 or The Cigar Galaxy). The former is a spiral galaxy and the latter a starburst galaxy both in the consellation of Ursa Major**
Skywatcher EQ6 Pro Mount
Scope: APM TMB 114mm @ F/7 Native Focal Length
Main Camera: Starlight Xpress SXVR-H18
Filters: Astronomik Luminance/Red/Green/Blue/6nm Hydrogen Alpha
Guidescope: Williams Optics Zenithstar 66mm @ F/5.9
Guidecamera: Starlight Xpress SX Guidecamera
Luminance Integration Total Time: 4.3 hours
RGB Integration Total Time: 1.5 hours
Ha Integration Total Time: 50 minutes

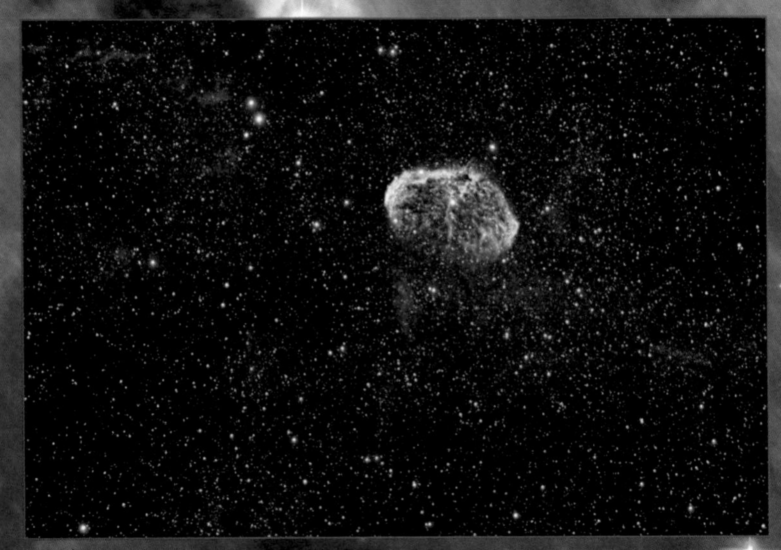

NGC 6888 Crescent Nebula
Skywatcher EQ6 Pro Mount
Scope: APM TMB 114mm @ F/5.6
Main Camera: Starlight Xpress SXVR-H18
Filters: Astronomik Hydrogen Alpha 6nm + Astronomik Oxygen III 12nm
Guidescope: Williams Optics Zenithstar 66mm @ F/5.9
Guidecamera: Starlight Xpress SX Guidecamera
Ha Total Integration Time: 15 hours
Oiii Total Integration Time: 6.7 hours

N90/NGC 602 star cluster: a tight cluster of new stars in the Small Magellanic Cloud, their light and sub-atomic particles blasting a cavity in the surrounding cloud.

Epilogue

It was the big questions of wonder: What am I? What am I doing here, alive, now, and on this planet? What is out there? What's the purpose of it all?

I have indeed found my place, our place, and while it is shockingly small and insignificant, it is also tremendously comforting. The very fact that we are all part of something so much bigger and grander than anything immediately around us and the fact that, as small as we are, we can know our circumstance, is itself a source of wonder, empowerment and deep spiritual fulfilment.

Knowing that in me, and through me, the principles governing everything in the Universe exist and flow, is all I need.

Professor Carolyn Porco, Leader of NASA's *Cassini* Imaging Team, excerpt from an interview, 2006

APPENDIX 1
THE 88 CONSTELLATIONS

Latin name	Pronunciation	Description
Andromeda	an-DROH-me-duh	The Chained Princess
Antlia	ANT-lee-uh	The Air Pump
Apus	APE-us	The Bird of Paradise
Aquarius	ah-KWAIR-ee-us	The Water Bearer
Aquila	uh-KWI-luh	The Eagle
Ara	AR-uh	The Altar
Aries	AIR-eez	The Ram
Auriga	oh-RYE-gah	The Charioteer
Bootes	boh-OH-teez	The Herdsman
Caelum	SEE-lum	The Chisel
Camelopardalis	ka-mel-o-PAR-da-lis	The Giraffe
Cancer	CAN-ser	The Crab
Canes Venatici	KAH-nez ve-NAT-eh-see	The Hunting Dogs
Canis Major	KAH-niss MAY-jer	The Great Dog
Canis Minor	KAH-niss MY-ner	The Little Dog
Capricornus	kap-reh-KOR-nuss	The Sea Goat
Carina	ka-RYE-nah	The Keel
Cassiopeia	kass-ee-oh-PEE-uh	The Queen
Centaurus	sen-TOR-us	The Centaur
Cepheus	SEE-fee-us	The King
Cetus	SEE-tus	The Whale (Sea Monster)
Chamaeleon	ka-MEE-lee-un	The Chameleon
Circinus	SUR-sin-us	The Drawing Compass
Columba	koh-LUM-bah	The Dove
Coma Berenices	KOH-mah bear-eh-Nee-seez	Berenice's Hair
Corona Australis	kor-OH-nah os-TRAH-lis	The Southern Crown
Corona Borealis	kor-OH-nah bor-ee-AL-is	The Northern Crown
Corvus	KOR-vus	The Crow
Crater	KRAY-ter	The Cup
Crux	KRUKS	The Southern Cross
Cygnus	SIG-nus	The Swan
Delphinus	del-FIE-nus	The Dolphin
Dorado	doh-RAH-doh	The Dolphinfish
Draco	DRAY-koh	The Dragon
Equuleus	eh-KWOO-lee-us	The Little Horse
Eridanus	eh-RID-an-us	The River
Fornax	FOR-nax	The Furnace
Gemini	JEM-eh-nye	The Twins
Grus	GROOS	The Crane
Hercules	HER-kyu-leez	Hercules
Horologium	hor-oh-LOH-jee-um	The Clock
Hydra	HY-dra	The Sea Serpent
Hydrus	HY-drus	The Male Water Snake
Indus	IN-dus	The Indian

Latin name	Pronunciation	Description
Lacerta	lah-SIR-tah	The Lizard
Leo	LEE-oh	The Lion
Leo Minor	LEE-oh MY-ner	The Little Lion
Lepus	LEE-pus	The Hare
Libra	LEE-bra	The Scales (Balance)
Lupus	LOO-pus	The Wolf
Lynx	LINKS	The Lynx
Lyra	LYE-rah	The Lyre (Harp)
Mensa	MEN-sah	The Table, Table Mountain
Microscopium	my-kro-SKO-pee-um	The Microscope
Monoceros	moh-NO-ser-us	The Unicorn
Musca	MUSS-kah	The Fly
Norma	NOR-muh	The Carpenter's Square
Octans	OCK-tanz	The Octant
Ophiuchus	oh-fee-U-cuss	The Serpent Bearer
Orion	oh-RYE-un	The Hunter
Pavo	PAH-voh	The Peacock
Pegasus	PEG-a-sus	The Winged Horse
Perseus	PURR-see-us	The Hero
Phoenix	FEE-nicks	The Phoenix
Pictor	PIK-tor	The Painter's Easel
Pisces	PIE-seez	The Fish
Piscis Austrinus	PIE-sis OSS-trih-nuss	The Southern Fish
Puppis	PUPP-iss	The Stern
Pyxis	PIK-sis	The Compass
Reticulum	reh-TIK-u-lum	The Reticle
Sagitta	sa-JIT-ah	The Arrow
Sagittarius	sadge-ih-TAIR-ee-us	The Archer
Scorpius	SKOR-pee-us	The Scorpion
Sculptor	SKULP-tor	The Sculptor
Scutum	SKU-tum	The Shield
Serpens	SIR-penz	The Serpent
Sextans	SEX-tanz	The Sextant
Taurus	TORR-us	The Bull
Telescopium	tel-eh-SKO-pee-um	The Telescope
Triangulum	tri-ANG-gyu-lum	The Triangle
Triangulum Australe	tri-ANG-gyu-lum os-TRAH-lee	The Southern Triangle
Tucana	too-KAN-ah	The Toucan
Ursa Major	ER-suh MAY-jer	The Great Bear
Ursa Minor	ER-suh MY-ner	The Little Bear
Vela	VEE-lah	The Sail (of Argo)
Virgo	VER-go	The Maiden (Virgin)
Volans	VOH-lanz	The Flying Fish
Vulpecula	vul-PECK-you-lah	The Fox

APPENDIX 2
THE MESSIER OBJECTS

M	NGC	Type
1	1952	Crab nebula in Taurus
2	7089	Globular cluster in Aquarius
3	5272	Globular cluster in Canes Venatici
4	6121	Globular cluster in Scorpius
5	5904	Globular cluster in Serpens
6	6405	Open cluster in Scorpius (Butterfly Cluster)
7	6475	Open cluster in Scorpius
8	6523	Lagoon nebula in Sagittarius
9	6333	Globular cluster in Ophiuchus
10	6254	Globular cluster in Ophiuchus
11	6705	Open cluster in Scutum (Wild Duck Cluster)
12	6218	Globular cluster in Ophiuchus
13	6205	Globular cluster in Hercules (Hercules Cluster)
14	6402	Globular cluster in Ophiuchus
15	7078	Globular cluster in Pegasus
16	6611	Open cluster in Serpens (Eagle Nebula)
17	6618	Omega nebula in Sagittarius (also known as Swan or Horseshoe)
18	6613	Open cluster in Sagittarius
19	6273	Globular cluster in Ophiuchus
20	6514	Triffid nebula in Sagittarius
21	6531	Open cluster in Sagittarius
22	6656	Globular cluster in Sagittarius
23	6494	Open cluster in Sagittarius
24	–	Star field in Sagittarius
25	IC4725	Open cluster in Sagittarius
26	6694	Open cluster in Scutum
27	6853	Dumbbell nebula in Vulpecula
28	6626	Globular cluster in Sagittarius
29	6913	Open cluster in Cygnus
30	7099	Globular cluster in Capricornus
31	224	Andromeda spiral galaxy
32	221	Elliptical galaxy in Andromeda
33	598	Pinwheel spiral galaxy in Triangulum
34	1039	Open cluster in Perseus
35	2168	Open cluster in Gemini
36	1960	Open cluster in Auriga
37	2099	Open cluster in Auriga
38	1912	Open cluster in Auriga
39	7092	Open cluster in Cygnus
40	–	Double star in Ursa Major
41	2287	Open cluster in Canis Major
42	1976	Great Orion nebula
43	1982	Diffuse nebula in Orion
44	2632	Praesepe or Beehive open cluster in Cancer
45	–	Pleiades open cluster in Taurus
46	2437	Open cluster in Puppis
47	2422	Open cluster in Puppis
48	2548	Open cluster in Hydra
49	4472	Elliptical galaxy in Virgo
50	2323	Open cluster in Monoceros
51	5194-5	Whirlpool spiral galaxy in Canes Venatici
52	7654	Open cluster in Cassiopeia
53	5024	Globular cluster in Coma Berenices
54	6715	Globular cluster in Sagittarius
55	6809	Globular cluster in Sagittarius

M	NGC	Type
56	6779	Globular cluster in Lyra
57	6720	Ring nebula in Lyra
58	4579	Barred spiral galaxy in Virgo
59	4621	Elliptical galaxy in Virgo
60	4649	Elliptical galaxy in Virgo
61	4303	Spiral galaxy in Virgo
62	6266	Globular cluster in Ophiuchus
63	5055	Spiral galaxy in Canes Venatici
64	4826	Black Eye spiral galaxy in Coma Berenices
65	3623	Spiral galaxy in Leo
66	3627	Spiral galaxy in Leo
67	2682	Open cluster in Cancer
68	4590	Globular cluster in Hydra
69	6637	Globular cluster in Sagittarius
70	6681	Globular cluster in Sagittarius
71	6838	Globular cluster in Sagitta
72	6981	Globular cluster in Aquarius
73	6994	Group of four faint stars in Aquarius
74	628	Spiral galaxy in Pisces
75	6864	Globular cluster in Sagittarius
76	650-1	Planetary nebula in Perseus
77	1068	Spiral galaxy in Cetus
78	2068	Diffuse nebula in Orion
79	1904	Globular cluster in Lepus
80	6093	Globular cluster in Scorpius
81	3031	Spiral galaxy in Ursa Major
82	3034	Irregular galaxy in Ursa Major
83	5236	Spiral galaxy in Hydra
84	4374	Elliptical galaxy in Virgo
85	4382	Elliptical galaxy in Coma Berenices
86	4406	Elliptical galaxy in Virgo
87	4486	Elliptical galaxy in Virgo
88	4501	Spiral galaxy in Coma Berenices
89	4552	Elliptical galaxy in Virgo
90	4569	Spiral galaxy in Virgo
91	4548	Barred spiral galaxy in Coma Berenices
92	6341	Globular cluster in Hercules
93	2447	Open cluster in Puppis
94	4736	Spiral galaxy in Canes Venatici
95	3351	Spiral galaxy in Leo
96	3368	Spiral galaxy in Leo
97	3587	Owl nebula in Ursa Major
98	4192	Spiral galaxy in Coma Berenices
99	4254	Spiral galaxy in Coma Berenices
100	4321	Spiral galaxy in Coma Berenices
101	5457	Spiral galaxy in Ursa Major
102	–	Duplicate of M101 spiral galaxy in Ursa Major
103	581	Open cluster in Cassiopeia
104	4594	Sombrero galaxy in Virgo
105	3379	Elliptical galaxy in Leo
106	4258	Spiral galaxy in Canes Venatici
107	6171	Globular cluster in Ophiuchus
108	3556	Spiral galaxy in Ursa Major
109	3992	Barred spiral galaxy in Ursa Major
110	205	Elliptical galaxy in Andromeda

APPENDIX 3
USEFUL RESOURCES

MAGAZINES

Monthly magazines (and accompanying interactive CDs offering videos, TV programmes, virtual planetariums, and software) offer comprehensive, user-friendly monthly sky charts (and almanacs) detailing constellations and deep-sky objects. Additionally, they provide in-depth information on topical discoveries and celestial events (visiting comets and asteroids, solar and lunar eclipses, meteor showers, occultations, etc.), as well as regular data on lunar phases and planetary/satellite visibility. They are invaluable for astronomical equipment and accessories, with handy 'how-to' guides; lists of suppliers; information on national and local clubs, societies, and organisations; planetariums; and museums and lecture events. They are also an informative read, providing book reviews and regular features, written by experts in various fields or experienced amateurs.

Periodicals currently available include *Astronomy* (USA); *Astronomy Now* (UK); *Sky & Telescope* (USA); *Sky & Telescope* (Australia); and *Sky at Night* (UK).

WEBSITES

In addition to mainstream media, the advent of the Internet and iPhone, iPod and iPad technology has utterly revolutionised the world of professional and amateur astronomy on a scale as illuminating as astronomy itself. Stunning 'hot off the press' information is easily accessible, with interactive co-operation between amateurs and professionals bringing about progression, achievement, and public outreach on a truly global scale. Ever-changing, and ever-increasing, there are now thousands of astronomy websites. Some may become obsolete during this book's production process, so utilise the Google search engine too. The websites given below should be invaluable for enthusiasts at all levels, whether amateur or professional. Emphasis is placed on the first five, since these are excellent entry points for anyone starting out; they offer interactive sky charts, observing tools, news blogs, newsletters and updates at local, national, and international level, astroalerts, podcasts, event calendars, hobby Q&As, astro glossaries, product news, equipment supplier and dealer indexes, test reports, classifieds, software products and free downloads, astrophotography for all, shopping ... and much more. In addition, each hosts substantial links to other equally invaluable sites.

- Astronomy.com
- Astronomynow.com
- SkyandTelescope.com
- Austskyandtel.com.au
- Skyatnightmagazine.com

The following websites are also excellent for general use or interest:

britastro.org	The British Astronomical Association.
darksky.org	The site of the International Dark-Sky Association, with information on the campaign against light pollution.
esa.int	The European Space Agency
exoplanet.eu	The Extrasolar Planets Encyclopaedia.
heavens-above.com	Offers up-to-date star charts of the entire sky for your viewing location for any preferred date or time. Also offers planetary and lunar information: where to look for Earth-orbiting artificial satellites, as well as the International Space Station, Hubble Space Telescope, iridium flares, and more.
jpl.nasa.gov	The Jet Propulsion Laboratory, covering planetary exploration and space telescopes.
lunar-occultations.com	The International Occultation Timing Association for information on occultations and grazing occultations.
nasa.gov	The National Aeronautics and Space Administration. The superlative site of the American Space Agency, with up-to-the-minute information on past and present space exploration, including mission photographs and video footage. Also details ground- and space-based telescope discoveries and current data on all things astronomical. With links to many other stunning sites, this is an invaluable source for anyone interested in astronomy.
spacetelescope.org	The site of the US and European centres for the Hubble Space Telescope, offering fabulous images and animations.
spaceweather.com	For predictions of solar and auroral activity, as well as a host of images of sky phenomena.

■ RECOMMENDED READING

Abell, George O.; Morrison, David; and Wolff, Sidney C. *Exploration of the Universe* (Saunders College Publishing, 1991) – although dated, this remains excellent.

Arditti, David. *Setting-up a Small Observatory*, Patrick Moore's Practical Astronomy Series (Springer, 2009).

Clark, Dr Stuart. *Galaxy* (Quercus Publishing plc, 2008).
— *The Sun Kings* (Princeton University Press, 2007).
— *The Big Questions* (Quercus Publishing plc, 2010).
— *The Unknown Universe* (Head of Zeus, 2015).

Cox, Brian; Cohen, Andrew. *Wonders of the Universe* (Collins, 2011).

Dickinson, Terence; and Dyer, Alan. *The Backyard Astronomer's Guide*, 3rd edition (Firefly Books Ltd, 2008).

Dunlop, Storm, Tirion, Wil, *2016 Guide to the Night Sky* (Collins 2015).

Ellyard, David; and Tirion, Wil. *The Southern Sky Guide* (Cambridge University Press, 2008).

Hicks, John. *Building a Roll-Off Roof Observatory*, Patrick Moore's Practical Astronomy Series (Springer, 2009).

May, Brian; Moore, Patrick; and Lintott, Chris. *Bang! The Complete History of the Universe* (Carlton Books Ltd, 2006/7).

Mobberley, Martin. *Lunar and Planetary Webcam User's Guide*, Patrick Moore's Practical Astronomy Series (Springer, 2010).

Mollise, Rod. *Choosing and Using a New Cat*, Patrick Moore's Practical Astronomy Series (Springer, 2009).

Moore, Patrick. *Exploring the Night Sky with Binoculars*, 4th edition (Cambridge University Press, 2000).

North, Gerald. *Observing the Moon*, 2nd edition (Cambridge University Press, 2014).

O'Meara, Stephen James. *Observing the Night Sky with Binoculars* (Cambridge University Press, 2008).
— *Deep-Sky Companions: Hidden Treasures* (Cambridge University Press, 2007).

Ratledge, David. *Digital Astrophotography: The State of the Art*, Patrick Moore's Practical Astronomy Series (Springer, 2010).

Ridpath, Ian. *The Monthly Sky Guide*, 9th edition (Cambridge University Press, 2012).
— *Stars and Planets* (Collins, 2011).

Sagan, Dr Carl. *Cosmos* (Random House Inc., 1980).

Sinnott, Roger W. *Sky & Telescope's Pocket Sky Atlas* (Sky Publishing, 2013).

Stoyan, Ronald. *Atlas of the Messier Objects: Highlights of the Deep Sky* (Cambridge University Press, 2008).

Szymanek, Nik. *Shooting Stars - The Ultimate Guide to Photographing the Universe* (Pole Star Publications Ltd, 2015)

STAR CHARTS AND ATLASES

MacEvoy, Bruce; *The Cambridge Double Star Atlas* (Cambridge University Press, 2015).

Tirion, Wil. The Cambridge *Star Atlas*, fourth edition (Cambridge University Press, 2011).
— and Sinnott, Roger W. *Sky Atlas 2000.0* (Sky Publishing Corporation, 1998) – 26 big charts, 81,312 stars to magnitude 8.5, 2,700 deep-sky objects.

Tirion, Wil; Rappaport, Barry; Remaklus, Will. *Uranometria, All Sky Edition* - with Stars to Visual 9.5 Magnitude and 30,000 + Non-Stellar Objects. (Willmann Bell, 2012)

■ ASTROPHOTOGRAPHY SOFTWARE FOR THE PC

DSLR Camera Control:

APT (Astro Photography Tool)
AstroArt
Backyard EOS
Images Plus Camera Control
IRIS
Maxim DL
Palm DSLR

Software-Assisted Focusing:

APT (Astro Photography Tool)
AstroArt
Backyard EOS
Focus Max
Images Plus Camera Control
Maxim DSLR
Nebulosity

Image Acquisition Automation:

APT (Astro Photography Tool)
AstroArt
Backyard EOS
DSLR Shutter
Images Plus Camera Control
Maxim DSLR
Nebulosity

Image Calibration, Aligning and Stacking:

AIP (Astro Photography Tool)
AstroArt
Deepsky Stacker
IRIS
Images Plus
Maxim DSLR
Nebulosity
PixInsight
Regim
RegiStax

Image Correction and Enhancement:

AIP (Astro Photography Tool)
AstroArt
GIMP
Images Plus
IRIS
Maxim DSLR
Nebulosity
Paint Shop Pro
Photoshop
Photoshop Elements
Picture Window Pro
PixInsight

Autoguiding Software:

AstroArt
GuideDog
Guidemaster
K3CCD
Maxim DSLR
Megaguide
PHD (Push Here Dummy)

■ ASTROPHOTOGRAPHY SOFTWARE FOR THE MAC:

DSLR Camera Control and Focusing:

Astro IIDC
DSLR Shutter
iAstrophoto
Nebulosity

Image Processing Programs:

Astrostack
Lykenos
Nebulosity
PixInsight

Image Correction/Enhancement Programs:

Photoshop CS
Photoshop Elements
GIMP

Photo Utilities:

FITS Liberator (file format image handling software)
iPhoto (image viewer, database and image editing)
StarStax (stacks frames for star trail image)

REMOTE ASTROPHOTOGRAPHY

Remote telescopes for amateur astro-imaging around the world:

Bradford Robotic Telescope	Mount Teide, Tenerife	telescope.org
Mytelescope.com	New Brunswick, Canada	mytelescope.com
Slooh	Tenerife, Chile, Australia	slooh.com
Lightbuckets	New Mexico, Australia	lightbuckets.com
Global Rent-a-Scope	New Mexico, USA, South Australia	global-rent-a-scope.com

PRACTICAL RESOURCES

Amateur Telescope Makers	atmsite.org	Amateur telescope making
AstroGazer	astrogizmos.com	Portable PVC observing tents
Astromart	astromart.com	Classified ads/reviews
Cloudy Nights	Cloudynights.com	Astronomy gear peer reviews
Fluxtimator	leonid.arc.nasa.gov	Meteor shower info
Home Observatory UK	homeobservatoryuk @hotmail.co.uk	Build/install rolling roof models
Kendrick Observing Tents	kendrickastro.com	Portable observing tents
Pulsar Observatories	pulsarobservatories.com	Fibreglass dome observatories
Scope Test	Scopetest.com	Astronomy gear peer reviews
Telescope House	Telescopehouse.com	Premium UK dealer

APPENDIX 4
GLOSSARY OF TERMS

Aberration – A lens or mirror defect causing image distortion or, as with chromatic aberration, coloured fringes around an object.

Absolute magnitude – The apparent magnitude (brightness seen by an observer) a celestial object would have if 10 parsecs (two million AU) from Earth. Used to compare the brightness of many stars at different distances.

Absorption nebula – A cloud of dust and gas sufficiently dense to block visible light from background objects. Also known as dark nebula.

Accretion disc – Disc of interstellar material surrounding a celestial object, formed by matter caught in its gravitational pull.

AFOV – Apparent field of view: the diameter of the circular field of an eyepiece expressed in degrees (a 50-degree 25mm eyepiece reveals around 1-degree of sky).

AGN – Active galactic nucleus: the central region of a galaxy from where prodigious energy is released. See quasars, blazars and radio galaxies.

Albedo – The fraction of sunlight reflected by a planet or exoplanet.

Alt-azimuth mount – A telescope mount yielding simultaneous movement about the vertical (altitude) and horizontal (azimuth) axes for tracking objects.

Angular momentum – a measure of the momentum associated with motion about an axis or fixed point.

Angular size – The apparent diameter of an object in the Earth's sky, measured in degrees of arc.

Apparent magnitude – the brightness of an object as seen from Earth. The object's distance is not taken into account.

Aphelion – The point of a comet's or planet's orbit when it is farthest from the Sun. The opposite to perihelion.

Apochromatic refractor – A telescope using three or more lenses to bring red, green, and blue light to focus at the same point.

Apparition – The period during which a planet is best placed or visible for observation.

Arcminute – A small unit of measurement equal to one-60th of a degree. Astronomers measure the separation of stars in terms of degrees.

Arcsecond – A tiny unit of measurement equal to one-60th of an arcminute.

Aspherical – A lens or mirror surface that does not form part of a sphere: used to reduce aberration.

Asterism – A pattern formed by stars that are not necessarily within a particular constellation.

Asteroid – A rocky remnant from the formation of the Solar System, ranging from tens to thousands of metres in diameter. Most exist in a belt between Mars and Jupiter.

Astronomical Unit (AU) – The average distance between the centre of the Earth and the centre of the Sun – 93,000,000 miles (150 million km).

Blazar – Powerful jet of radiation emitted from the heart of a quasar towards an observer's line of sight.

Blueshift – The shifting of light to shorter wavelengths caused by low speeds of recession or gravitational fields. The shifting of light to longer wavelengths is known as redshift.

Brown dwarf – A type of stillborn star, formed from cool clouds of hydrogen gas but too small to ignite normal nuclear reactions.

Celestial equator – the great circle which is the projection of Earth's equator onto the celestial sphere.

Celestial sphere – The name given to the projection of the night sky on to an imaginary sphere around Earth upon which the right ascension and declination coordinates are mapped.

Centrifugal force – A fictitious force that can be thought of as acting outwards on any body that rotates or moves along a curved path.

Chandrasekhar Limit – The upper limit to the mass of a white dwarf star (equal to 1.4 times the mass of the Sun).

CMB – Cosmic microwave background: electromagnetic radiation observed uniformly from every direction in space (to one part in 10,000). Thought to be the redshifted remnant radiation emitted around 300,000 years after the Big Bang.

Collimation – The alignment of telescope mirrors with the centre of the focuser for precise viewing.

Condensation – to change, or cause to change, from a gaseous to a liquid to a solid state.

Conduction – The direct transfer of energy from one atom, or electron, to another.

Conjunction – The configuration of a planet when it has the same celestial longitude as the Sun, or the configuration when any two celestial objects have the same celestial longitude (right ascension).

Constellation – One of the 88 recognised groups of stars into which the night sky is divided.

Convection – The transfer of energy by moving currents of a fluid containing that energy.

Cosmic rays – atomic nuclei (predominantly protons) striking Earth's atmosphere with enormously high energies.

Dark matter – Non-luminous mass whose presence is inferred by its gravitational effect on luminous matter.

Daylight Saving Time (DST) – When local time is one hour ahead of Greenwich Mean Time (GMT).

Declination (Dec) – The angular distance of a body north or south of the celestial equator: a positive value indicates north, negative indicates south.

Diffraction grating – A system of tightly spaced equidistant slits or reflecting strips that, by diffraction and interference, create a spectrum.

DSO – Deep-sky object: the common name for faint distant objects such as galaxies and star clusters, used to differentiate from objects in the Solar System.

Eccentricity – Orbits are almost circular or elongated; the more elongated, the more 'eccentric' it is.

Eclipse – When one celestial body passes in front of another, dimming or blocking its light.

Ecliptic – The apparent path of the Sun around the celestial sphere.

Electromagnetic radiation – energy in the form of electromagnetic waves: oscillating electric and magnetic fields at right angles to each other and the direction of transmission. Examples are light and radio waves.

Electromagnetic spectrum – The entire array of electromagnetic emissions.

Electron – A negatively charged subatomic particle that normally moves around an atomic nucleus.

Ellipse – The oval closed passage followed by a celestial object moving under gravity, such as a planet around a star.

Elongation – The angle in the sky between the Sun and a planet.

Emission nebulae – Gaseous nebulae that derive their visible light from the UV light of nearby stars.

Equinox – The two occasions in a year when the Sun crosses the celestial equator: the spring (vernal) equinox occurs around 21 March (Sun crosses from south to north), and the autumnal equinox occurs around 23 September (Sun crosses from north to south) – the hours of daylight and night are equal.

Escape velocity – The velocity a body must have to permanently escape the gravity of another body.

Event horizon – The largest surface surrounding the singularity of a black hole from within which no signal can escape. At the event horizon, the escape velocity from the black hole is equal to the speed of light.

Exomoon – Moon of an exoplanet.

Exoplanet – a planet orbiting a star beyond the Solar System.

Faculae – Bright regions near the limb of the Sun.

FOV – Field of view (also called actual field): the area of sky visible through the eyepiece expressed in angular degrees.

Geocentric – any system using the centre of the Earth as its reference point.

Gravitation – the mutual attraction of material bodies or particles.

Heliocentric – any system using the centre of the Sun as its reference point.

Heliopause – The boundary of the Sun's magnetic influence.

Inferior planet – A planet whose distance from the Sun is less than Earth's.

Inflation – The hypothetical rapid expansion (inflation) of the Universe after the Big Bang.

Infrared radiation – The electromagnetic radiation with a wavelength longer than the longest (red) wavelengths; perceivable by the eye but shorter than radio wavelengths.

Interferometer – An instrument that uses two or more detectors to collect radiation from a single source: the signals are combined and analysed for greater resolution.

Intergalactic medium – The sparse distribution of gas and dust between galaxies.

Interstellar medium (IMS) – The material between the stars in a galaxy, comprising 99% gas and 1% dust. Hydrogen forms 90% of the gas, the remainder contains helium and other elements. The dust is carbon and silicates.

Ionisation – the process whereby an atom or molecule is converted to an ion or by which an ion is converted to another ion by the loss or gain of electrons. An atom is completely ionised when all electrons are lost.

Km/s – Kilometres per second.

Lagrange points – Locations in space where the combined gravitational forces of two large bodies, such as Earth and the Sun or Earth and the Moon, equal the centrifugal force felt by a much smaller third body.

Light year (ly) – The distance travelled by light, or any other form of electromagnetic radiation, in a vacuum over one year: 6 trillion miles (9.5 trillion km); 63,240AU.

Limb darkening – Where a star is less bright near its limb than near the centre of the disc.

Luminosity – The amount of energy radiated by a celestial object independent of distance: valued in terms of 'solar luminosities' – how many times more luminous they are than the Sun.

Magnetic field – the area of space near a magnetized body wherein magnetic forces can be detected.

Magnetosphere – The region around a planet where the magnetic field dominates the interplanetary field carried by the solar wind, enabling charged particles to be trapped.

Magnitude – A measure of the brightness of an astronomical object (intrinsic or observed): the lower the number, the brighter the object. Magnitudes brighter than zero are represented with a negative value, eg -2. Magnitude 1 is 100 times brighter than magnitude 6.

Main Sequence – The band on the Hertzsprung-Russell diagram running from upper left to lower right, in which most stars lie whilst burning hydrogen in their cores.

Mass – A measure of the total amount of material in a body; defined by its inertial properties or its gravitational effect on other bodies.

Meteor – a streak of light observed as a 'shooting' or 'falling star' as a result of tiny interplanetary dust burning up in Earth's upper atmosphere.

Meteorite – A meteor that has survived the burning passage through Earth's atmosphere and impacted the ground.

Meteoroid – Rocky debris in space smaller than an asteroid.

Milky Way – The spiral galaxy to which the Sun belongs.

Mi/s – Miles per second.

Molecule – a combination of two or more atoms bound together; the smallest particle of a chemical compound or substance exhibiting the chemical properties of the substance.

Nanometre – A billionth of a metre.

NEA – Near-Earth asteroid: a class of asteroids with orbits that come closs, or cross, the orbit of Earth

Neutron – A subatomic particle with no charge and a mass almost equal to a proton.

Nucleus – Generally the core of an object, such as an atom, comet, or galaxy.

Occultation – When a celestial object of larger angular size passes in front of a smaller object.

Open cluster – A group of gravitationally bound stars formed from the same cloud of gas, typically numbering a few thousand.

Opposition – The position of a superior planet when it is opposite the Sun as viewed from Earth.

Orbital period – The time interval taken for a body to revolve around another body or point.

Parallax – The apparent displacement, or difference in the apparent position, of a target caused by change (or difference) at the point of observation. There are two kinds: diurnal, when a target is observed from opposite sides of Earth's surface; and annual, when observed from opposite points in Earth's orbit.

Parsec – A distance of 3.26ly

Perihelion – The point of a comet's or planet's orbit when it is nearest to the Sun. The opposite to aphelion.

Photon – a discrete 'packet' of electromagnetic radiation travelling in waves, but interacting with matter as though composed of particles, at the speed of light.

Planetesimal – Hypothetical small body that formed in the Solar Nebula as an intermediate step between the tiny grains and larger planets we see today.

Plutoid – A term for dwarf planets beyond the orbit of Neptune.

Precession – The gradual circular motion of the Earth's axis of rotation: one cycle takes 25,800 years to complete.

Proton – A heavy subatomic particle with a positive charge; one of the two constituents of an atomic nucleus.

Proto-stellar (or -star or -planet or -galaxy) – the original material from which a star has condensed.

Pulsar – A rapidly rotating neutron star emitting jets of radiation from its magnetic poles in the form of X-rays or gamma rays. These jets are observed in the form of pulses as a result of the rotation.

Quasar – A highly energetic distant galaxy emitting enormous electromagnetic radiation – thought to result from central massive black holes accreting material on to a surrounding disc.

Radial velocity – the component of velocity that lies in the line of sight, acting towards or away from the observer.

Radiation – The transmission of energy through a vacuum: also the transmitted energy itself.

Radio galaxy – an active galactic nuclei where the observed central black hole is obscured by the galactic torus (ring doughnut) of gas and dust. Strong sources of radio emission.

Radius – A straight line joining the centre of a circle or sphere to any point on the circumference or surface.

Red dwarf – Small cool stars that predominate in the Universe. They have 10% the mass of the Sun and surface temperatures between 2,200°C and 3,700°C.

Redshift – The shifting of light to longer wavelengths caused by high speeds of recession or gravitational fields. The shifting of light to shorter wavelengths is known as blueshift.

Reflection nebula – A dense dust cloud illuminated by starlight.

Resolution – the size of the smallest detail made visible by imaging systems.

Resonance – A condition where one object is subjected to periodic gravitational disturbance by another, commonly found when two objects orbit a third.

Retrograde motion – An apparent westward motion of a planet on the celestial sphere or with respect to the stars.

Right ascension (RA) – One of the two coordinates used in the equatorial coordinate system, the other being Declination (Dec). It is the angular distance measured eastwards along the celestial equator from the vernal equinox and is expressed in hours, minutes, and seconds, from 0 to 24 hours.

RR Lyrae star – A type of pulsating variable star, usually found in globular clusters, that varies from around 0.2 to 2 magnitudes in less than a day.

Seeing – A measure of the steadiness of Earth's atmosphere: the greater the turbulence, the more an astronomical image is distorted.

Sidereal period – The period of revolution of one body around another with respect to the stars.

Small Magellanic Cloud (SMC) – An irregular galaxy in the Local Group, 210,000ly distant and containing several hundred million stars.

Solstice – Either of the two occasions each year when the Sun reaches its most northerly or southerly declination (23.5° north or south): summer solstice is around 21 June, and winter around 21 December.

Spectra/spectrum – The array of colours or wavelengths resulting from light being dispersed through a prism or diffraction grating.

Superior conjunction – The configuration of a planet in which it and the Sun have the same longitude, but the planet is more distant than the Sun.

Superior planet – A planet more distant from the Sun than Earth.

Supernova – The death of a high-mass star that has exhausted its fuel: unable to support itself against gravity, it contracts and explodes leaving a 'remnant' dead neutron star, pulsar, or black hole with an expanding gas cloud.

Terminator – The line dividing the illuminated and dark part of a moon or planet.

Trans-Neptunian object (TNO) – Any object orbiting beyond the orbit of the planet Neptune.

Transit – When a small celestial body passes in front of a larger one.

Universal Time (UT) – Essentially Greenwich Mean Time (GMT) – local time at Greenwich, UK.

Variable star – A star that varies in light output over time, whether caused by internal processes, such as pulsations (intrinsic), or external influences, as in binary systems (extrinsic).

White dwarf – A star that has exhausted most or all of its nuclear fuel and has collapsed to a very small size.

Zenith – The point on the celestial sphere directly above the observer.

APPENDIX 5.1
STAR MAP OF NORTHERN HEMISPHERE

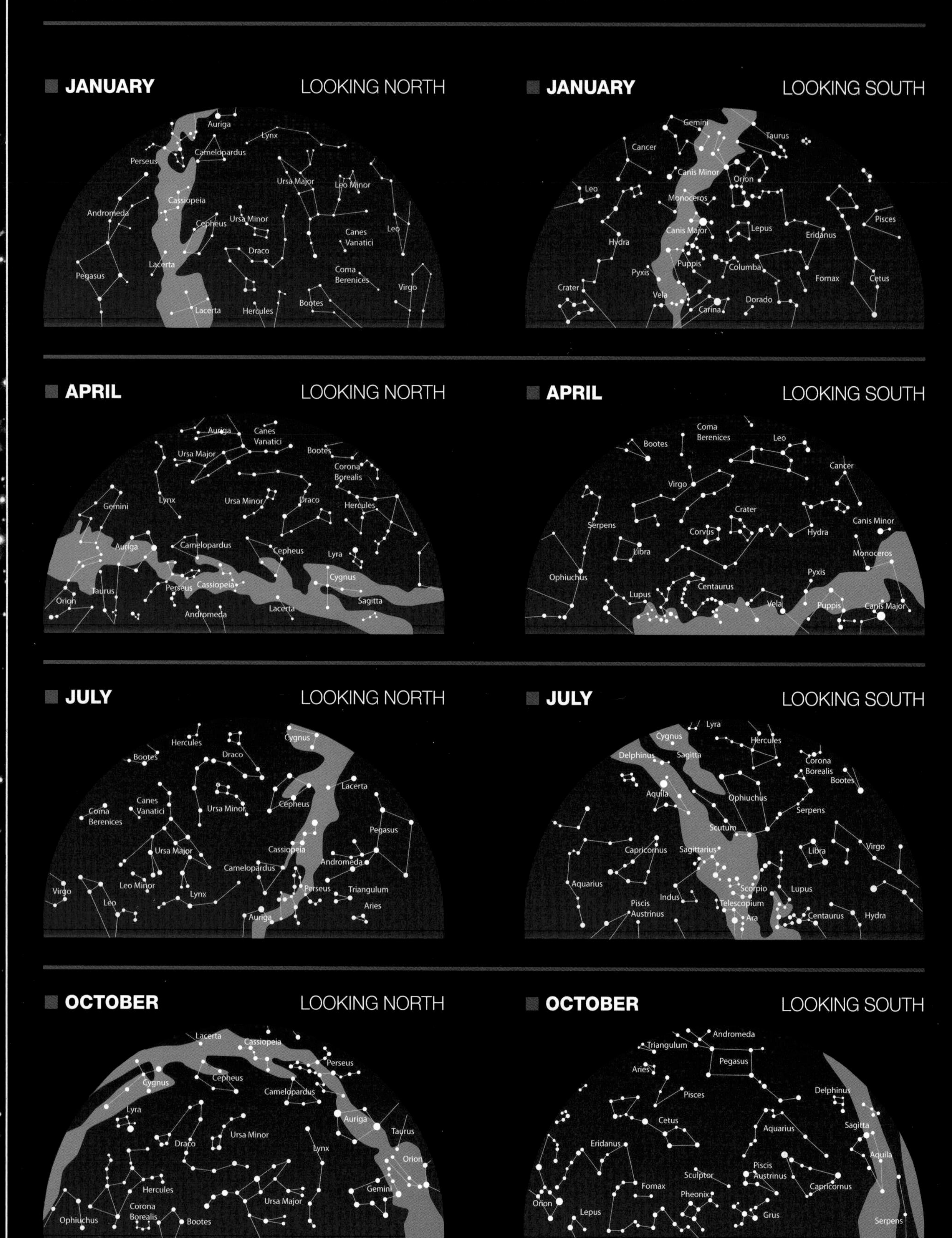

APPENDIX 5.2
STAR MAP OF SOUTHERN HEMISPHERE

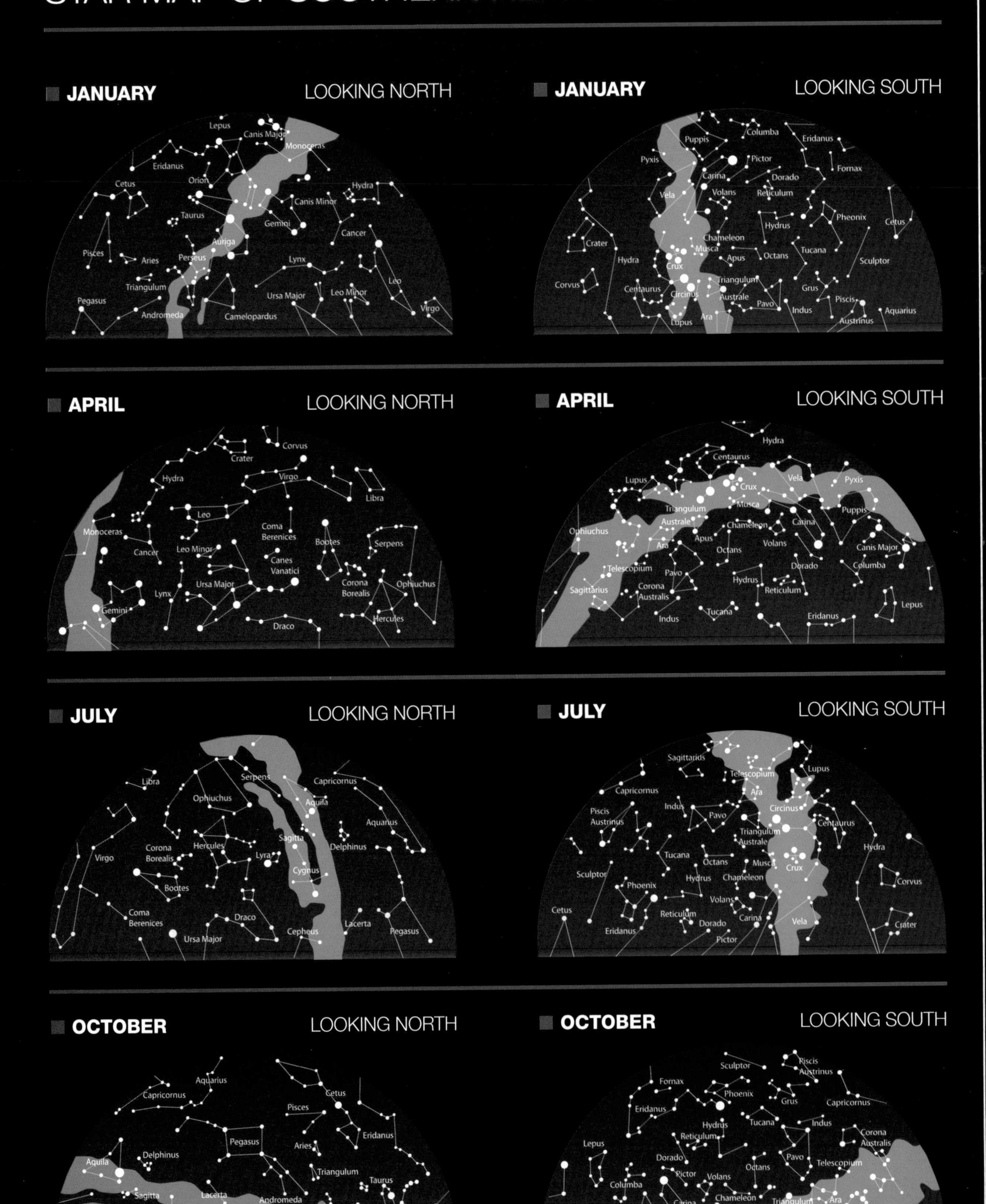

APPENDIX 6
FEATURES OF THE MOON

1. Apennine Mountains
2. Archimedes
3. Aristarchus
4. Aristillus
5. Aristoteles
6. Autolychus
7. Billy
8. Bullialdus
9. Caucasus Mountains
10. Copernicus
11. Eudoxus
12. Gassendi
13. Goclenius
14. Grimaldi
15. Hercules
16. Hevelius
17. Kepler
18. Lacus Mortis
19. Langrenus
20. Manilius
21. Mare Crisium
22. Mare Fecunditatis
23. Mare Frigoris
24. Mare Humorum
25. Mare Imbrium
26. Mare Nectaris
27. Mare Nubium
28. Mare Serenitatis
29. Mare Smythii
30. Mare Tranquillitatis
31. Mare Vaporum
32. Menelaus
33. Oceanus Procellarum
34. Pitatus
35. Plato
36. Plinius
37. Posidonius
38. Proclus
39. Schiller
40. Sinus Iridum
41. Theophilus
42. Tycho
43. Vitello

Apollo landings:

- 11 Apollo 11
- 12 Apollo 12
- 14 Apollo 14
- 15 Apollo 15
- 16 Apollo 16
- 17 Apollo 17

Acknowledgements

I owe an enormous debt of thanks to so many:

- David Wragg: who always believed in me and whose love and kindness made this entire project possible.
- My Editor, Derek Smith, and Editorial Director, Mark Hughes: thank you for such an opportunity and for your support and encouragement throughout.
- My Project Manager, Louise McIntyre and Designer, John Loasby; thank you for your patience, kindness and professionalism throughout the revisionary process.
- Dr Stuart Clark: whom I admire immensely and who made time, in a hectic schedule, to read the manuscript and offer a much-needed final vote of confidence.
- Keith Cooper, Editor of the UK's *Astronomy Now*: who offered invaluable material and support and without whom I would still be stuck on the launch pad!
- Nik Szymanek: a superb astrophotographer – without him Chapter 5 would not have been written.
- Sir Patrick Moore CBE FRAS and Dr Brian May: it is an indescribable honour to include their contributions.
- Leif J Robinson, Editor Emeritus, *Sky & Telescope* – a constant inspiration.
- Simulation Curriculum Corp: Mike Goodman and, especially, Brenda Shaw, for their excellent Starry Night images.
- Brent Kikawa, Ian Lauer, Meade Instruments Corporation, and Steve Collingwood, Telescope House: for invaluable high-resolution images.
- Chris Baker and Phil Sokell, Optical Hardware.
- Daniel Cerda, Explore Scientific USA: for your alacrity and beautiful images.
- Peter Carboni, Tele Vue Optics, Inc.
- All those kind personnel who sourced, or gave permission for, various images, but especially Eli Slawson, Charles Blue, Dr John M. Hill, Daniel Lewis PhD, Martha Lineham, and Mark Vivian.
- And last, but never least, special friends who have been part of this wonderful journey: Tom Blakeley, John Buchanan, Nicola Clark, Claire Evans, Debra and Mike Gervais, Ross Howard, Kathy Jones, Mark Matthews (Seaford Bay Music), Raymond and Toni Morgan, Leigh O'Brien, Mick Pickford, Captain Chris Sample, Marc Thompson, Burt Trowell and Carol White.

Picture credits

All images courtesy **nasaimages.org** apart from the following: **ALMA(ESO/NAOJ/NRAO)** 167 top **ATIK Cameras/Jo Surzyn** 184-185, 186 btm left & right **Author** 12 btm, 15 top, 110 top, 120, 139-142, 143 top left, 148 btm **Bridgeman Art Library** 10 btm **British Library Board** 11, 12 top, 13 **Carnegie Observatories** 165 btm right **Canon** 108 top, 174, 175 top, 176 top & mid right, 177 mid left & btm right **Celestron** 112 btm left, 116 top & btm left, 180 mid & btm right **CSIRO ASKAP** 168 btm, 169 **Mary Evans Picture Library** 9 top, 10 top, 36 btm, 147 **Mount Wilson Observatory** 14 **Hinode JAXA/NASA** 29 mid left **ESA** 20 top, 60 top, 61 top, 63 mid, 80 btm, 161 top right, **ESA/C.Carreau** 62 top, 162 btm **ESA/Rosetta/NAVCAM** 47 mid right **European Southern Observatory** 23 top inset, 64, 65 mid, 77 left, 150, 151 btm, 152, 166 mid right, 167 btm **ESO/S. Brunier** 66/67 btm **ESO/Stéphane Guisard** 86, Evryscope 163 mid right **Explore Scientific USA/Daniel Cerda** 112 top **Fermilab Visual Media Services/SDSS** 166 top **Getty Images** 7 **Gnixon/Wikimedia** 79 btm **J.M. Hill/LBTO** 153 top **Haynes Publishing** 23 top & mid, 26 mid & btm, 32 top, 34, 35 top, 42 btm, 45 top, 46 btm, 47 btm, 48 btm, 49 top, 55 btm, 57, 58 top, 60 btm, 62 btm, 67 top, 81, 83 top, 84 btm, 85 top, 88-89, 90 top, 91, 93 btm, 94-95, 97 right, 98 right, 99 right, 100 right, 101 right, 102, 107 btm, 110 btm, 111 mid & btm, 114, 121 right, 122, 126 top, 146, 151 top, 182, 202-203 **Haynes Publishing/Wikimedia** 72-73 **Rochus Hess/Wikimedia Commons** 190 btm left **David Hinds Ltd** 112 btm left **Institute for Astronomy/PAN-STARRS** 171 top left **iStock** 22 btm, 23 btm, 106 top, 108 left, 178 btm left **LIGO** 165 top left **LOFAR** 168 top left **Richie Jarvis** 141 top left, 191 **LSST Corp./NOAO** 171 btm right **Lunar and Planetary Institute** 25 top, 27 btm **Marc-Andre Besel & Wiphu Rujopakarn/LBTO** 153 btm **Meade Instruments Corporation** 93 top, 104 top, 107 top right, 113, 115, 116, 121 top, 123, 126 btm, 127 btm, 129 btm right, 130 mid right & btm right, 131 top left, 132 btm left & right, **MIT** 162 top left **Julian Monaghan 5 Mount Wilson Observatory** 154 left **mytelescope.com** 190 right **NASA/ESA** 65 btm **NASA/ESA/A. Schaller** 63 btm **NASA/Goddard Space Flight Center Scientific Visualization Studio** 20 mid, 51 top **NASA/JPL-Caltech** 18 top, 33 mid, 36 mid, 48 mid, 51 mid, 78, 159 top left **NASA/JPL-Caltech/Cornell University** 50 top **NASA/JPL/Space Science Institute** 40 btm **NASA/Messenger** 28 top & btm left **Marie-Lan Nguyen/Wikimedia Commons** 188 btm right **NRAO/AUI/NSF/O. Eilek & Kassim** 149 btm left **Orion** 180 top **Philips** 188 top right, **Howard Robinson** 175 btm **Sky-Watcher USA/Jeff Simon** 117, 118, 119 top left & btm left **Nik Szymanek** 48 top, 176 left, 178 top, mid left & mid right, 179 left, 179 left & top right, 181 btm right, 183 btm, 187, 204 **Telescope House** 127 top, 129 top right, 130 top left, 131 top right & btm right, 132 mid right **Telescope House/Astrozap** 131 btm left, **Telescope House/Baader** 130 btm left, 132 mid left **Telescope House/Celestron** 132 top right **Telescope House/Teirad** 130 top right **Tele Vue Optics Inc/Peter Carboni** 124 btm **The Sky/Nik Tring Astro** 119 btm right, **Szymanek** 181 btm right, **Zoltan Trenovszki/365 Astronomy** 189 **Skyshed** 138 btm, 143 top right **TMT Observation Corporation** 166 top left, mid & btm **Voyager 1 Team, NASA** 30 btm **Wikimedia** 23 mid inset, 35 btm, 92 btm, 96 top & btm, 97 btm left, 98 mid left, 100 top left, 104 btm, 105, 106 btm, 107 top left, 109, 110 mid, 111 top, 124 top, 125, 129 top left, 132 top right, 138 top, 148 top

Index